# Intelligent Estimation

*Intelligent Estimation: A Soft Computing Paradigm* presents a unified treatment of a new soft computing framework using AI-based approaches for system identification, parameter estimation, and filtering. This Neuro-Fuzzy-GA-based methodology, succinctly referred to as Intelligent Estimation (IE), integrates these estimation and filtering tasks within a single paradigm.

Offering a thorough understanding of soft computing-based estimation concepts and theory, the book discusses a modelling-control-system approach with numerous practical applications in solving mathematical modelling problems for industrial and aerospace engineering systems. It delves into theory, concepts, and various ramifications of neural networks, fuzzy logic, and genetic algorithms for modelling, system identification, filtering, and estimation.

This book is intended for upper-level undergraduate and graduate engineering students studying soft computing, intelligent systems, and advanced control systems in industry applications.

Instructors will be able to utilize a solutions manual and figure slides for their course.

**Jitendra R. Raol** received his BE and ME degrees in electrical engineering from M.S. University (MSU) of Baroda, Vadodara, in 1971 and 1973, respectively, and a PhD (in electrical and computer engineering) from McMaster University, Hamilton, Canada, in 1986. He joined the National Aeronautical Laboratory (NAL) in 1975. At CSIR-NAL, he was involved in human pilot modelling in fix- and motion-based research flight simulators. He re-joined NAL in 1986 and retired in July 2007 as Scientist-G (and Head of Flight Mechanics and Control division at CSIR-NAL). He was a fellow of the IEE/IET and a senior member of the IEEE. He is a life-fellow of the Aeronautical Society of India and a life member of the Systems Society of India. Dr. Raol has published nearly 160 research papers and written several books and numerous technical reports. His main research interests include data fusion, system identification, state/parameter estimation, flight mechanics-flight data analysis, H-infinity filtering, nonlinear filtering, artificial neural networks, fuzzy logic systems, genetic algorithms, soft technologies for robotics, and data-systems analytics (DaSyA).

**Sara Mohan George** holds a BTech in electronics and communication from Mahatma Gandhi University, Kerala, India; MTech and PhD degrees from Visvesvaraya Technological University, Karnataka, India. She is currently an Assistant Professor in Ramaiah Institute of Technology, Bangalore, India. Her research interests include nonlinear filtering, signal processing, and embedded system design. She has published nearly 20 technical papers in various conferences and journals. She is also the co-author of *Advances in State and Parameter Estimation: Theory and Practice* (CRC Press 2025).

# Intelligent Estimation

## A Soft Computing Paradigm

Jitendra R. Raol and Sara M. George

CRC Press
Taylor & Francis Group
Boca Raton London New York

CRC Press is an imprint of the
Taylor & Francis Group, an **informa** business

Designed cover image: Shutterstock



First edition published 2026
by CRC Press
2385 NW Executive Center Drive, Suite 320, Boca Raton FL 33431

and by CRC Press
4 Park Square, Milton Park, Abingdon, Oxon, OX14 4RN

*CRC Press is an imprint of Taylor & Francis Group, LLC*



ISBN: 978-1-041-01823-0 (hbk)
ISBN: 978-1-041-01838-4 (pbk)
ISBN: 978-1-003-61658-0 (ebk)

DOI: 10.1201/9781003616580

Typeset in Times
by Newgen Publishing UK

Access the Instructor Resources: www.routledge.com/9781041018230

# Contents

# Preface

The modern field of system identification, parameter estimation, and filtering (SIPEST) has been in vogue for the last seven, or more, decades, as a collective discipline to recon with, supported by extensive theoretical, analytical, and experimental work, research, and strengthened by numerous practical applications to solve mathematical modelling problems for many industrial and aerospace engineering systems. Concurrently, there has been a considerable and very significant progress in the field of AI (artificial intelligence) augmented by the new paradigms of artificial neural networks, fuzzy logic, and genetic algorithms, the latter collectively called soft computing in computer science/engineering and information sciences. There is reasonably a good literature that covers the various individual topics and ingredients of AI as studied and investigated with some emphasis on their use and applications to many science and engineering problems.

The title of the present book should be read as: *Intelligent Estimation: An Integrated Soft Computing Paradigm-Theory and Practice*; and here an attempt is made, perhaps for the first time, to collect several AI-based approaches from the open literature, and our own work, applicable to SIPEST, and to provide a unified treatment for this new paradigm of 'soft computing' (SC), highlighting the merits and limitations for their practical use. We call this Neuro-Fuzzy-GA-Based System Identification, Parameter Estimation, and Filtering, and think of it succinctly as Intelligent Estimation (IE), *if not directly so, but definitely signifying it.* Thus, the theory, concepts, and various ramifications of artificial neural networks, fuzzy logic, and genetic algorithms for mathematical modelling, system identification, filtering, state estimation, and parameter estimation are covered in this book.

Where appropriate we use numerical simulation examples coded in MATLAB (MATLAB is the trade mark of MathWorks Ltd. USA), or other software tools (like Python) to illustrate the concepts of IE. The user should have access to PC-based MATLAB software and other toolboxes: signal processing, control system, system identification, neural networks, fuzzy logic, genetic algorithms, statistics, and optimization; none of these tools are provided as part of the book. However, the programs/scripts written by the authors for the illustrative examples may be obtained from the URL of the book.

There are some good books on system identification, filtering, state estimation, and parameter estimation, although almost all miss out the new methodologies of IE.

The treatment in the present volume is comprehensive and covers several related disciplines which will help in understanding the AI-based SIPEST concepts and methods from a practical point of view. An engineering approach is used rather than a pure mathematical approach (it is not at all a Theorem-Proof approach), of course without losing sight of essential mathematics. Where appropriate, some novel methods, approaches, techniques, and/or algorithms, and analytical stability results are presented.

Also, some novel relationships amongst various recurrent neural network structures and amongst fuzzy operators/fuzzy implication functions are brought out, and where feasible asymptotic convergence results are obtained for some parameter estimation methods based on Lyapunov's normalized energy functional; these are novel aspects in the present book.

The AI-based ANN, FL, GA (SC) paradigm for application to modelling of dynamic systems is a very important, exciting, and crucial area for research, development, and teaching. The end users of this integrated technology will be systems-aero-mechanical-civil-chemical engineers, educational institutions, R&D laboratories, aerospace and other industries, space mechanics and transportation/automation industry, environmental science/engineers, as well as economics system researchers.

***Disclaimer****:* Although care has been taken in the presentation of the theoretical and practical/simulation results, their usage and any implications would be the users/readers own responsibilities, and the authors and publisher would not be answerable for any eventuality caused by the usage of the results of this book.

Acknowledgements

**Jitendra R. Raol:**
I am very grateful to Drs. Girija G., Jatinder Singh, and Sudesh Kashyap for their moral and technical support.

I am also thankful to all the past and present doctors of NAL for my reasonable health for all these years.

I am too grateful to my mother, my wife Virmati, my daughter Harshakumari, my son, Mayur, and our daughter-in-law Amina for their continual support for all these years.

**Sara M. George:**
I thank the Almighty for His blessings, guidance, and grace, without which the completion of this work would not have been possible.

I would like to express my deepest gratitude to my mentor, Dr. J. R. Raol, whose exceptional guidance, insightful suggestions, and constant encouragement have been instrumental in shaping these chapters. His depth of knowledge, clarity of thought, and unwavering support have inspired me throughout this work. I consider it a privilege to have had the opportunity to learn from him.

I am profoundly grateful to my parents and sisters, who have consistently believed in me and given me the freedom and confidence to pursue my interests and aspirations. Their faith has been the foundation of my journey. My heartfelt appreciation goes to my husband, James, for his unconditional love, patience, and understanding. His steady support and encouragement have sustained me through every phase of this effort. I am equally thankful to my daughter, Joanna, whose affection, warmth, and joyful presence have been a constant source of motivation.

This work is a reflection of the support, love, and guidance I have received from all these individuals and I remain truly indebted to them.

We are also grateful to Dr. Sudesh Kumar Kashyap (Deputy Head, FMCD, CSIR–NAL, Bangalore) and Dr. Lakshmi Shrinivasan (Associate Professor, Department of Electronics and Communication Engineering, Ramaiah Institute of Technology, Bangalore) for their technical discussions and for reviewing certain chapters of the book. Their thoughtful inputs and attention to detail have been invaluable.

We remain deeply grateful to Ms. Kyra Lindholm for her dedicated coordination toward the book project and for her continuous technical support during its progression. Our thanks also go to Ms. Sonia Tam, whose diligent efforts in keeping track of the book project ensured smooth coordination throughout.

# About the Authors

**Jitendra R. Raol** received BE and ME degrees in electrical engineering from M. S. University (MSU) of Baroda, Vadodara, in 1971 and 1973 respectively and a PhD (electrical and computer engineering) from McMaster University, Hamilton, Canada, in 1986; at both places he was also a post graduate research and teaching assistant. He joined the National Aeronautical Laboratory (NAL, Bangalore) in 1975. At CSIR-NAL he was involved in human pilot modelling in fix- and motion-based research flight simulators. He re-joined NAL in 1986 and retired in July 2007 as Scientist-G (and Head, flight mechanics and control division at CSIR-NAL).

He has visited Syria, Germany, the United Kingdom, Canada, China, the United States of America, and South Africa on deputation/fellowships to i) work on research problems on system identification, neural networks, parameter estimation, multi-sensor data fusion, and robotics; ii) present technical papers at international conferences; and/or iii) deliver guest lectures. He has given several guest lectures at many Indian colleges & Universities, and Honeywell (HTSL, Bangalore).

He was a fellow of the IEE/IET (United Kingdom) and a senior member of the IEEE (United States). He is a life-fellow of the Aeronautical Society of India and a life member of the Systems Society of India. During his studies at the MSU he had received the Dean and Professor Suba Rao memorial prize and M. C. Ghia charitable fellowship. In 1976, he won the K. F. Antia Memorial Prize of the Institution of Engineers (India) for his research paper on nonlinear filtering with time delay. He was awarded a certificate of merit by the Institution of Engineers (India) for his paper on parameter estimation of unstable systems. He received the best poster paper award from the National Conference on Sensor Technology (New Delhi) for a paper on sensor data fusion. He also received a gold medal and a certificate for a paper related to target tracking (from the Institute of Electronics and Telecommunications Engineers, India). He was also one of the five recipients of the CSIR (Council of Scientific and Industrial Research, India) prestigious technology shield for the year 2003 for leadership and contributions to the development of Integrated Flight Mechanics and Control Technology for Aerospace Vehicles in the country; the shield was associated with a plaque, a certificate, and a project-grant-prize of INRs. 30,00,000 for the project work. He was one of the five recipients of the Chellaram Foundation Diabetes Research Award 2018 for the best paper (presented at the 2nd International Diabetes Summit, March 2018, Pune, India; which carried a prize of 100,000 INRs).. He received the Sir Thomas Ward memorial prize of the Institution of Engineers (India) in 2019 (jointly) for the paper on Image Centroid Tracking with Fuzzy Logic…, and it carried a gold medal and a certificate. He featured in the Stanford University list as one of the world's top 2% scientists/researchers for the years 2019 and 2021.

He has published nearly 160 research papers and numerous technical reports. He has guest-edited two special issues of Sadhana (an engineering journal published by the Indian Academy of Sciences, Bangalore) on i) advances in modelling, system identification, and parameter estimation (jointly with Late Prof. Dr. Naresh Kumar Sinha), and ii) multi-source, multi sensor information fusion. He has also guest-edited two special issues of the Defense Science Journal (New Delhi, India) on i) mobile intelligent autonomous systems (jointly with Dr. Ajith K. Gopal, CSIR-SA), and ii) aerospace avionics & allied technologies (jointly with Prof. A. Ramachandran, MSRIT).

He has co-authored an IEE/IET (London, UK) Control Series book *Modelling and Parameter Estimation of Dynamic Systems* (2004), and the CRC Press (Taylor & Francis, Florida, USA) books, *Flight Mechanics Modelling and Analysis* (2009, and the revised edition in March 2023), *Nonlinear Filtering: Concepts and Engineering Applications* (2017), *Control Systems: Classical, Modern, and AI-based Approaches* (2019), and *Situation Assessment in Aviation: Bayesian Network and Fuzzy Logic Approaches* (2024).

He has authored the CRC Press books *Multi-sensor Data Fusion with MATLAB* (2010), and *Data Fusion Mathematics – Theory and Practice* (2015), and with Ajith K. Gopal (CSIR-SA) edited the CRC press book *Mobile Intelligent Autonomous Systems* (2012).

His recent jointly authored CRC books are (2025): *Mathematical Modelling of Aerospace Dynamic Systems with Practical Applications*, *Advances in State and Parameter Estimation-Theory and Practice*, and *Data Fusion Mathematics-Theory and Practice*, 2nd Edn.

He has served as a member/chairman of numerous advisory-, technical project review-, and doctoral examination committees. He has also conducted sponsored research and worked on several projects from Industry as well as other R&D organizations to CSIR-NAL with substantial budgets. Under his formal technical guidance, 11 doctoral and 8 master research scholars have had received their degrees successfully; he has also guided jointly/informally one dozen research students for their doctoral studies. He is a reviewer of a dozen national/international journals, and has evaluated several MTech./Doctoral theses (from India and overseas). He has been with: i) MSRIT (M. S. Ramaiah Institute of Technology, Bengaluru) as emeritus professor for five years; ii) the Govt. college of engineering, Kunnur (Kerala) as a senior research advisor; and iii) the dept. of aerospace engineering (IISc., Bangalore) as a consultant on modelling and parameter estimation for Type I diabetes-patients' data for a period of three months. He was also a consultant to a private company in Bangalore that provides software and algorithmic solutions to target tracking, navigation, and data fusion problems. He is a member of the governing body of the Centurion University of Technology and Management (CUTM), Andhra Pradesh, India.

His main research interests have been and are data fusion, system identification, state/parameter estimation, flight mechanics-flight data analysis, H-infinity filtering, nonlinear filtering, artificial neural networks, fuzzy logic systems, genetic algorithms, and soft technologies for robotics.

On the literary front, he has also authored a few books as a collection of his 320 (free-) verses on various facets closely related to science, philosophy, evolution, and life itself. He has also contributed 80 articles/stories and 1200 'bites' (long quotes) on matrubharti.com (#1 Indian Content Community) and seven ebooks on Amazon KDP (kindle direct publishing) covering social, philosophical, science, and human life-related aspects and issues. He was featured as one of the top 100 downloaded English authors of the matrubharti.com for the years 2021–2024. His new area of study and research is data-systems analytics (DaSyA).

**Sara Mohan George** holds a BTech in electronics and communication engineering from Mahatma Gandhi University, Kerala, and MTech and PhD degrees from Visvesvaraya Technological University (VTU), Karnataka. She is currently an Assistant Professor at the Ramaiah Institute of Technology, Bangalore.

Her areas of expertise span nonlinear filtering, signal processing, system identification, neural networks, and embedded system design. In addition to her core research interests, she actively explores interdisciplinary applications of AI, including intelligent sensing systems, drone-based data acquisition, wearable healthcare technologies, and image-based diagnostic tools. She has guided numerous student projects in embedded systems and machine learning. She has received a research grant of ₹10 lakhs for her project that monitors and manages gait in Parkinson's patients. She has also been recognized multiple times by FAER (Foundation for Advancement of Education and Research) for innovative student projects under her mentorship, reflecting her strong contribution to practice-oriented engineering education and research.

Over the years, she has published nearly 20 technical papers in reputed national and international conferences and journals. Her research contributions integrate strong theoretical foundations with practical implementation, addressing real-world problems in engineering and applied science.

She is also the author of the book *Advances in State and Parameter Estimation – Theory and Practice*, published by CRC Press, Taylor & Francis, Florida, USA, in February 2025.

# 1 Introduction
## *Intelligent Estimation*

We briefly present certain aspects and features of artificial intelligence (AI), intelligent systems (IS), and intelligent control (IC); and then signify and specialize to the aspects of intelligent estimation (IE), the latter mainly based on soft computing paradigm. A very brief note is presented on intellect-intelligence (I-I) in Appendix 1A. A brief discussion on parameter estimation and need of soft computing is given in Appendix 1B. In Appendix 1C Sub-components of AI are briefly explained.

We are in fact an intelligent species, (an outcome of long drawn evolutionary process), and hence we would like to eventually build intelligent systems-based technologies around us to help us in carrying out increasingly complex and sophisticated tasks to facilitate our useful and happier living on this planet.

We have invented and used highly sophisticated computers, mainly digital, and we further want these machines to think, since we are also thinking (natural-) biological machines; we would like to have the automation of activities that are associated with our own thinking, learning, and decision-making. An attempt in the realm of AI is the art of creating intelligent machines (mostly computer based) that perform functions that require human intelligence. To build AI-based systems, we need to study our own mental faculties, may be through the use of computational models. Hence, one important facet of the pursuit of AI is the study of computations that make it possible to perceive, reason, and act. Figure 1.1 depicts several aspects of AI and areas of AI: search (from presented data), logic (deductions), knowledge representation, machine learning, planning, visualization, building expert systems, and utilization for a desired application, e.g. robotics; this mimics certain key components of the human brain's functionality.

## 1.1 CONCEPT OF ARTIFICIAL INTELLIGENCE (AI)

AI is a field of study that seeks to explain, and emulate (human's) intelligent behaviour in terms of computational processes, it is also considered to be a branch of computing (computer) science that is concerned with the automation of intelligent behaviour [1]. The quad-elements model (QEM) concept of AI is depicted in Figure 1.2.

Let us pick up the element QEM(2,1) which signifies: systems that act like humans. Here, by systems, we mean: i) an algorithm (like an agent), ii) a software (SW) program, iii) hardware (HW) system (a smart device/sensor/computer), and/or iv) any switching mechanism that has a logic. The main aspect here being how to make the computers (in fact any system that has some logic and computing mechanism) to do things at which, in the present/current times, humans are better (in some qualitative or quantitative manner) than other species. When certain tasks are performed by the machines, then we want them to perform the way we perform these tasks using our intelligence. This is highlighted by the so-called Turing test, Figure 1.3; assume you have entered

DOI: 10.1201/9781003616580-1

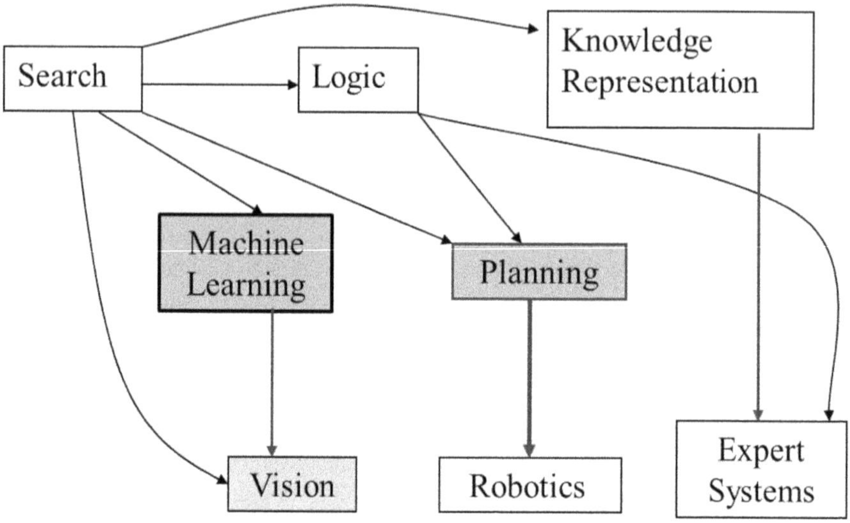


**FIGURE 1.1** Key components of Artificial Intelligence (AI)-mimicking human intelligence functionality process

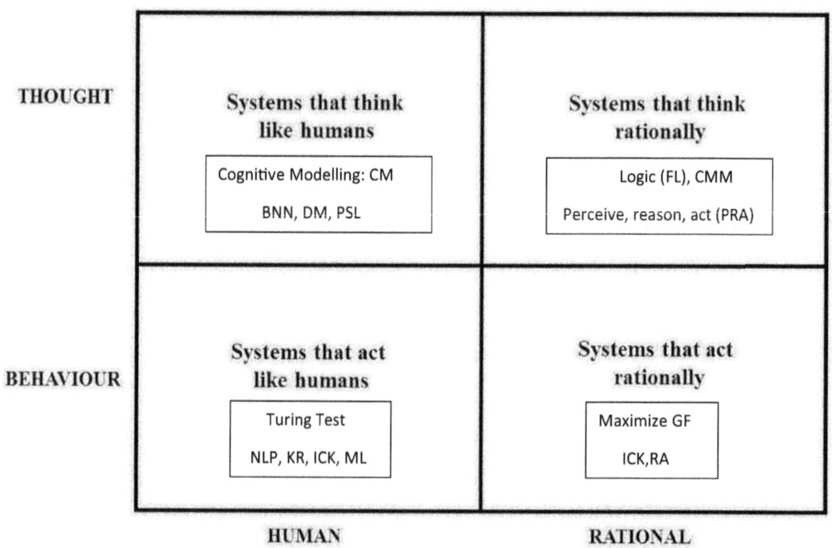


**FIGURE 1.2** Quad-elements model (QEM) of AI: An element QEM(i,j) signifies an important aspect of AI (the abbreviations are explained in the text of Chapter 1)

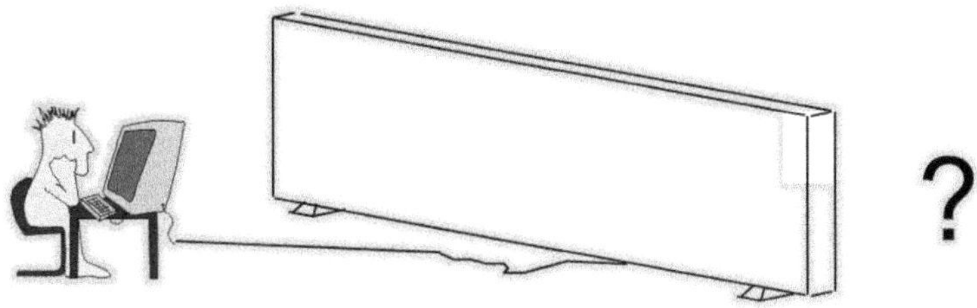


**FIGURE 1.3** A system, for example a computer, that thinks like a human

into a room that has a computer terminal, and within a fixed period of time, you can type what you want, and study the replies. At the other end, there is either a human, or a computer. If it is a computer, and at the end of the period of time, you cannot reliably ascertain whether it is a system/computer, or a human, then the system can be deemed to be an intelligent one. Here, the idea is to achieve human-level performance in all cognitive tasks. These include: i) natural language processing (NLP) for communication with human, ii) knowledge representation (KR) to store processed information effectively and efficiently, iii) automated reasoning to retrieve and answer questions using the stored information-cum-knowledge (ICK), and iv) machine learning (ML) to adapt to new circumstances. The complete Turing test includes two more aspects: i) computer vision to perceive objects, e.g., seeing, and ii) robotics-to move objects, i.e., acting.

Let us now pick the element QEM(1,1) which signifies: systems that think like humans. Here, the emphasis for the systems is to think like we think. This requires cognitive modelling (CM), that is the humans as observed from 'inside' (mainly this involves study of our biological neural networks, BNNs). We can do this by introspection vs. psychological experiments. The idea, thus in cognitive science is to make computers think, i.e., the machines with minds, relatively, in a nearly complete and literal sense. Here, we call upon the automation of activities that humans associate with their own thinking: decision-making (DM), and problem solving, and learning (PSL).

Let us now pick the element QEM(1,2) which signifies: systems that think rationally. Here, the emphasis for the systems is to think rationally. Ironically, humans are not always 'rational' in their thinking. As we know from the evolutionary nature, that all (or almost all) biological species are neither kind nor cruel, they are just indifferent, i.e., often we observe that these are cruel to the members of other species, and kind to their own members. In fact, these species do not think rationally or irrationally. But, with the highest level of intelligence, humans can think irrationally also, since the very fact of the ability of rational thinking in humans, can produce an irrational thought as a by-product (under existential pressure, difficult times, under fear, over-sentimentalism, over-reactions, and knee-jerk responses). The rationality can be defined in terms of logic, but logic cannot express every aspect of act. The uncertainty is difficult to rationalize. Of course, now fuzzy logic (FL) can be used to define a kind of uncertainty that is called vagueness. The logical approach would need more computational time, and programming, since it also needs 'guidance'. The demand here is to study the mental faculties by using computational models, and evolve the computational mechanisms (CMM) that make machines perceive, reason, and act (PRA).

Let us now pick the element QEM(2,2) which signifies: systems that act rationally. Here, the emphasis for the systems is to act rationally, which signifies doing the right thing. This has to be in terms of maximizing the goal function (GF, i.e., achievement), given certain amount of information. Here, it is not very important whether the AI-system replicates a human thought process, makes the same decisions as humans do, or uses purely a logical reasoning. The latter is a part of the rational agent, and does not necessarily cover all of the rationality. This is because, sometimes a logic cannot reason a correct inference. In such cases, a specific in-domain human ICK is used; and suggests that one should study the AI as a rational agent, because it is logic+ domain knowledge. The rational agent (RA) is an entity that perceives and acts. In an abstract sense, RA is a function from percept histories (past actions, experiences) to actions: f: P→A. So, here we seek an agent with best performance for a given environment and task, e.g., design a best program for given machine resources.

So, AI can be in plain terms stated as: produced by human art or an effort rather than originally done naturally, and is the ability to acquire knowledge and use it. Thus, in AI, we develop the ideas that enable computers to be intelligent.

## 1.2 BASICS OF AI

The foundation of AI has the major constituents: i) philosophy, ii) psychology, iii) computer engineering, iv) control theory and cybernetics, and v) linguistics. The origin of AI could be that the study of human intelligence began with no formal expression, and the ideas were initiated about mind that can pave the way for the machines to act as humans, and perform their internal operations as well. Mathematics imparts a formalism to these three main aspects of AI: i) computation, ii) logic, and iii) probability (which can be replaced or aided by fuzzy logic (FL), and associated possibility theory). The computation leads to analysis of the problems that can be made amenable to computing. The logic provides some reasoning, like If,…, Then,…, etc. as fuzzy rule/s; and thought experiments. Probability assigns a degree of belief to handle uncertainty in AI. Then, the decision theory combines probability and utility theory to arrive at some final actions. The FL assigns a degree of 'belongingness' and permits incorporation of human experts' knowledge into the rule-base. In psychology, we ask questions: how do humans think and act?; it is a study of human reason(-ing) and acting, and thus provides reasoning models for AI. Thus, in this field of study, humans and other animals can be thought of as information processing/gathering/giving machines to support the developmental ideas for AI. In the areas of computer engineering, we work to develop efficient computers, algorithms, time-sharing concepts, object-oriented programs, and concurrent processing, and even build parallel computing machines/algorithms based on the architectures of recurrent neural networks (RNNs). The control theory and cybernetics try to answer the question of how artefacts can operate under their own control?; these artefacts tend/try to adjust their actions to do better for the environment over the process-time based on an objective function/performance criterion, and feedback from the environment. These aspects are also applicable to language, vision, and planning. As for the linguistics, for understanding natural languages for the purpose of use in AI, different approaches can be adopted/adapted from the linguist work: syntactic, semantic, and knowledge representation.

Some merits of AI are: i) more powerful, more useful computers have/can come out, ii) the (human–machine) interfaces have improved, iii) very difficult and new problems are being solved, and iv) information is handled in better ways, and relieves the information overload, because it is converted into knowledge (compact mathematical and functionality process models, like the data fusion process model). The demerits of AI are: i) costs are high/increased, ii) the SW development is slow/expensive, and iii) experienced programmers are few.

## 1.3 REPRESENTATION OF KNOWLEDGE AND LEARNING FOR/IN AI

Acquiring information (and data), learning (from the data and experience), knowledge representation (shaping the gained/learned knowledge), and reasoning (providing logic) (KRR) are the most important aspects in AI. The point is that if we are going to act in a rational manner, then we should be able to describe that environment and draw some inferences by representing the acquired information in a useable form. Questions in the domain of KRR are: i) how to describe the existing knowledge, ii) how to do it concisely, iii) how do we do it right, iv) how do we generate new knowledge, and v) how do we deal with the uncertainty (or uncertain knowledge, i.e., the one with large dispersion, large covariance, and lower Fisher's information matrix—FIM). Since learning from the presented data/information is very important in AI, we dwell more on this facet.

### 1.3.1 Learning

The idea here is to generate new facts/concepts/inferences from the old/previous facts. We must be able to distinguish different situations in the new environments. So, if a system is going to (rationally) act, then it must be able to change its actions with new experience, i.e., act like a learned person. This requires that to generate an intelligent behaviour, the system has to interact with the environment. The properly developed intelligent systems are expected to: i) accept inputs from sensors (sensory inputs, sound, vision,…), ii) interact with humans (to understand language, recognize speech, generate text/speech/graphics), and iii) modify the environment. Learning: i) is essential for unknown environment (since the designer does not have omniscience/all-pervading knowledge), ii) is useful as a method of constructing/developing a system (an agent is exposed to reality, new environment), and iii) modifies the agents' decision-making approaches to improve the performance of the AI systems. Thus, the learning brings about: i) a persistent change in performance due to interaction with the environment, ii) a permanent change in knowledge/behaviour due to experience, and iii) an enduring change in capacity due to practice.

### 1.3.2 Theory and Approaches of Learning

Theory of learning encompasses: i) behaviourism, ii) cognitivism, iii) social learning aspects, iv) social constructivism, v) multiple intelligences, and vi) brain-based learning. Learning is defined as an outward expression (showing off/announcing, displaying skills), and focuses on observable behaviour. It is context dependent, and is based on classical/operant conditioning: reflexes, and feedback/reinforcement.

Cognitivism grew in response to behaviourism, based on the concept that knowledge is stored cognitively as symbols. Thus, learning becomes a process of connecting symbols in a meaningful and memorable way. So, the study of learning focused on the mental processes that facilitate symbol-connections. The cognitive learning theory is based on: i) discovery learning, and ii) meaningful verbal learning. The features of the discovery learning are: i) one can learn anything, if it is presented in terms that he/she can understand, and ii) confront the learner with problems and help find the solutions. The meaningful verbal learning has a feature: any new material is presented in a systematic way, and then, it is connected to existing structures (mathematical modelling) in a meaningful way, the latter aspect also has to be defined. If the learner has difficulty with a new presented fact/data, then one needs to go back to the concrete anchors, i.e., advance organizers, and then provide a discovery approach to the learner, and the results of the learning will be enhanced. Cognitive learning has some criticism: i) I/O models are mechanistic/deterministic, ii) it does not account adequately for individuality, and iii) there is little emphasis on affective characteristics.

The aspects/features in/of/for brain-based learning (BBL) that grew out of neuroscience, and constructivism, are: i) the brain is a parallel processor (biological neural networks, BNNs, are massively parallel and naturally adaptive circuits/systems), ii) the entire body is learning (from five sensors: seeing/vision, hearing/acoustic, taste, smell, and touch), iii) it is a search for meaning and patterning, iv) emotions are critical, v) the processing is in parts as well as in wholes, vi) focused attention and peripheral perception, vii) it is conscious as well as unconscious (in the sense of unawareness) process, and viii) each brain is unique, and has several types of memories. The BBL provides opportunities for group learning and for self-expression and making personal connections to the content; it has to take into account the regular effect of (multi-sensory) environmental changes.

The main problems of BBL are: i) research is conducted by neuroscientists, and less by the educational/teaching researchers, and ii) lack of understanding of the brain itself makes BBL questionable.

As far as knowledge in learning is concerned, for logical learning, we do not need a priori knowledge, but then we need to incorporate any available knowledge, otherwise the effort of the previous knowledge is a waste. So, we need to have the approach of inductive learning in the logical setting. The approach is to find a hypothesis that explains the classifications of the examples given their descriptions: that is, the hypothesis maps descriptions to classifications.

In a learning process one can use feedback: i) supervised learning, correct answers for each example; ii) unsupervised learning, correct answers not given; and iii) reinforcement learning, occasional rewards. Function learning is a learning of a function from the examples, say ‘f’ is a target/desired function: for pair (x,f(x)), find a hypothesis ‘h’ such that ‘h’ is nearly ‘f’, given a training set of examples (data), the curve fitting, as seen from Figure 1.4, if ‘h’ agrees with all the examples presented, then ‘h’ is consistent.

In computational learning theory we are seeking general laws that govern learning, and the features are: i) sample (examples) complexity, how many sets we need to learn a hypothesis successfully (performance), ii) computational complexity needed to learn (time and efficiency), and iii) mistakes in learning, errors of fitting the

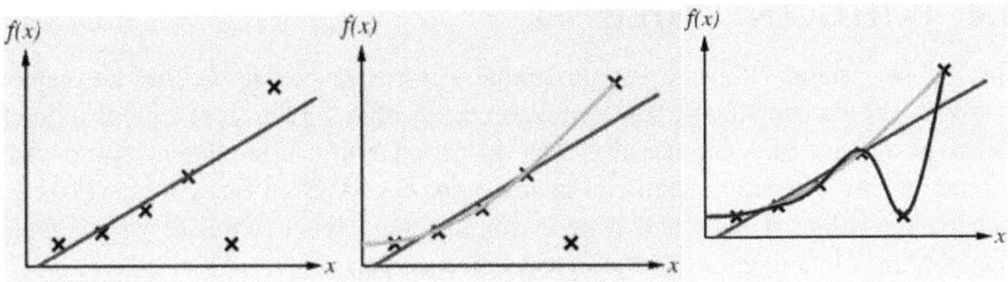


**FIGURE 1.4** Learning by induction: Curve fitting, learning (that is fitting) a function from (to) the presented/empirical data (for the third case a better/nonlinear hypothesis/model is required for the given data)

hypothesis to the function (performance metrics). In the case of statistical learning we need to include the agents to handle uncertainty by using probability/decision theory/fuzzy logic, since these uncertainties abound in nature and hence, in the data/examples presented to the learner.

Bayesian learning calculates probability of each hypothesis given the data, and makes predictions based on this, i.e., computes the a posteriori probability, from which some recursive solutions are required to be found. This learning is based on all the hypotheses that are weighted by their probabilities, and then the predictions are made. Since, the problem space is very large, one resorts to approximations. The approaches are: (i) maximum a posteriori (MAP), which requires prior probability information, and (ii) maximum likelihood (ML), in which the prior probability is assumed to be uniform and therefore need not be considered explicitly. As a result, when a large amount of data is available, ML provides a good fit to the observed data. For this reason, ML is often regarded as a non-Bayesian statistical learning method, since prior probability is not taken into account, yet it produces reasonable results when all hypotheses are of similar complexity. The ML approach is the most widely used method for parameter estimation, especially in applications involving aircraft and helicopters.

The full Bayesian learning gives the best prediction, but it is a complex process (like the filter error method for parameter estimation) and often intractable. The MAP selects a single best hypothesis and the prior is still used. The ML assumes a uniform prior; the process is: i) choose parameterized family of mathematical models to describe the data, ii) write down the likelihood of the data as a function of parameters, iii) take the derivative of the log-likelihood with regard to each parameter, and iv) determine the parameters values such that the derivatives are zero; leading to ML estimates.

Most applications of AI are/could be for: i) autonomous planning and scheduling (autonomous rovers, telescope scheduling), ii) analysis of data from space-based missions, iii) image-guided surgery, iv) image analysis and enhancement, v) autonomous vehicle control, vi) pedestrian detection, vii) games, viii) bioinformatics (gene expression data analysis, prediction of protein structures), ix) text classification/document sorting (web pages, e-mails, news articles), and x) video image classification (music composition, picture drawing, natural language processing).

## 1.4 INTELLIGENT SYSTEMS (IS)

Intelligent systems (ISs) are computationally intensive paradigms that are nature-inspired, yet mathematically sound and are increasingly finding applications in many fields of science and engineering. The so called artificial-intelligent approaches, including soft computing (artificial neural networks (ANNs), fuzzy logic (FL), and genetic algorithms (GAs)), and systems employ computers to emulate various faculties of human intelligence and biological functions [2].

The intelligent systems are suited for: i) search and optimization, ii) pattern recognition and matching, iii) planning and uncertainty management, and iv) control, and adaptation.

In essence, the intelligence is often identified/specified/signified in terms of: i) competence (ability to do a task in a better way), ii) expertise (real good working knowledge of carrying out a task), iii) talent (much more than average ability in doing many things), iv) schooling (training), v) IQ (intelligent quotient-cognitive intelligence), and vi) social interaction (mutual behavioural aspects).

In an alternative way, it is signified as: i) cognitive intelligence, ii) emotional intelligence, and iii) spiritual intelligence. However, from the computational point-of-view, the intelligence is characterized by its: i) flexibility (expanding and accommodating ability), ii) adaptability (meet the changing environment), iii) memory (storage capacity, remembering from the experience), iv) learning (new strategies), v) temporal dynamics (adopting to time dimension, time varying aspect), vi) reasoning (thinking and decision-making ability), and vii) the ability to manage uncertain/imprecise information; it is in the latter aspect in which soft computing, especially fuzzy logic technology, plays a very significant role.

Many important features of such ISs are: i) information gathering (collecting data and acquiring the knowledge), ii) understanding (assimilating), iii) making inferences (deciphering), and iv) applying it to understand and solve new problems efficiently. In AI we have: i) to rationally think as humans do, and ii) to rationally act as humans act. An IS includes intelligent behaviour as seen in nature as a whole: i) biological genetics (distributed information), ii) evolution (survivability), iii) chaos (structured randomness), and iv) natural adaptation (sufficability). The idea in IS, supported by AI is to find solutions that are better than their predecessors, and not necessarily the optimal ones. In a simpler way the IS should mimic some aspects of intelligence exhibited by nature: i) learning (from the presented data/information), ii) adaptability (meet the demands of changes in the system), iii) robustness across problem domains (assured performance in the presence of the uncertainty), iv) improving efficiency (more useful output than the input over time and/or space), v) information compression (data to knowledge), and vi) extrapolated reasoning [2].

Another new or novel perceptive for the AI should be (as defined by J.R. Raol): make an attempt to build algorithms, software, machines, and/or systems that perceive, think, reason out, make decisions, and act rationally in the sense of achieving 'probably approximately correct algorithm/performance/perfect behaviour' in the areas of problem solving and learning [3]. Then gradually approach towards perfection by ever refining these artefacts. This perspective of AI is practically achievable and would give a definite measure of evaluating the performance of the AI systems and one can calibrate the nearness of it to the human's thinking and acting; that it gives a means of measure in the sense of achieving the higher and higher probability in the sense of/for lesser and lesser approximation, higher probability for more and more correct computational algorithm: for learning, fuzzy logic-based inferences, and decision-making.

### 1.4.1 ISs in Aerospace Engineering

Some novel applications of IS are: i) spacecraft autonomy, ii) aircraft control (fault diagnosis, fuzzy gain scheduling), iii) math modelling, iv) airfoil design, v) satellite operations, vi) missile design, and vii) vehicle health management. Using ISs, complex/sophisticated ideas can be implemented and tested with rapid development cycles. The role of ISs in aerospace engineering is: i) function as intelligent assistants to augment human expertise (e.g., pilots' situation assessment), and ii) act as a substitute for human expertise in endeavours that save cost, time, and life.

#### 1.4.1.1 ISs for Modelling

Intelligent systems provide two aspects for modelling: i) generalization implies that the model could be used not only to represent just the data gathered but the knowledge that the data represents, and ii) robustness is the system's ability to perform within certain bounds of its nominal performance in spite of uncertainty, that is however bounded. The techniques of ANNs, FL, GA, and expert systems can be used by aerospace engineers for modelling. Knowledge representation, in general contains syntax and semantics, and includes [2]:

- Mathematical equations (ARMA, state space models)
- Rule-based/expert systems/inferencing system (based on reasoning) consists of: a knowledge base (a set of rules/known facts), acquired data (derived facts/data), and an inference engine.
- Fuzzy modelling, consisting of: i) system I/O variables, ii) fuzzy membership functions, iii) fuzzy implication functions, iv) aggregation functions, and v) defuzzification methods. Attractive features of this are: i) reduced design complexity, ii) rapid prototyping, iii) flexibility, iv) simplicity, and v) cost effectiveness.
- Neural models: ANNs are brain/BNNs-inspired connectionist models that consist of many similar linear and nonlinear computational elements called neurons connected in parallel and complex patterns. These simple neurons have the ability to perform tasks such as memory recall, pattern recognition, and learning.
- Tree structure: an algorithm for placing and locating objects in a database; it finds data by repeatedly making choices at decision points called nodes.

Intelligent modelling tools are used for operator behaviour modelling, flight test data analysis/modelling, and aerodynamic modelling for design.

#### 1.4.1.2 Improved Robustness Using Fuzzification

The technique of the fuzzy granularization combined with the knowledge available through an existing linear dynamic controller provides an outer fuzzy shell to existing control techniques.

The fuzzification of the inputs and outputs might lead to better robustness to system uncertainties via the manipulation of the distance between the membership functions.

#### 1.4.1.3 Intelligent Systems for Search

Search techniques are employed in combinatorial optimization, and consist of: i) an initial state, ii) a set of operators, iii) application of the operators produces a sequence of new states called the path, and iv) a goal test is defined and tested to determine when a new state becomes the desired goal state. A general combinatorial task involves deciding: i) the specifications of those entities (what type of aircraft, the opponent's aircraft descriptions, identification number of aircraft), and ii) the way in which those entities are brought together (various elementary tactics formations with their relative positions). A few search techniques used by the ISs are [2]:

- Golden section: the interval is reduced by throwing out regions that do not contain the optimum.
- Breadth-first search: search a state space by constructing a tree consisting of a set of leaves and branches, the algorithm defines a way to move through the tree structure.
- Depth-first search: before going deeper, the algorithm follows a single branch of the tree down as many levels as possible until a solution/dead-end is reached.
- Heuristic dynamic programming (HDP): based on an attempt to approximate the Bellman equation, variations of HDP can be used in solving combinatorial optimization problems.
- A*Search: combines the cost estimate $J(n)$ of traversing from step $n$ to the goal state and $U(n)$ which is the known path cost from the start node to step $n$ [3].
- Genetic algorithm: a type of evolutionary computation that uses a genetic/evolutionary metaphor.

For aerospace applications, intelligent search techniques are used for: i) searching a design space, ii) searching for an optimal scheduling and planning schemes, and iii) solving combinatorial optimization problems.

#### 1.4.1.4 Search for Optimal Air Combat Tactics

In the case of a four-airplane division of two elements, each element consists of two aircraft, treated as a unit, and this hierarchical concept can be used to develop a GA-based approach to search for optimal air-combat tactics. The tactics implementation proceeds as:

- Specify a set of commonly used element, the division formations, the underlying tactical manoeuvres, and attack tactics.
- Formulate a set of principles for aggregating the small formation tactics for large engagements, and implement a method for doing this in the GA algorithm.
- Use the resultant formation tactics to drive the engagement, and use the performance metric generator, and search for the best engagement tactics so as to optimize these metrics.

### 1.4.2 IS for Design

In aerospace systems, the design is interdisciplinary: wing (aerodynamic), propulsion (engine control), and automatic control (autopilot, stability augmentation) designs might need extensive redesign (design iterations) for arriving at an integrated design. ISs can help in many ways: i) neural networks and fuzzy logic for representation of known data/designer expertise (mathematical modelling), and ii) genetic algorithms as optimization procedures for searching through large spaces with several constraints (of systems' dynamics and stability), and subsystem interactions. For example, to maximize engine performance under degraded conditions, a fault tolerant engine control scheme can be applied, for which an engine performance estimator using a combination of a genetic algorithm (GA) and RBN for the implementation can be designed. The usual input selection, executed by simple inspection of data files, can be replaced by automatic inspection using GAs. The performance can be evaluated by an objective function (PI, performance index), the error is defined as the difference between desired performance and the current design's performance.

## 1.5 INTELLIGENT CONTROL (IC)

The two important aspects to intelligent control (IC) are: i) the 'intelligence' to analyze the changing environment, i.e., the analytical ability to comprehend/react to the changing environment; and ii) the resources to respond to the changing environment, i.e., the physical components of the system that are necessary to react to the environment. One application is: maintain performance under any of the situations: i) loss of control due to failure, ii) aircraft characteristics change due to damage (C.G., or mass/inertia), and/or iii) environmental effects due to wind and turbulence. The IC encompasses: i) conventional control-optimal, robust, stochastic, linear, and nonlinear, and ii) fuzzy, genetic, and neuro-control technologies.

### 1.5.1 Levels of Intelligent Control

An IC design is stated as: given the dynamic system

$$x(t + 1) = f(x(t),u(t),t) + w; \tag{1.1}$$

In (1.1), w is an unknown disturbance. A set of goals is generated:

$$x_g(t +1) = g(x_g(t), x(t), t) \tag{1.2}$$

A performance measure is defined as:

$$J(t+1) = \sum_{t=0}^{T} F(x_g(t), x(t), u(t), t) \tag{1.3}$$

A planning function is specified as

$$P(t+1) = p(x(t),P(t),t,\nu) \tag{1.4}$$

In (1.4), v is system faults and emergencies. The intelligent controller needs to arrive at a control u(t), such that the system: i) is locally stable (includes requirements of handling quality of the system), ii) follows closely the desired path (closeness defined by a PI), iii) constantly optimizes long-term and short-term goals, and iv) reacts to changing environments by properly adapting the planning functionality [2]. A practical way to accommodate the IC is to have various levels of capabilities for self-improvement (using some adaptive mechanism, and switching based on some logic, say fuzzy logic), which is also an important goal of human intelligence. The IC is defined with various levels of intelligence (control) so that it will accommodate easily the contributions from: cognitive science, computer hardware, sensors/actuators, learning theory, and newer control architectures. For various tasks in the levels of self-improvement, the designer of control systems can use the combination of the existing control methods augmented by the soft-computing-based paradigms.

### 1.5.2 Level Self-improvement of Description

Some main levels for IC and self-improvement are briefly stated here [2], in all or some of which ANNs, FL, and/or GA-based approaches can be used appropriately:

*Level 0 IC–a robust controller*: Tracking Error (TE) with robust feedback control tends to zero, self-improvement of TE is an important goal of many control techniques. To achieve this, one designs robust feedback controllers with constant gains that improve the error (i.e., reduce the errors) as time goes to infinity.

*Level 1 IC–an adaptive controller*: TE + Control Parameters (CP), robust feedback control with adaptive control parameters (error tends to zero for non-nominal operations); feedback control is self-improving of control parameters towards the goal of achieving better tracking error or some error oriented goal, is the next level in intelligent control.

*Level 2 IC–an optimal controller*: TE+CP+Performance Measure (PM), optimal control: robust, adaptive feedback control that minimizes or maximizes a utility function over time. Self-improvement of an estimate of the performance error towards the goal of optimization of an utility function over time (error tends to zero and a PI is optimized) is the next level in IC.

*Level 3 IC-a planning controller*: TE+CP+PM+ Planning function: Level 2 + the ability to plan ahead of time for uncertain situations, simulate, and model uncertainties. This Level 3 IC includes self-improvement of planning functions for contingency, emergencies, and faults. These planning functions could be static for Level 2 but should be self-improving for Level 3.

## 1.6 INTELLIGENT FLIGHT CONTROL – A BRIEF NOTE

In most flight control systems for aerospace vehicles one finds extensive/extended application of the classical control method, and/or the so called modern control approaches in selective cases. The aim of design and development of complex need-based/sophisticated control for flight (-dynamics) is: i) to improve the otherwise poor static stability or low damping in dynamics for a particular axis, ii) to provide enhanced stability to inherently unstable (or relaxed stability) aircraft/dynamics, iii) to improve the pilot-aircraft interactions/coupling, iv) to improve the safety and reliability of the aircraft's functions (in presence of certain expected/limited failures), and v) to reduce the workload of the pilot in handling the complex flight missions/ operations and tasks, e.g., use of an autopilot.

The conventional methods have been very successfully applied to the problems of flight control, guidance, and navigation for variety of aircrafts in the last several decades. However, recently there has been an upward trend of applications of artificial neural networks (ANN), fuzzy logic/system (FL/S), and GAs to the design of flight control systems for aircraft, missiles, UAVs (unmanned aerial vehicles) and MAVs (micro/mini air vehicles), along with use of allied and supportive technologies: i) parameter/state estimation, ii) H-infinity control/filtering, iii) object-oriented programming (OOP), and iv) handling quality evaluation of pilot-aircraft interaction. We feel that some more rigorous blending and cross-fertilization of these approaches with conventional/classical approaches can emerge as do-able, useful, robust, and yet intelligent methods for the design, evaluation, and synthesis of flight control systems [3,4]. The ANNs, FLSs, and GAs that are called the major components of the soft computing (SC) are crucial subsystems/ingredients of a successfully built AI system. An AI system would use all the SC paradigms in an integrated way of providing: i) the structure (ANNs), ii) incorporation of knowledge (FLSs), and iii) optimization/ robustness (GAs) of the strategies.

### 1.6.1 ANN-based Systems for Flight Control

ANNs are emerging paradigms for solving complex/involved problems in sciences and engineering and have salient features: i) they mimic the behaviour of the human brain, ii) they have massively parallel architecture/structures, iii) they can be represented by adaptive circuits with input channel/s, weights (parameters/coefficients), one or two hidden layers and output channel/s with some nonlinearities, iv) the weights can be tuned to obtain the optimum performance of the ANN – be it modelling of a dynamic system or nonlinear curve fitting, v) it requires the use of a training algorithm to determine the weights (where the acquired knowledge is stored), vi) they can have a feedback type arrangement within the neuronal structure, vii) the trained network can be used for predicting the behaviour of the dynamic system that is modelled using the ANN, viii) the ANN can be easily coded/validated using standard software procedures, ix) optimally structured ANNs can be hard-wired/firm-wired and embedded into a chip for practical applications – this will be the generalization of the erstwhile analogue circuits-cum-computers, and x) then the neural network-based system can be truly termed as a new generation powerful and massively parallel computer.

An aerodynamic model (used for design of flight control laws) could be highly nonlinear and dependent on many physical variables, which might be difficult to model and to have an explicit math-model; the difference/s between the math-model/s and the real system/s' data would cause performance degradation of the flight control system. To alleviate this problem, an ANN can be used and the weights of the NW are adjusted so as to compensate for the effect of the (initial) modelling errors. The ways in which ANNs can be used for control augmentation are: i) the normal control can be aided by a RBFNN controller for online learning to represent the local inverse dynamics of an aircraft, ii) to compensate for uncertainty without explicitly identifying changes in the aircraft model, iii) the ANN nonlinearity, the sigmoidal function can be made adaptive and used in the desired dynamics block of the flight controller (say MRAC), iv) the learning ability can be incorporated into the gain scheduling of the controller, and v) sensor and/or actuator failure detection and management. Major merits would be: i) the controller becomes more robust and more insensitive to the plant parameter variations, and ii) the online learning ability would be useful in handling certain unexpected behaviour.

### 1.6.2 Fuzzy Logic Systems (FLSs) for Flight Control

The salient features of FLSs are: i) they are based on multivalued logic, ii) they do not have any specific architecture like ANN, iii) they are based on certain rules which need to be a priori specified, iv) fuzzy logic is a machine intelligent approach in which desired behaviour can be specified in rules in which an expert's (design engineer's) experience can be captured/incorporated, v) it deals with approximate reasoning in uncertain situations where truth is a matter of degree, and vi) it is based on the computational mechanism (algorithm) with which decisions can be inferred, in spite of incomplete knowledge, the process of inference engine can be used for decision-making and control.

FL-based control is suited to multivariable and non-linear processes and can achieve desired control action. The heuristic fuzzy control does not require deep knowledge of the controlled process, but, the control policy should be known a priori; however, an improvement to this type of controller can be made by an application of (math-) model-based fuzzy system, wherein the model of the controlled process is assumed to be known and used. This will require assuring the stability of the FL-based flight control system. FLS can be used to augment the flight control system in certain ways: i) FL will approximately duplicate some of the ways a pilot might respond to an aircraft (useful in case of a damage or failure), ii) the rules can be used to create a nonlinear response with low damping for large errors and high damping for small error, iii) to incorporate the complex nonlinear strategies based on pilot or system design engineer's experience and intelligence within the control law, iv) adaptive fuzzy gain scheduling (AGS), and v) FL-based adaptive tuning of Kalman filter (KF) for adaptive estimation/control.

A brief application review is presented in Table 1.1, and is just an indicative, and not exhaustive of various possibilities [5]-[9].

### 1.6.3 Associated Technologies for Flight Control

Certain associated technologies (for which certain paradigms of soft computing can be used) for flight control also play a key role in understanding the behaviour of flight control systems and are briefly highlighted next:

**TABLE 1.1**
**A brief description on applications of ANNs and FLS to flight control**

| Approach | Aircraft and/or problem addressed | Inferences drawn | Ref. |
|---|---|---|---|
| FFNN-online learning | F-8; an autopilot mode to track the pilot's throttle/ pitch rate command | | [5] |
| RBFNN (radial basis function NN-variable Gaussian functions | F-8; online learning to represent the local inverse dynamics | Compact ANN structure; small tracking error | [6] |
| ANN | F-16; to compensate for the effect of the aerodynamic modelling error | The errors exponentially converge to a compact set | [7] |
| FL | Guided missile; adaptive fuzzy gain scheduling | High degree of automation; better terminal guidance performance; less control effort; a smaller miss distance. | [8] |
| ANN; adaptive nonlinearity | U.S. Navy; U.S. AF & NASA program on X-36/ Xv-15 | Compensates an uncertainty without explicitly identifying changes in the aircraft model | [9] |
| FL | E-6A aircraft in a limited way; simulation mode for F-18 automatic carrier landing system | Allows the incorporation of complex nonlinear strategies based on pilot's and/or design engineer's intelligence in the control law | [9] |

***System modelling, identification, and parameter estimation*:** Math-modelling of aerospace vehicles is very important since many applications require such information in the form of aerodynamic (stability and control) derivatives that appear in the model: i) to explain aerodynamic, stability and control behaviour of the flight vehicle, ii) for design of flight control systems, and iii) for high-fidelity flight simulators. Determination of aerodynamic derivatives of an aircraft with a fly-by-wire control system (FBWCS for unstable aircraft) poses new situations for identification and parameter estimation problem: when identification experiments are performed with an aircraft operating in the closed loop, the feedback introduces correlations among the I/O variables. This correlation causes the identifiability problems that would render estimates of some derivatives uncertain and unclear. Also, due to the feedback action, that would be constantly trying to generate controlled responses, the measured responses would not display the modes of the (original) vehicle, causing an identifiability issue. Also, the complexity of the problem increases when the basic aircraft is unstable, due to the fact that the numerical integration of the state model (in the estimation program) leads to numerical divergence. The problem is further compounded if these data are noisy. This is a fertile field for application of ANNs and FLSs-based methods for system identification, parameter estimation, and state estimation for: i) modelling pilot control activity and behaviour in several mission-oriented tasks to quantitatively evaluate his/her workload and supervisory effectiveness in a FBW aircraft, and ii) modelling of the aircraft from a control requirement point of view can also be done using the system identification methods.

***H-infinity control and filtering:*** In the HI (H-infinity) concept, the minimum of the worst case error (i.e., the least of the maximum error) is sought in order to arrive at the robust (sub-optimal) solution to control/filtering problem. Specific studies useful to flight control are: i) concept of gain and phase margin, ii) H-infinity filtering for generalizations of the conventional/KF algorithms, and iii) combination of H-2 (Kalman filter) and HI filtering algorithms.

***The object-oriented programming (OOP) in control system design:*** The OOP has a good potential application to the design of flight control system, and the merits are: i) use of data modelling concept, ii) encapsulation of data and functions, and iii) OOP makes software decomposition like the decomposition of a dynamical system. The 'objects' provide a much more natural basis for a good CACSD (computer-aided control system design/SW tool) than the conventionally used simple data types and functions. The aspect is that the OO-paradigm can be used in the field of CACSD, i.e., the package would provide a set of data structures adequate to describe the 'objects' that are found in control system design; for example, the fundamental 'objects' are 'system' and 'results': a model in the form of a set of equations is a 'system' and the time and frequency responses are a 'result'.

A synergy of a typical use of ANN-FL-based systems for flight control with some associated technologies is depicted in Figure 1.5.

### 1.6.4 Recent Intelligent Flight Control System

NASA's intelligent flight control system (IFCS) [10] is, perhaps, one of the latest developments in this direction; i.e., to effectively optimize aircraft performance in normal as well as failure conditions. It is designed to incorporate self-learning ANN into the flight control software that enables the pilot to maintain the control and to safely land the aircraft, in case the test aircraft suffers a failure of a control surface/damage to the airframe. The IFCS is an integration of the ANN technology and the state-of-the-art control strategies/algorithms that identify the changes in the aircraft stability/control characteristics in case of failure; it responds to such changes and immediately adjusts to maintain the best possible performance of the aircraft-flight. The adaptive ANN-SW learns the new changed flight characteristics in real time, and helps the pilot regain the control of the aircraft and/or prevent an accident; the ANN strategy is directly applied to FCS-feedback errors. The ANNs are also used/validated to organize/map the aerodynamic changes and these are then fed to FCS, which in turn uses this information to stabilize the aircraft. In some cases, the ANN-based strategy can be used in the direct adaptive mode to continuously apply the required

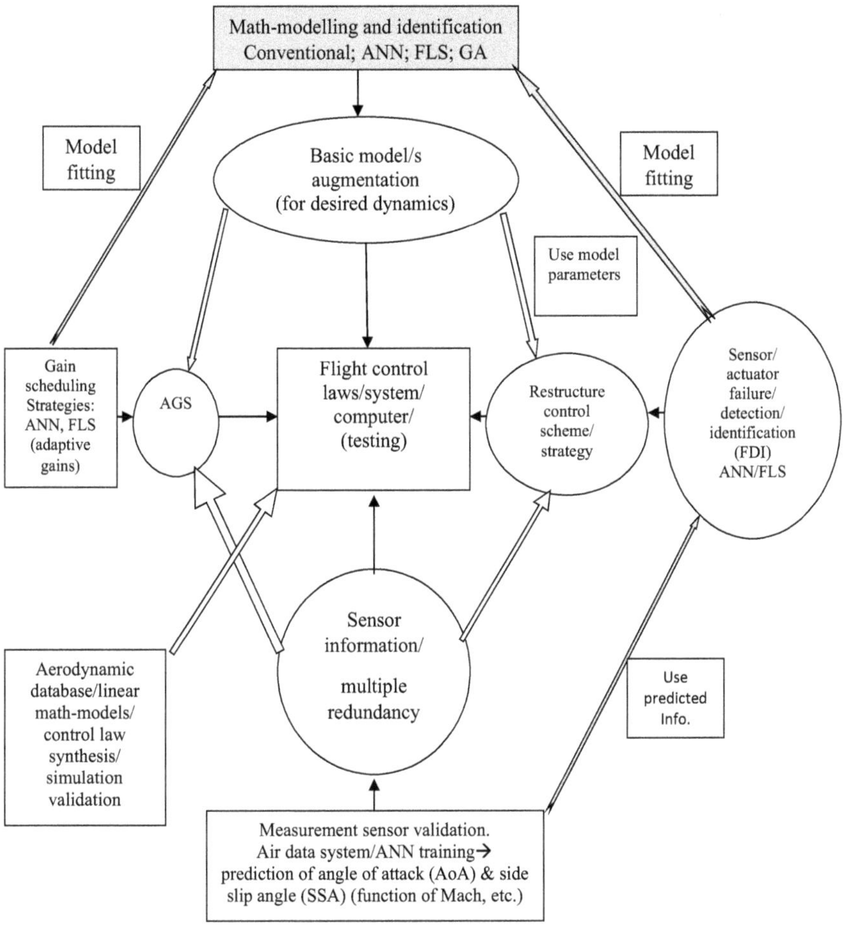


**FIGURE 1.5** A concept of ANN-FLS(-GA) augmented, aided intelligent-flight control system: simulation and evaluation (adapted/modified from [13])

corrections to FCS and also learn by observing the effects on the performance of the aircraft behaviour.

## 1.7 CLASSICAL VS INTELLIGENT CONTROL

We can now summarize some important differences between the classical and intelligent control. For developing a classical control system (CCS, including modern control) to control a plant (process, dynamic system, or an agent), the designer constructs a *mathematical model* (MM) of the system. This MM contains all (or should contain most) of the dynamics of the plant that affect controlling the very plant [11]. This form of control can be called the mathematicians' approach, because the designer should mathematically model the plant that is to be controlled. Whereas, for developing an intelligent control system (ICS) to control an agent (plant), the designer *inputs the system behaviour* and the ICS *abstractly models* the plant. This type of control is called the lazy man's approach, because the designer doesn't need to know the internal dynamics of the plant. It must be emphasized here, that in many cases, the plant may be too complex to be modelled. Of course, for general systems and intelligent systems (based on AI), there are many more differences than highlighted here from the control-point-of-view.

One can use the notion of 'shifting' intelligence: i) for CCS, the designer must model the system to be controlled and therefore the intelligence/knowledge lies with her/him; the intelligence is shifted towards the designer, and ii) for ICS, the software abstractly models the system and therefore the intelligence/knowledge lies with it; the intelligence is shifted towards the software. However, it must be noted here that the designer must have some knowledge of the plant, since he is ultimately designing the system; it is just that s/he does not need enough knowledge to develop an accurate model of the plant.

Here we just capture main aspects of ICS; which is that the system that has to be controlled does not have to be rigidly modelled. The designer inputs the appropriate stimuli to the ICS, and evaluates its performance (on its outputs), the ICS itself develops a model (now it is not a MM in the sense of the classical sense) of the plant to be controlled; of course, for the ICS the modelling is accomplished by ANNs and FL.

Humans perform many complex tasks without knowing exactly how they do these, often in a very simple way. The types of ICS approaches popularly studied and applied are: i) ANNs, ii) fuzzy logic, iii) genetic algorithms/programming, iv) reinforcement learning, and v) support vector machines. Now, let us dwell on how some of these are called the ingredients of the intelligent systems, and help build AI-based machines/computers and control systems.

First, let us take the ANNs. These are constructed so as to mimic some (functional) behaviour of the so-called biological neural networks (BNNs). The functional structure of an ANN is such that several simple neuronal units/circuits are connected in parallel, and these are in millions; the knowledge is stored in the input weights (synapses) of the neurons, and adjusting these weights gives the neurons the ability to store different information. The ANN is presented with the data/examples and trained to learn, say, patterns. Once, it is trained well, it is able to recognize the patterns and works or acts as humans act for the same job. Similarly, ANNs can be trained to do many complex tasks, and the learned ANNs can then perform these tasks by themselves. In this way ANNs are relatively close(-er) to human's functionality, and can perform the tasks that humans do, and thus the ANNs are called an ingredient of AI.

Now, let us take the case of FL which is modelled on the basis of how the reasoning goes on in the human brain; the human's reasoning is approximate, non-quantitative, and non-binary, i.e., it is multi-valued. The FL's inference process is devised such that it is able to incorporate the heuristic knowledge of the human expert (about the process/plant/environment), and also fuzzifies the input variables and takes into account the ranges of the variables rather than the crisp values, i.e., it assigns a degree of 'belongingness' of a variable to certain (fuzzy) sets. Thus the FL works with imprecise information, due to uncertainty, and still makes decisions like humans do in the similar situations.

Genetic programming (GP) is derived from GA, except that the output of GP is another program. The GA is based on the evolutionary strategy of mating-crossover-reproduction-retaining of the fittest offsprings and propagation, and repeating the process; over a period of time (from a few seconds/days to thousands of years, depending up on the type of species), the process produces evolutionary stable strategies (ESS) for the survival of the species. Now, in this process of EP (evolutionary process/programming), there is no 'explicit' intelligence seen working, but as we can conjecture, there should be some 'implicit' intelligence working or playing a role, without which such an awesome task cannot be performed. Thus, GA/GP are safely considered as a part of AI. The GA in fact is very simple, and yet very powerful in solving optimization problems, and can lead to robust solutions, if not optimal ones. The idea is to create new (GP) programs to solve a control problem based on the ones that work best now. The steps to implementing GP are : i) to generate a random group of functions and terminals, each random group will constitute a computer program, functions are operators such as +, -, *, /, etc. Terminals are the I/Os to the problem; ii) to execute each program and assign a number (called a fitness value) to each program to see how well it did in solving the problem; iii) to create a new population via mutation, crossover, and the most fit programs, mutation involves randomly changing functions and terminals of a program.

Reinforcement learning involves presenting an agent (a robot) with a set of action choices, (initially equally rewarding). The agent makes choices, based on a behavioural policy, and receives rewards or punishments via a reward function based on the choices made. In the process, the rewards/punishments affect how the agent will make its choices in future encounters, and thus it learns what choices are rewarding and which are not.

## 1.8 INTELLIGENT ESTIMATION (IE)

Intelligent estimation (IE) refers to a method of making estimations that leverages advanced data analysis techniques, often utilizing machine learning (ML) or artificial intelligence (AI) to produce more accurate and informed approximations, particularly in situations with complex or incomplete information (here soft computing techniques are used), going beyond simple calculations or rule-based estimations; essentially, using intelligent algorithms to improve the quality of estimations. The main points about intelligent estimation are:

i) Data-driven: It relies on large datasets to train models that can learn patterns and relationships, leading to better predictions;
ii) Adaptive: It can adjust its estimations based on new data and changing circumstances, making it more dynamic than traditional methods; and
iii) Complex scenarios: It can handle situations with multiple variables and uncertainties, which might be difficult to estimate manually.

Applications of intelligent estimation are:

i) Software development: Estimating project timelines and costs with greater accuracy by analyzing historical data and project complexity;
ii) Supply chain management: Predicting demand and inventory levels based on market trends and past sales data;

iii) Manufacturing: Estimating production costs and potential risks based on machine performance and material usage; and
iv) Healthcare: Predicting disease progression or treatment outcomes based on patient data.

The much earlier introduction and discussion on this aspect was in Chapter 10 of [12], wherein parameter estimation methods using FFNNs, RNNs, and GA were discussed, and illustrated using MATLAB-based simulated data/examples. Similarities and contra-distinctions among four RNN structures were also brought out.

In this book we present methods for system identification, parameter estimation, and state estimation (including filtering which is a special case of general estimation theory) based on soft computing techniques of AI: ANNs, FL, and GAs. This

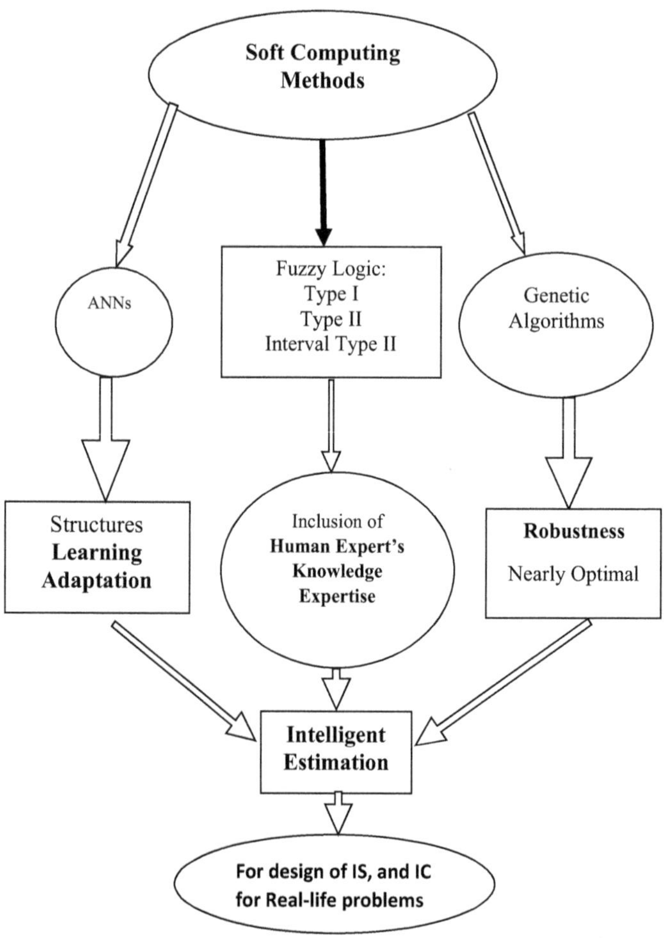


**FIGURE 1.6** Soft Computing: Intelligent Estimation (IE) → Intelligent Systems (IS)/ Intelligent Control (IC)

approach of math modelling would form a definite ingredient of AI systems: i) intelligent systems, and ii) intelligent control as described in the previous sections of this Introductory part of the book, as depicted in Figure 1.6.

## REFERENCES


1. Russell, S.J., and Norvig, P., *Artificial Intelligence-A Modern Approach*, Prentice Hall, Englewood Cliffs, New Jersey, USA, 1995. https://people.engr.tamu.edu/guni/csce625/slides/AI.pdf, accessed April, 2026
2. Krishna Kumar, K. "Intelligent Systems for Aerospace Engineering-An Overview", https://ti.arc.nasa.gov/m/pub-archive/364h/0364%20(Krishna).pdf
3. Raol, J.R., and Ajith, K. Gopal (Eds.). *Mobile Intelligent Autonomous Systems*, CRC Press (Taylor & Francis Informa Group), Florida, USA, 2013.
4. Raol, J.R. "Intelligent Flight Control-A Brief Survey", In *Control and Data Fusion e-Journal, CADFEJL*, (Bangalore, India), Vol. 1, No. 1, pp. 2–9, Jan-Feb. 2017.
5. Sadhukhan D., and Feteih, S. "F8 Neurocontroller Based on Dynamic Inversion", *Journal of Guidance, Control and Dynamics*, Vol. 19, No. 1, pp. 150–156, 1996.
6. Li, Y., Sundararajan, N., and Saratchandran, P., "Stable Neuro-Flight Controller Using Fully Tuned Radial Basis Function Neural Networks", *Journal of Guidance, Control and Dynamics*, Vol. 24, No. 4, pp. 665–674, July-Aug. 2001.
7. Lee, T., and Kim, Y., "Non-Linear Adaptive Flight Control Using Backstopping and Neural Networks Controller", *Journal of Guidance, Control and Dynamics*, Vol. 24, No. 4, pp. 675–682, July-Aug. 2001.
8. Lin, C.L., and Su, H.W., "Adaptive Fuzzy Gain Scheduling in Guidance System Design", *Journal of Guidance, Control and Dynamics*, Vol. 24, No. 4, pp. 683–692, July-Aug. 2001.
9. Steingerg, M.L., "Comparison of Intelligent, Adaptive and Non-Linear Flight Control Laws", *Journal of Guidance, Control and Dynamics*, Vol. 24, No. 4, pp. 693–699, July-Aug. 2001.
10. Gibbs, Y. (Ed.), *NASA Armstrong fact sheet: Intelligent Flight Control System,* March 2014. www.nasa.gov/wp-content/uploads/2021/09/120321main_fs-076-dfrc.pdf, accessed April 2026.
11. Smith, B., *Classical vs Intelligent Control. Special Topics in Robotics*, Memorial University, Newfoundland, Canada, 2002. www.scribd.com/document/375086649/Classical-vs-Intelligent-Control, accessed April 2026.
12. Raol, J.R., Girija, G., and Singh, J., *Modelling and Parameter Estimation for Dynamic Systems. IEE/IET Control Series Books*, Vol. 65, London, UK, 2004.
13. Raol, J. R. and J. Singh. *Flight Mechanics Modelling and Analysis.* CRC Press, FL. USA 2009.


## APPENDIX 1A WHAT IS INTELLECT-INTELLIGENCE (I-I)?

The I-I has varied facets: intrinsic, cognitive intelligence/intelligent quotient (CI/IQ), emotional intelligence/emotional quotient (EI/EQ), contemplative/meditative intelligence (CMI), spiritual intelligence/spiritual quotient (SI/SQ), and now evolution-conscious(-ness) intelligence (ECI). The body and mind (BM) are related by psychosomatic links, and the mind and intellect (MI) by thoughts, to make the total BMI (body-mind-intelligence).

The evolutionary nature (EN) is intelligent without being too vocal about its intelligence, it is beyond the grasp of our own intelligence. The tremendous I-I is so vividly exhibited automatically that there is not a single aspect that can be said to be devoid of even a little I-I.

What does intelligence mean?; it is: i) the memory, the repeatability, the learning; and ii) the recalling, the adaptation, the logic, and the thinking; all these work almost concurrently, and are intricately woven; the intellect/intelligence has to summon the wisdom for choosing right from wrong, and good from bad; and this wisdom is a mature, and highly evolved process of the I-I itself. Where does the wisdom come from?; from the past experiences of the self, from constant learning, total alertness, and being continually aware of the deeds and acts in this worldly forest wherein the dangerous predators roam freely to devour our own personalities.

The intrinsic intellect (often thought as given by 'God', or the 'Creator'; inherited in our biological brains) is being recognized in three ways: i) the basic physiological process(-ing) capacity: intellect governing the intelligent functioning of the body logistics; ii) the intrinsic thought process(-ing) intellect: governing our logic functioning; and iii) the intelligence governing our acquiring of information and knowledge from the external world.

All living humans have all these three ways of thinking: in different measures; low to high, with varying quality and quantum.

The intelligence further acts in three ways: i) the cognitive one governing the routine skill function and scientific/technological investigations and understandings; ii) the emotional one governing the family relations/management and human resource development/management; and iii) the spiritual one beaconing the self-development, and self-realization.

## APPENDIX 1B PARAMETER ESTIMATION AND SOFT COMPUTING

We give here, a brief summary of various parameter estimation approaches, and the needs of soft computing methods.

### 1B.1 Parameter Estimation

The maximum likelihood-output error method (MLE-OEM) utilizes output error-related cost function. It utilizes the maximum likelihood principle and Fisher's information matrix (FIM). The inverse of this matrix gives the covariance measure and hence the uncertainty bounds on the parameter estimates. Maximum likelihood estimation has nice theoretical properties [1].

The MLE-OEM, popularly known as output error method is a batch-iterative procedure. In one shot, all the measurements are handled and parameter corrections are computed. Subsequently, a new parameter estimate (a vector of parameters) is obtained. This process is again repeated with a new computation of residuals, etc. The OEM has two limitations: i) it can handle only measurement noise, and ii) for unstable systems, it might diverge. The first limitation is overcome by using a Kalman filter (KF) type formulation within the structure of the maximum likelihood output error method to handle process noise; this leads to the filter error method (FEM). In this approach, the cost function contains filtered/predicted measurements (obtained by KF) instead of the predicted measurements based on just the state-equation integration. This makes the method more complex and computationally intensive. The FEM can compete with the extended Kalman filter (EKF), which can handle process as well as measurement noises and also estimate parameters as additional states. One major advantage of the Kalman filter/extended Kalman filter is that it is a recursive technique and very suitable for on-line/real-time applications. For the latter application, a factorization filter might be very promising. One major disadvantage of Kalman filter is the filter tuning, for which the adaptive approaches need to be used. The second limitation of the OEM for unstable systems can be overcome by using the so-called stabilized output error methods (SOEM), which use measured states. This stabilizes the estimation process. Alternatively, the EKF or the extended factorization filter can be used, since it has some implicit stability property in the filtering equation. The FEM can be efficiently used for unstable/augmented systems.

Since the output error method is an iterative process, all the predicted measurements are available and the measurement covariance matrix R can be computed in each iteration. The EKF for parameter estimation could pose some problems since the covariance matrix part for the states and the parameters would be of quite different magnitudes. Another major limitation of the KF type approach is that it cannot determine the model error, although it can get good state estimates. The latter is achieved by process noise tuning. This limitation can be overcome by using the model error estimation method (MEEM); this approach provides estimation of the model error, i.e., model discrepancy with respect to time. However, it cannot handle process noise. In this sense, the model error estimation can compete with the output error method, and additionally, it can be a recursive method. However, it requires tuning like the KF. The model discrepancy needs to be fitted with another model, the parameters of which can be estimated using recursive least squares method, and then two models in cascade can be taken as a complete model of the dynamic system.

Another approach, which parallels the model error estimation, is the estimation before modelling approach (EBM); this approach has two steps: i) the EKF to estimate the states (and scale factors and bias related parameters), and ii) a regression method to estimate the parameters of the state model or related model.

The model error estimation also has two steps: i) state estimation and discrepancy estimation using the invariant embedding method, and ii) a regression method to estimate the parameters from the discrepancy time-history. Both, the EBM, and MEEM can be used for parameter estimation of non-linear system; also the OEM, and FEM can be used.

The FFNN-based approach somewhat parallels the two-step methodologies, but it is quite distinct from these: i) it first predicts the measurements, ii) then the trained network is used repeatedly to obtain differential states/measurements, and iii) the parameters are determined by the Delta method and averaging.

The RNN-based approach looks quite distinct from many approaches, but a closer look reveals that the equation error method (EEM) and the output error method (OEM)-based formulations can be solved using the RNN-based structures. In fact, the equation error method and the output error method can be so formulated without invoking RNN theory and still will look as if they are based on certain variants of the RNNs. This revealing observation is important from practical application of the RNNs for parameter estimation, especially for on-line/real-time implementation using adaptive circuits/VLSI, etc. Of course, one needs to address the problem of convergence of the RNN solutions to true parameters.

Interestingly, the parameter estimation procedure using recurrent neural network differs from that based on the feed forward neural network. In the recurrent neural network, the so-called weights (weighting matrix W) are pre-computed using the correlation like expressions between $\dot{x}, x, u$ etc. The integration of a certain expression, which depends on the sigmoid non-linearity, weight matrix and bias vector, and some initial guesstimates of the states of the recurrent neural network results in the new states of the network. These states are the estimated parameters (of the intended state-space model). This contrasts with the procedure of estimation using the feed forward neural network. In the feed forward neural network, the weights of the network are not the parameters of direct interest. In a recurrent neural network also, the weights are of no direct interest, although they are pre-computed and not updated as in a feed forward neural network. In both methods, we do not get to know more about the statistical properties of the estimates and their errors.

The genetic algorithms provide yet another alternative method that is based on direct cost function minimization and not on the gradient of the cost function. This is very useful for types of problems where the gradient could be ill-defined. However, the genetic algorithms need several iterations for convergence and stopping rules are needed. One limitation is that we cannot get parameter uncertainties, since these are related to second order gradients. In that case, some mixed approach can be used, i.e., after the convergence, the second order gradients can be evaluated.

## 1B.2 Soft Computing (SC)

Soft computing differs from the conventional computing (CC) in the following ways:

i) The CC can be called as a ‘hard’ computing, or better firm computing! (because a SW or an algorithm/program resides in the HW);
ii) The role-model for soft computing is as such the human mind;
iii) The human mind does not always take hard decisions; and
iv) The mind takes soft or often vague decisions.

SC is so called because it is tolerant of imprecision, and vagueness. The directive principle of SC is: i) imprecision: exploit the tolerance for imprecision (of definitions/expressions of the views); ii) uncertainty: in partial decisions/data/information; iii) partial truth: often available; and iv) approximation: of computing new decisions/estimations.

The aim of the SC methods is to achieve tractability/solvability, robustness, and low cost solutions to many complex problems in science and engineering [2].

Biological neural networks (BNNs) and many decision-making abilities work ‘as if’ on the principles of imprecision and uncertainty. BNNs and our (humans) decision-making approaches always try to make sense out of such data and in such environments; and in such cases, we hardly use formal mathematics: be it learning, adaptation, and/or data fusion.

Thus, the SC paradigms in fact then model our behaviour and activities and provide computational tools so that we can formally and legitimately use these techniques for solving difficult problems, and develop/build artificial (intelligent-AI-like) systems that work/act like humans.

Such AI-based systems (HW/SW/algorithms/procedures) can also work in hazardous situations and areas, and can also perform tasks of data/image processing, decision-making and data fusion. The SC paradigms in fact are complementary (in fact supplementary) rather than competitive (with the existing methods) to handle complex problems that require the applications/building of SC-based AI systems. Also, SC is viewed as a foundation-component for the conceptual and computational (artificial) intelligence/CAI.

In parameter estimation using the ANNs, FL, and the GAs, the main ingredients or the methods of the SC are in an evolving state. New results on convergence, uniqueness, robustness, and parameter error-covariance need to be explored. Perhaps, such results could be obtained by using the existing analytical results of estimation and statistical theories. Theoretical limit theorems are needed to obtain more confidence in these approaches.

## REFERENCES


1. Raol, J. R., Girija, G., and Singh, J., *Modelling and Parameter Estimation for Dynamic Systems. IEE/IET Control Series Books*, The Institution of Electrical Engineers (IEE), Vol. 65, London, UK, 2004.
2. Zadeh, L. A., Soft computing and fuzzy logic. *IEEE Software*, 11(6), pp. 48–56, 1994.


## APPENDIX 1C SUB-COMPONENTS OF AI

There are four components of/in AI-Artificial Intelligence (see Chapter 1, Figure 1.2, Quad-Elements Model-QEM):

i) Human-Behaviour (HB), element (2,1)→ Systems that act like humans (SALH).
ii) Human-Thought (HT), element (1,1)→ Systems that think like humans (STLH).
iii) Rational-Behaviour (RB), element (2,2)→ Systems that act rationally (SAR).
iv) Rational-Thought (RT), element (1,2)→ Systems that think rationally (STR).

Within each of these components, there are certain sub-components that are either processes, actions, or techniques (PAT) that help achieve the main task in AI as outlined in the QEM of AI, i.e., the tasks of the four sub-components [1]. These PAT are delineated as follows;

i) For HB-SALH: Turing test, NLP – Natural language processing, KR – Knowledge representation, ICK – Information-cum-knowledge, and ML – Machine learning.

ii) For HT-STLH: CM – Cognitive modelling, BNN – biological neural network, DM – Decision-making, and PSL – Problem solving language.
iii) For RB-SAR: Maximize GF-Goal function, ICK, and RA – Rational agents.
iv) For RT-STR: FL – Fuzzy logic, CMM – Cognitive methods mechanisms, and PRA – Perceive, reason and act, Now we describe these subcomponents briefly.

## A1C.1 For HB-SALH

The PATs that are instrumental in achieving the tasks in SALH are next described.

a) **Turing test:**
It is a test for checking if a system (device, digital computer, or SW) acts like a human. It's a method to assess whether a machine can exhibit intelligent behaviour that is indistinguishable from that of a human, usually through a conversation where a human evaluator tries to identify if they are interacting with a machine or another person.
It was developed by Alan Turing, a mathematician and computer scientist, in his 1950 paper 'Computing Machinery and Intelligence'. A human evaluator engages in a text-based conversation with both a human and a machine, without knowing which is which; and if the evaluator cannot reliably tell/decipher the machine apart from the human, then the machine is considered to have passed the test; this means that the machine acts like the human. It is argued that the Turing Test only measures the ability to mimic human conversation and not true intelligence or consciousness.
b) **Natural language processing (NLP):**
It is a technology that enables computers to understand human language. It's a branch of artificial intelligence (AI) that uses machine learning, statistical modelling, and computational linguistics. Examples of NLP are: i) Email filters: NLP can identify spam, promotions, and social media-related emails; ii) Smart assistants: NLP can recognize speech patterns and respond to voice commands; Siri, Alexa, and Google Assistant; iii) Search results: NLP can provide relevant search results based on user intent; iv) Chatbots: NLP can understand what a person is typing and respond appropriately; v) Translation apps: NLP can decipher foreign languages; vi) Sentiment analysis: NLP can determine whether the meaning behind data is positive, negative, or neutral; vii) Document summarization: NLP can summarize documents; and viii) Grammar/spell checking: NLP can check for grammatical and spelling errors. NLP is used in many everyday products and services.
c) **Knowledge representation (KR):**
It is the process of structuring information so that computers can use it to solve problems. It's a key part of AI that allows machines to reason and act intelligently. KR uses techniques like ontologies and knowledge graphs to represent information about the world. KR allows AI to interpret data, make inferences, and solve problems. KR can be used in expert systems and to make complex software easier to define and maintain. What kinds of knowledge are represented?: i) Objects: Facts about things in the world that can be

observed, touched, or manipulated; ii) Events/Facts: Actions or occurrences/truths that take place in the world; iii) Performance: The behaviour of agents or systems that perform a task; and iv) Meta-knowledge: Knowledge about knowledge, including its structure, sources, and reliability.

KR helps AI systems: i) make rational decisions, ii) solve complex problems, and iii) exhibit human-like intelligence.

d) **Information-cum-knowledge (ICK):**

Information is the foundation for knowledge, which is created by analyzing, interpreting, and applying information.

Information: i) Processed data that is passed from one person to another, and ii) Provides a structured understanding of a topic.

Knowledge: i) Information that has been analyzed, synthesized, and refined; ii) Involves applying information to solve problems or make decisions; iii) Is unique to each person and is shaped by past experience and insight

How information becomes knowledge: i) Information is processed and interpreted to create knowledge; ii) Knowledge is built by adding context, interpretation, and judgment to information; and iii) Knowledge is the result of a continuous learning process.

Examples: i) A news article is information, but the knowledge gained from reading it is the result of processing and interpreting the information, and ii) A bottle of ketchup is information, but the knowledge gained from using it/eating the ketchup, is the result of processing and interpreting the information, viz. determining the taste of the ketchup.

e) **Machine learning (ML):**

It is a subset of AI that allows machines to learn and improve over time. ML uses algorithms to analyze data, learn from it (improves its performance and accuracy), and then make decisions; can perform tasks autonomously.

Examples of ML: i) Speech recognition: Computers can identify words in spoken language; ii) Natural language processing: Computers can understand and communicate with human language; iii) Image and video processing: Computers can recognize faces, objects, and actions in images and videos; iv) Sentiment analysis: Computers can identify and categorize positive, neutral, and negative attitudes in text.

ML techniques are: i) Deep learning: A subset of ML that uses neural networks to process and analyze information, and ii) Linear regression: A statistical technique that finds the relationship between variables. Benefits of ML are: i) Better data integrity, ii) Faster decision-making, and iii) Increased operational efficiency.

## A1C.2 For HT-STLH

The PATs that are instrumental in achieving this task are described next.

a) **Cognitive modelling (CM):**

It is a technique that simulates human thought processes to create computer models. The goal is to make AI systems behave more human-like.

How it works: i) Understand human thought: Use introspection, psychological experiments, and brain imaging to understand how humans think; ii) Create a model: Use the understanding of human thought to create a computational model; and iii) Apply the model: Use the model to improve decision-making, task completion, and research

It is used in: i) Expert systems; ii) Natural language processing (NLP) to help computers understand and respond to human language; iii) Robotics; iv) Virtual reality (VR); and v) Neural networks.

The CM can improve: i) human-computer interaction, ii) research, iii) task-critical decision-making, and iv) task completion. Cognitive modelling is also used in psychology, and cognitive sciences.

b) **Biological neural network (BNN)**: This is described in Chapter 2.

c) **Decision-making (DM):**

Decision-making is used to develop systems that think like humans to analyze data, identify patterns, and make choices based on complex information, mimicking the way humans process information and make decisions, often incorporating elements like context awareness, emotional reasoning, and learning from experience to achieve a more human-like decision-making process. Key aspects of decision-making in human-like AI systems are:

i) **Data analysis and pattern recognition:** AI systems analyze large datasets to identify patterns and trends, similar to how humans gather information and make inferences from observations;
ii) **Probabilistic reasoning:** Incorporating uncertainty and probability into decision-making, reflecting the human ability to weigh different possibilities and make decisions even with incomplete information;
iii) **Heuristics and biases:** Implementing cognitive shortcuts and biases that humans often exhibit in decision-making to create more realistic behaviour;
iv) **Context awareness:** Taking into account the surrounding situation and relevant details when making decisions, just like humans do in real-world scenarios;
v) **Reinforcement learning:** Allowing the system to learn from feedback and adjust its decision-making strategies over time, similar to how humans learn through experience; and
vi) **Multi-criteria evaluation:** Considering multiple factors and their relative importance when making a decision, mirroring how humans weigh different aspects of a situation.

How decision-making is used in human-like AI systems: i) Chatbots: Using natural language processing and decision-making algorithms to understand user intent and respond in a natural, conversational way; ii) Medical diagnosis support: Analyzing patient data to suggest potential diagnoses based on medical knowledge and decision rules; iii) Self-driving cars: Making real-time driving decisions based on sensor data, traffic conditions, and road rules, considering potential risks and uncertainties; and iv) Recommendation systems: Suggesting products or services to users based on their past behaviour and preferences, taking into account factors like context and user psychology.

d) **Problem solving language (PSL):**
PSL is used to develop systems that think like humans by enabling the system to represent complex problems in a way that mirrors human thought processes, allowing it to break down problems into smaller, manageable steps, explore potential solutions through reasoning and logic, and ultimately arrive at solutions that are not just accurate but also intuitive and adaptable, much like how humans solve problems in real-world situations.

Key aspects of problem solving language are: i) Breaking down complexity: Similar to how humans analyze a problem, PSL allows systems to identify key components, relationships, and constraints within a problem, enabling them to break it down into smaller, more manageable parts; ii) Heuristics and reasoning: By incorporating heuristics, which are mental shortcuts based on experience, systems can make informed decisions and navigate complex scenarios without needing to exhaustively explore every possibility, mimicking human intuition; iii) Natural language processing: Integrating natural language capabilities allows systems to understand and respond to queries in a way that is similar to human conversation, facilitating more natural interactions; and iv) Learning and adaptation: Machine learning techniques, combined with problem solving language, enable systems to learn from past experiences and adapt their approach to new problems, just as humans do.

How problem solving language is used in AI development: i) Planning algorithms: Used to create a sequence of actions to achieve a goal, taking into account potential obstacles and constraints, much like humans plan their day; ii) Decision support systems: Employing problem solving language to analyze various options and recommend the most optimal course of action based on defined criteria; iii) Expert systems: Incorporate knowledge bases with rules and reasoning mechanisms to mimic the decision-making process of a human expert in a specific domain; and iv) Chatbots and virtual assistants: Utilize natural language processing and problem solving algorithms to understand user queries and provide relevant responses in a conversational manner.

## A1C.3 For RB-SAR

The PATs that are instrumental in achieving this task are described next.

a) **Maximize GF – Goal function**:
A GF refers to a mathematical equation used in optimization problems to represent the desired outcome or target that an algorithm is trying to achieve, essentially defining what best means in a given situation; it tells the algorithm which solution to prioritize by assigning a numerical value to each possible outcome, with the highest value representing the best solution.

b) **ICK:** This has been already discussed in section A1C.1.

c) **Rational agent (RA):**
A rational agent is a person, machine, or software that makes decisions based on information and premises. It aims to perform actions that are optimal for its situation.

How do rational agents work? i) Observe: The agent observes its environment; ii) Analyze: The agent analyzes the information it receives; and iii) Decide: The agent selects the best action to take based on its analysis. The examples of rational agents are: i) Chess players: Analyze the board to determine the best move; and ii) Autonomous vehicles: Use sensors, cameras, and radar to make decisions. The concept of rational agents is studied in many disciplines, including artificial intelligence, cognitive science, economics, ethics, and game theory. Rationality doesn't mean perfection. A rational agent may not have all the information it needs, or it may face uncertainty. However, it will always try to make the best decision it can with the information it has.

## A1C.4 For RT-STR

The PATs that are instrumental in achieving this task are described next.

a) **Fuzzy logic (FL)**: This will be discussed in Chapter 5.
b) **Cognitive methods mechanisms (CMM):** This has been discussed in Section A1C.2.
c) **Perceive, reason and act (PRA):**

Unlike traditional software, AI agents are not static, rule-based systems; they perceive their environment, learn from data (from observations and measurements), reason through complex problems, and take autonomous actions, and communicate seamlessly with humans and other systems.

## REFERENCE

1. Russell, S.J., and Norvig, P. *Artificial Intelligence-A Modern Approach*, Prentice Hall, Englewood Cliffs, New Jersey, USA, 1995. https://people.engr.tamu.edu/guni/csce625/slides/AI.pdf, Accessed April, 2026.

# 2 Concepts and Theory of Artificial Neural Networks

The research in the area of artificial neural networks (ANNs) has advanced at a very rapid pace in recent times. An ANN possesses good ability to learn adaptively, which is one of the key features of the artificial intelligence (AI). The decision process in an ANN is based on certain nonlinear operations, and linear aspects are special cases. Such nonlinearities are useful in: i) improving the convergence speed (of the algorithm), ii) providing more general nonlinear mapping between I/O signals, and iii) reducing the effect of outliers in the measurements. Much of the exposition in this chapter is influenced by [1]–[10].

ANNs have evolved from simple perceptron models to sophisticated architectures capable of solving complex real-world problems, spanning domains such as image recognition, natural language processing, and control systems.

## 2.1 BIOLOGICAL NEURAL NETWORKS

The human brain contains approximately 10–500 billion neurons and is organized into about 1000 major modules consisting of several neural networks (NWs). Each neural network may contain on the order of 100,000 neurons. A typical neuron consists of dendrites, a central cell body, and an axon, and the axon of each neuron connects to several hundred (or even thousands of) other neurons. Signal flow in a neuron is from the dendrites through the cell body and then along the axon. The connection between the axon of one neuron and the dendrite of another neuron is called a synapse, which is the point at which a nerve impulse is relayed; see Figure 2.1 which shows the conceptual evolution of an ANN's functionality from the biological neural system.

Signals can be transmitted unchanged or can be altered by synapses. A synapse is able to increase/decrease the strength of the connection from the neuron to another neuron and cause excitation/inhibition of a subsequent neuron; this is where the information is stored.

The information processing abilities of biological neural systems follow from highly parallel processes operating on representations that are distributed over many neurons. One motivation for ANN is to capture this kind of highly parallel computation based on distributed structural representations.

Broad functions in a Biological Neural System (BNS/BNN) are: i) dendrites – input branching tree of fibres – connect to a set of other neurons, and – receptive surfaces for input signals; ii) soma – cell body – all the logical functions of the

DOI: 10.1201/9781003616580-2

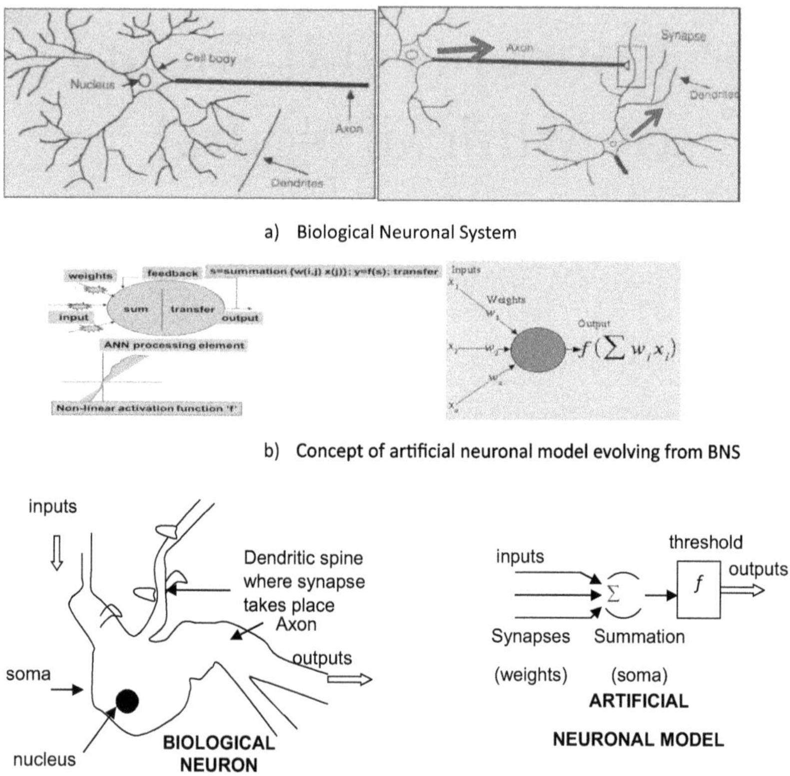


a) Biological Neuronal System

b) Concept of artificial neuronal model evolving from BNS

c) Simple elemental/module representation of functionalities of BNN and ANN

**FIGURE 2.1** Biological neural system/network (BNS-BNN) to artificial neural network (ANN) – a conceptual transformation from perceived reality to computational model: a) BNS; b) concept of artificial neuronal model evolving from BNS; c) simple elemental/module representation of functionalities of BNN and ANN

neurons are realized here; iii) synapse – specialized contacts on a neuron – interfaces some axons to the spines of the input dendrites – can increase/dampen the neuron excitation, and iii) axon – nerve fibre – final output channel – signals converted into nerve pulses (spikes) to target cells.

## 2.2 ARTIFICIAL NEURAL NETWORKS

The ANNs have some similarities to the BNS; it has massive parallelism and yet consists of very simple processing elements. The feedforward neural network (FFNN) is an information processing system of a large number of simple processing elements, Figure 2.2. These elements are called artificial neurons or nodes. These neurons are interconnected by links, which are represented by the so-called weights, and they cooperate to perform parallel distributed computing to carry out a desired computational task. The neural networks are so called because the early researchers in the area of ANNs were involved in the study of functioning of the human brain and modelling of the neuron system [1]. They should be more appropriately called as massively parallel adaptive (and switching) circuits or filters. This is because the ANNs have technical roots in the area of analogue circuits, computing, and signal processing.

ANNs are used for I/O subspace modelling because the basic neural network functions can adequately approximate the system behaviour in an overall sense. The FFNNs can be thought of as nonlinear black box model structures, the parameters (coefficients/weights) of which can be estimated by conventional optimization methods. These are more suitable for system identification, time-series modelling and prediction, pattern recognition/classification, sensor failure detection, and estimation of aerodynamic coefficients. Lately these have also been used for parameter estimation of dynamic systems. In this case, the FFNN is used for predicting the time histories of aerodynamic coefficients, via training-set data samples/signals, and then some regression method (based on the classical principle of Least Squares) is used to estimate the aerodynamic parameters (the aerodynamic stability and control derivatives of an aircraft or helicopter) from the predicted aerodynamic coefficients.

An illustration and comparison of biological neuron and artificial neuron are given in Figure 2.1 and Table 2.1

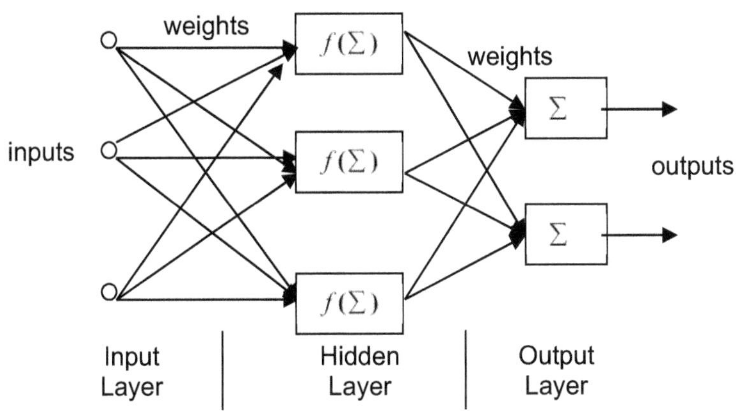


**FIGURE 2.2** Feedforward neural network (FFNN) structure with one hidden layer

**TABLE 2.1**
**Comparison of two neural systems**

| Biological neuron (of human brain) | Artificial neuron |
|---|---|
| Signals received by dendrites and passed on to neuron receptive surfaces | Data enters through input layer |
| Inputs are fed to the neurons through specialized contacts called synapse | Weights provide the connection between the nodes in the input and output layers |
| All logical functions of neurons are accomplished in Soma | Non-linear activation function operates upon the summation of the product of weights and inputs $f(\Sigma Wx_i)$ |
| Output signal is delivered by the axon nerve fibre | The output layer produces the network's predicted response |

### 2.2.1 Feedforward Neural Networks and Training Algorithms

One of the most successful ANNs is the so-called FFNN; it has found successful applications in the pattern recognition, nonlinear curve fitting/mapping, flight data analysis, aircraft flight mechanics modelling, adaptive control, and system identification.

The FFNNs have a non-cyclic and layered topology and hence can be considered to have structure-free (in the conventional polynomial model-sense) nonlinear mapping between I/O signals of a system. The chosen network is first trained using the training-set data and then the trained network is used for prediction of the output time history using a different input set (validation set), which belongs to the same class of the data; the process is similar to the one used as cross-validation in system identification literature. The weights of the network are determined using the so-called backpropagation/gradient-based procedure (basically the steepest descent method is used). Because of the sequential-layered disposition of the weights of the FFNN, the estimation of the weights requires propagation of the error of the output layer in a backward (logical) direction, hence the name backpropagation. Here, the training (and estimation) algorithms are described using the matrix/vector notation for the sake of clarity and ease of implementation in PC MATLAB. Even if one does not have the Neural Network toolbox of MATLAB, the simulation studies can be carried out easily and very efficiently using the available and newly formulated dot-em (.m) files.

The FFNN has the following variables: i) $u_0$ = input to (input layer of) the network, ii) $n_i$ =number of input neurons = number of inputs $u_0$, iii) $n_h$ = number of neurons of the hidden layer, iv) $n_o$ =number of output neurons = number of outputs $z$, v) $W_1 = n_h$ x $n_i$ weight matrix between input and hidden layer, vi) $W_{10} = n_h$ x 1 bias weight vector, vii) $W_2 = n_o$ x $n_h$ weight matrix between hidden and output layer, viii) $W_{20} = n_o$ x 1 bias weight vector, and ix) $\mu$ = learning rate or step size.

#### 2.2.1.1 Backpropagation Algorithm for Training

This algorithm is based on the steepest descent optimization (SDM) method. The forward pass signal computation is done using the following sets of equations, since $u_0$ is known and initial guesstimates of the weights are known:

$$y_1 = W_1 u_0 + W_{10} \tag{2.1}$$

$$u_1 = f(y_1) \tag{2.2}$$

Here, $y_1$ is a vector of intermediate values and $u_1$ is the input to the hidden layer. The function $f(y_1)$ is a nonlinear sigmoidal activation function and is given by

$$f(y_i) = \frac{1 - e^{-\lambda y_i}}{1 + e^{-\lambda y_i}} \tag{2.3}$$

Next, the signal between the hidden and output layers is computed:

$$y_2 = W_2 u_1 + W_{20} \tag{2.4}$$

$$u_2 = f(y_2) \tag{2.5}$$

Here, $u_2$ is the signal at the output layer. The learning rule is derived next. Often, an unconstrained optimization problem for parameter estimation is transformed into an equivalent system of differential/difference equations, which in turn constitutes a basic neural network algorithm to solve:

$$\frac{dW}{dt} = -\mu(t)\frac{\partial E(W)}{\partial W} \tag{2.6}$$

With the output error defined as $e = z - u_2$, and a suitable quadratic cost function based on it, the expression for the gradient is obtained as:

$$\frac{\partial E}{\partial W_2} = -f'(y_2)(z - u_2)u_1^T \tag{2.7}$$

Here, $u_1$ is the gradient of $y_2$ with respect to $W_2$. The derivative $f'$ of the node activation function $f$ is given by

$$f'(y_i) = \frac{2\lambda_i e^{-\lambda y_i}}{\left(1 + e^{-\lambda y_i}\right)^2} \tag{2.8}$$

The expression (2.7) follows directly from the quadratic cost function, also known as a criterion, defined as $E = \frac{1}{2}(z - u_2)(z - u_2)^T$ and using eqns. (2.4) and (2.5). The modified error of the output layer can be expressed as

$$e_{2b} = f'(y_2)(z - u_2) \tag{2.9}$$

Thus, the recursive weight update rule for the output layer is given as:

$$W_2(i+1) = W_2(i) + \mu e_{2b} u_1^T + \Omega[W_2(i) - W_2(i-1)] \tag{2.10}$$

The $\Omega$ is the momentum constant and is used to smooth out the weight changes and to accelerate the convergence of the algorithm. The backpropagation of the error and the weight update rule for $W_1$ are given as:

$$e_{1b} = f'(y_1)W_2^T e_{2b} \tag{2.11}$$

$$W_1(i+1) = W_1(i) + \mu\, e_{1b} u_0^T + \Omega[W_1(i) - W_1(i-1)] \tag{2.12}$$

The training-set data/signals (TSD) are presented to the network in a sequential manner and this process is called pattern learning in neural network literature. The TSD are presented again but with initial weights from the outputs from the previous cycle. This process is stopped when the convergence is reached. The entire process is called recursive-iterative. It must be noted here that, the values of $\mu$ in eqns. (2.10) and (2.12) need not be the same. Similar observation applies to $\Omega$ .

#### 2.2.1.2 Backpropagation Recursive Least Squares Filtering Algorithms

***Algorithm with nonlinear output layer***

During the forward pass training of the network, the signals $y$ and $u$ are computed for each layer as is done in the back-propagation algorithm. The filter gains $K_1$ and $K_2$ are computed for both the layers and the forgetting factors $f_1$ and $f_2$ are chosen. The formulation is the usual scalar data processing scheme: For layer 1, the updates for filter gain $K_1$ and covariance matrix $P_1$ are given as:

$$K_1 = P_1 u_0 (f_1 + u_0 P_1 u_0)^{-1} \tag{2.13}$$

$$P_1 = (P_1 - K_1 u_0 P_1)/f_1 \tag{2.14}$$

For layer 2, the updates for filter gain $K_2$ and covariance matrix $P_2$ are given as:

$$K_2 = P_2 u_1 (f_2 + u_1 P_2 u_1)^{-1} \tag{2.15}$$

$$P_2 = (P_2 - K_2 u_1 P_2)/f_2 \tag{2.16}$$

The modified output error is given as:

$$e_{2b} = f'(y_2)(z - u_2) \tag{2.17}$$

The backpropagation of the output error to inner/hidden layer gives inner layer error as:

$$e_{1b} = f'(y_1)W_2^T e_{2b} \tag{2.18}$$

And finally, the weight update rule for the output error is

$$W_2(i+1) = W_2(i) + (d - y_2)K_2^T \tag{2.19}$$

Here, $d$ is given by

$$d_i = \frac{1}{\lambda} \ln\left[\frac{1+z_i}{1-z_i}\right]; \text{and } z_i \neq 1 \tag{2.20}$$

For the hidden layer, the rule is:

$$W_1(i+1) = W_1(i) + \mu \, e_{1b} K_1^T \tag{2.21}$$

Here, the additional computation of Kalman gains is needed, otherwise the procedure for training is similar to the back-propagation algorithm. We note here that when the weight update rule of eqn. (2.21) is used, the range of values of $\mu$ would not generally be the same as that when the rule of eqn. (2.12) is applied.

***Algorithm with linear output layer***

In this case, the output layer does not have nonlinearities. Only the inner layer has nonlinearities. The linear Kalman filter concept, therefore, is directly applicable in this case. Since, the output layer block is linear, the output signal is computed as

$$u_2 = y_2 \tag{2.22}$$

The Kalman gain computations are as per the algorithm discussed above. Since, the output layer has no nonlinearity, the error for output layer is:

$$e_{2b} = e_2 = (z - y_2) \tag{2.23}$$

The back-propagation of the output error gives:

$$e_{1b} = f(y_1) W_2^T e_{2b} \tag{2.24}$$

Finally, the weight update rules are:

$$W_2(i+1) = W_2(i) + e_{2b} K_2^T \tag{2.25}$$

$$W_1(i+1) = W_1(i) + \mu \, e_{1b} K_1^T \tag{2.26}$$

Once the data are scanned (used/processed) and the convergence achieved, the estimated weights of the last iteration are used as inputs and presented again to the network to predict the output. This output is compared with the desired/available output in order to judge the network's ability for prediction.

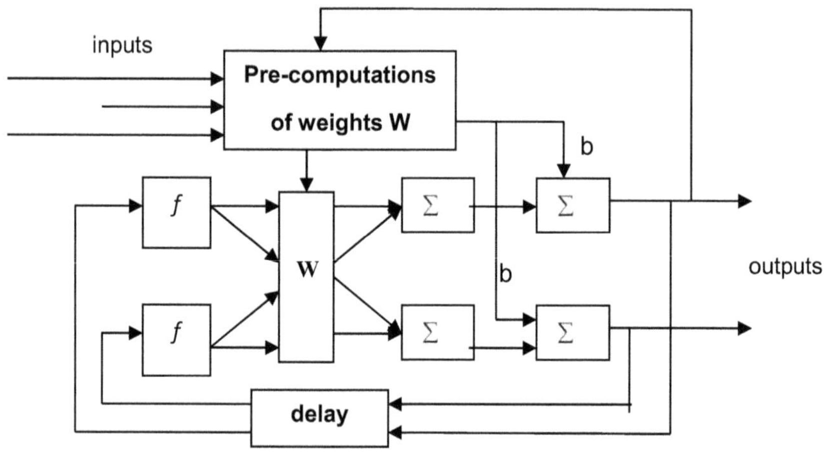


**FIGURE 2.3** Schematic of a recurrent neural network (RNN)

### 2.2.2 Recurrent Neural Networks (RNN)

Modelling of a system using ANNs has recently become popular with application to signal processing, pattern recognition, system identification, and control. Estimation of parameters using empirical data plays a crucial role in modelling and identification/estimation of dynamic systems. Often equation error and output error methods are used for parameter estimation of dynamic systems. These are generally batch iterative procedures wherein a set of data is processed to compute the gradient of a cost function and estimation error. The estimation of parameters is then refined using an iterative procedure based on the improved estimates (innovations sequence) of error and its gradients. Such methods can be termed as batch-iterative. The ANNs provide new/alternative paradigms to handle the problem of parameter estimation with potential application to on-line estimation. Especially recurrent neural networks are easily amenable to such possibilities due to their special structure: feedforward neural networks with a feedback feature (Figure 2.3). In order to obtain fast solutions, a system of parallel computers can be used. This will require the parallelization of the conventional parameter estimation algorithms. Since ANNs have massively parallel processing capacity, they can be easily adapted to parameter estimation problems for on-line applications. In particular, the RNNs can be considered as more suitable for the problem of parameter estimation of linear and nonlinear dynamic systems, compared to perhaps FFNNs. The RNNs are dynamic neural networks, and hence amenable to explicit parameter estimation in state-space models.

#### 2.2.2.1 Variants of RNNs – Unification of Structures

Here, four variants of RNNs are studied from the point of view of explicit parameter estimation. In literature, several variants of the basic Hopfield Neural Network (HNN)

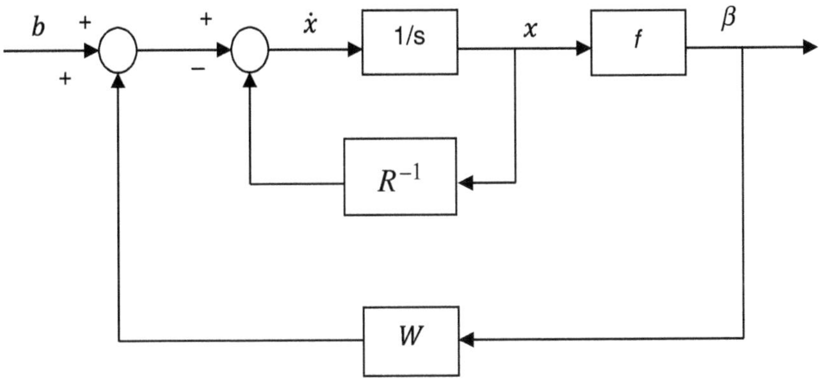


**FIGURE 2.4** Schematic of RNN-S structure

structure are available. The three variants are related to each other by affine or linear transformation of their states. The variants are classified by the way in which the sigmoid non-linearity operates: either on states (-S), weighted states (-WS), residual (-E) of the network signal, or forcing input (-FI).

**(a) RNN-S (HNN)**

This network is known as HNN; the Hopfield neural network model has a number of mutually interconnected information processing units called neurons. In this configuration, the outputs of the network are a non-linear function of the states of the network (and hence 'S'). The dynamic representation of the network is given as (Figure 2.4):

$$\dot{x}_i(t) = -x_i(t)R^{-1} + \sum_{j=1}^{n} w_{ij}\beta_j(t) + b_i; \quad j = 1,\ldots,n \tag{2.27}$$

Here, $x$ is the internal state of the neurons, $\beta$ is the output state, $\beta_j(t) = f(x_j(t))$, $w_{ij}$ are the neuron weights, $b$ is the bias input to the neurons, and $f$ is the sigmoid non-linearity. $R$ is the neuron impedance and $n$ is the dimension of the neuron state. Eqn. (2.27) can also be written as:

$$\dot{x}(t) = -x(t)R^{-1} + W\{f(x(t)\} + b \tag{2.28}$$

Eqn. (2.28) can be considered as a representation of "classical" neuro-dynamics. In comparison to biological neurons, eqn. (2.28) obtains a simple system retaining essential features: i) neuron as a transducer of input to output, ii) a smooth sigmoidal response up to a maximum level of output, and iii) feedback nature of connections. Thus, the model retains two aspects: dynamicity and nonlinearity.

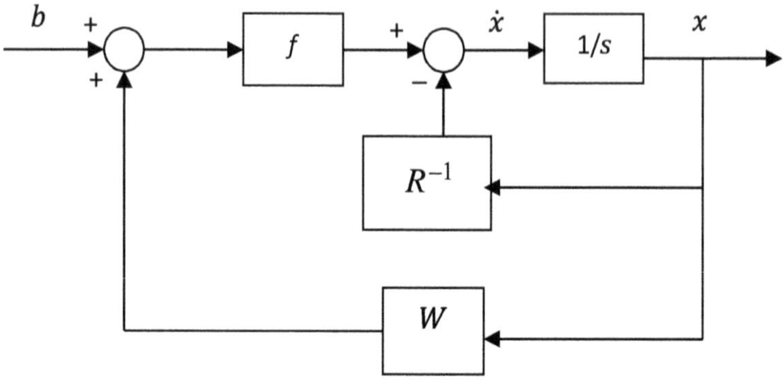


**FIGURE 2.5** Schematic of RNN-FI structure

**(b) RNN-FI**

In this configuration of the RNN, the non-linearity operates on the forcing input: FI= weighted states + input to the networks → modified input = $f(Wx+b)$. The dynamics of this network can be given as (Figure 2.5):

$$\dot{x}_i(t) = -x_i(t)R^{-1} + f\left(\sum_{j=1}^{n} w_{ij}x_j(t) + b_i\right) \tag{2.29}$$

Here, $f(.) = f(FI)$. This network is related to the RNN-S by affine transformation. Use $x_H(t) = Wx + bR$ to obtain the following equivalence:

$$\begin{aligned} W\dot{x} &= -(Wx+bR)R^{-1} + Wf(Wx+bR)+b \\ W\dot{x} &= -WxR^{-1} - b + Wf(Wx+bR)+b \\ \dot{x} &= -xR^{-1} + f(Wx+bR) \\ \dot{x} &= -xR^{-1} + f(FI) \end{aligned} \tag{2.30}$$

Here, *FI* is the modified input vector, due to the *bR* term. Here invertibility of *W* is a necessary condition. We see that eqn. (2.30) has exactly the same form as that of RNN-FI.

**(c) RNN-WS**

In this configuration, the non-linearity operates on the weighted states, hence WS. The dynamics of this neural network are described as (Figure 2.6):

$$\dot{x}_i(t) = -x_i(t)R^{-1} + f(s_i) + b_i \tag{2.31}$$

Here, $s_i = \sum_{j=1}^{n} w_{ij}x_j$. It can be seen that the network is related to RNN-S by linear transformation. Substitute $x_H(t) = Wx$ in eqn. (2.31) to obtain:

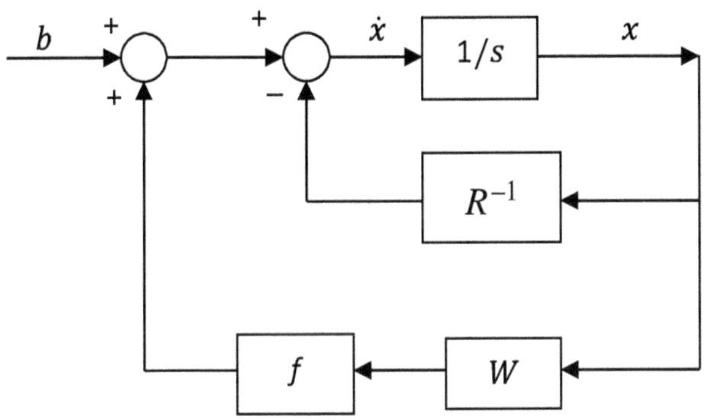


**FIGURE 2.6** Schematic of RNN-WS structure

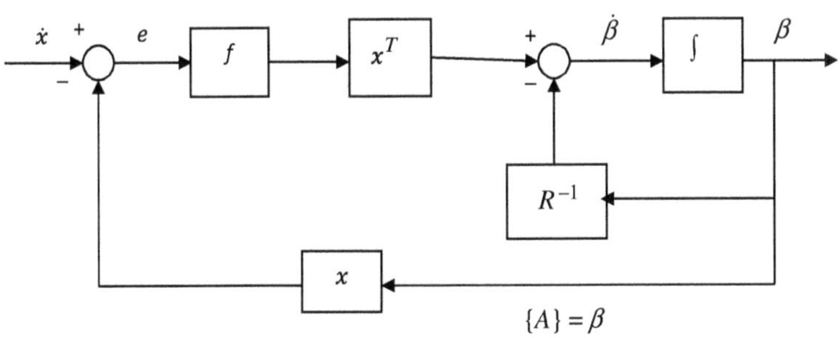


**FIGURE 2.7** Schematic of RNN-E structure

$$\begin{aligned} W\dot{x} &= -(Wx)R^{-1} + Wf(Wx) + b \\ \dot{x} &= -xR^{-1} + f(s) + W^{-1}b \end{aligned} \tag{2.32}$$

Here, we have a modified input vector. The matrix $W$ must be invertible.

**(d) RNN-E**

In this configuration, the non-linearity directly operates on the residual error or equation error. Hence, the function $f$ or its derivative $f'$ does not enter into the neuron dynamic equation. Yet, it does affect the residual by way of quantizing them and thereby reducing the effect of measurement outliers. The dynamics are given by (Figure 2.7):

$$\dot{x}_i(t) = -x_i(t)R^{-1} + \sum_{j=1}^{n} w_{ij}x_j(t) + b_i \tag{2.33}$$

For RNN-E, we say that the internal state $x_i$ is $\beta_i$, the parameters of the general dynamic system. In that case, the $x_i$ of eqn. (2.33) does not represent the state of this general dynamic system.

### 2.2.3 Radial Basis Function Neural Networks (RBFNN)

The radial basis function (RBF) NW is a part of two-layer ANN. In the ANN each hidden unit implements a radial activated function. Then, the output is a weighted sum of outputs from the hidden units. The input (unit) is nonlinear whereas the output is linear [9]. The weights are found by a suitable training algorithm using a mean square error (MSE) between the training set output and the NW-predicted output. The RBF NWs have been used for: i) modelling of chaotic time-series, ii) system identification, iii) control engineering, iv) modelling of electronic devices, v) (communications) channel equalization, vi) speech recognition, vii) image registration, viii) shape from shading identification, ix) 3-D object modelling, x) motion estimation and moving object segmentation, and xi) sensor data fusion.

#### 2.2.3.1 The NW Structure

Figure 2.8 depicts the RBF NW topology. The RBFs are wired into the 2-layer FFNN, with a set of I/Os. It has one hidden layer with certain units and each one is an RBF. Various functions are used as activation functions: i) thin-plate spline, ii) Gaussian function, and iii) mixture of Gaussian functions. The Gaussian activation function is given by:

$$\phi_i(u) = \exp\left\{-(u - m_i)^T P_i^{-1}(u - m_i)\right\} \tag{2.34}$$

Here, we have $u$ as the input signal-data/features, the hidden layer units as $i=1,2,...n_h$, and $m$ and $P$ are the mean and covariance matrix of the i-th Gaussian function, respectively. The RBFNN geometrically represents a bump in the multi-dimensional space, then the mean vector is the location and $P$ is the shape of the activation function,

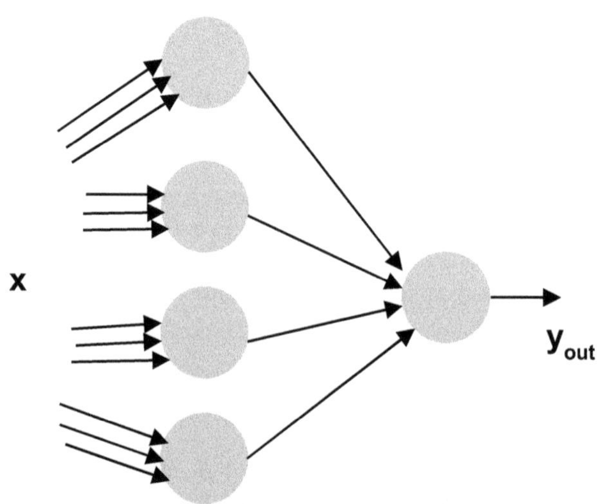


**FIGURE 2.8** Radial basis function neural network

and in effect the activation function models a pdf with *m* and *P* as the first and second order statistics. Thus, the output layer results in the following expression:

$$U_j(u) = \sum_{i=1}^{n_h} W_{ij}\phi_i(u) \tag{2.35}$$

The *W*s are the contribution of the hidden units to the output units, and we have as the number of output units: $j=1,2,\ldots\ n_0$.

In pattern classification the RBFNN used is a sigmoid function:

$$\phi_j(u) = \frac{1}{1+\exp\{-U_j(u)\}} \tag{2.36}$$

#### 2.2.3.2 RBFNN Properties

RBF NW is specified by its localization and an activation hyper surface, and if it is a Gaussian function, then these are characterized by the two parameters: mean and covariance (*P*). If this matrix *P* is diagonal, with all its elements equal, then the hyper surface is a hyper sphere, otherwise, it is hyper ellipsoid. In the latter case the influence of the activation function decreases according to the Mahalanobis distance (MD) from the centre of the ellipsoid. The normalized MD is given by the following expression:

$$d(u,v) = (u-v)P^{-1}(u-v)^T \tag{2.37}$$

Here, *P* is the covariance of the input data. What this simply means is that the samples located at a larger MD from the centre of the RBF will not be able to activate the basis function effectively. Hence, the maximum activation is achieved when the data coincide with the mean vector. The Gaussian basis functions are quasi-orthogonal. Also, the product of the two bases functions is almost zero if their centres are far away from each other. The RBFNNs can be considered as the potential functions. The hidden units with the output weights *W*s having the same sign for a certain output unit, will have their activation fields joined together, in the same way as the electrical charges of the same sign form the electrical fields, and the weights with the different signs will form the activation fields like the electrical fields of the opposite signs. The RBFNNs are found to be suitable for interpolation, functional modelling, for modelling the pdfs (probability density functions), and even for implementing the Bayesian rule.

#### 2.2.3.3 RBFNN Training Algorithm

For the training, the cost function is defined as:

$$J = \sum_{n=1}^{N}\left(U_j(u_n) - V(u_n)\right)^T\left(U_j(u_n) - V(u_n)\right) \tag{2.38}$$

Here, $N$ is the total number of the vectors from the training-set data (TSD), $U$ is the RBFNN output vector, and the $V$ is the vector associated with the data samples, $u_n$, the TSD. The cost function $J$ is minimized with the RBF NW parameters. Several techniques can be used to obtain the optimal training algorithm: i) orthogonal LS method with the Gram-Schmidt algorithm, ii) SD (steepest descent) algorithm, iii) expectation maximization (EM) algorithm, and iv) learning vector quantization. The centres of the RBFNN are initialized randomly. The Euclidean distance is used for adaptation to the closest centre. The centre is updated as:

$$\tilde{m}_i = \tilde{m}_i + \eta\left(u_n - \tilde{m}_i\right) \tag{2.39}$$

The training rate in eqn. (2.39) can be taken as the reciprocal of the total number of the data samples. Then, the centre is the classical 1st order statistical estimation. Then, the output weights are estimated by using the LS method.

## 2.3 NEURAL NETWORK WEIGHTS AND TRAJECTORY LEARNING ALGORITHMS

ANNs operate based on the interaction of neurons interconnected by weighted connections. These weights are the core parameters that guide how the input signals transform as they propagate through the network. By modifying these weights during the training process, ANNs adapt their behaviour to perform tasks like classification, prediction, and regression with improved accuracy.

Network weights determine the strength and direction of the relationship between two neurons in adjacent layers. Each weight signifies the influence that a neuron's output in one layer has on a neuron in the next. The weighted sum of inputs, combined with a bias term, is typically passed through an activation function to produce the neuron's output. The adjustment of these weights during training enables the network to approximate the desired function, whether it involves recognizing images, interpreting language, or predicting trends.

The process of weight adjustment follows a specific trajectory, representing a path through the optimization landscape. This landscape, defined by the loss function, includes peaks, valleys, and plateaus corresponding to various configurations of weights. The objective of training is to traverse this high-dimensional surface efficiently, converging on a configuration where the loss is minimized and the network performs optimally.

Trajectory learning algorithms govern this process by providing systematic methods for updating weights during training. These algorithms define how the network explores the optimization landscape to identify the best set of weights for the task at hand. The trajectory of learning is not just a simple downhill path; it often requires strategies to escape local minima and navigate challenging regions in the loss surface.

The choice of trajectory learning algorithms significantly affects the network's performance, convergence speed, and stability. Classical algorithms like Stochastic Gradient Descent (SGD) and its variants – such as momentum-based methods and adaptive learning rate techniques – are widely used to guide the trajectory. They rely on gradient information to determine the direction and magnitude of weight updates, ensuring the network moves toward a better configuration with every iteration.

In practice, weight initialization, learning rate, and regularization play critical roles in determining how effectively a network navigates its learning trajectory. Improper initialization or poor choice of hyperparameters can lead to vanishing or exploding gradients, overfitting, or convergence to suboptimal solutions. Advanced techniques, such as learning rate scheduling and batch normalization, are often employed to address these challenges, ensuring the learning trajectory remains efficient and effective.

This understanding of network weights and their learning trajectories serves as the foundation for the subsections that follow, where detailed algorithms for supervised and unsupervised learning will be discussed. These algorithms, built on the principles of trajectory optimization, enable neural networks to solve complex tasks across diverse application domains.

### 2.3.1 Supervised Learning

Supervised learning is one of the most widely used paradigms in artificial neural networks. In this approach, the network is trained using labelled data, where each input is paired with a corresponding target output. The network's task is to learn a mapping from inputs to outputs by adjusting its weights and biases such that the predicted outputs closely match the actual targets. This is achieved by minimizing a predefined loss function that quantifies the error between the predicted and desired outputs.

The training process in supervised learning involves feeding input-output pairs into the network and propagating the errors backward to update the weights. This mechanism, commonly referred to as error backpropagation, relies on the gradient of the loss function with respect to the weights. By iteratively updating the weights along the direction of steepest descent, the network converges toward an optimal solution.

In supervised learning, the trajectory of weight updates depends not only on the magnitude and direction of gradients but also on the specific rules employed to guide these updates. Different update rules and techniques have been developed to improve the efficiency, speed, and stability of the learning process. These include methods that leverage gradient history, use adaptive learning rates, or restructure the network dynamically during training.

To gain a deeper understanding of supervised learning in neural networks, it is essential to examine the various strategies used for error computation and weight updates. The subsections that follow will cover the nuances of these methods in detail, starting with approaches based on the error derivatives from previous and present epochs. Advanced optimization techniques such as QuickProp and Resilient Propagation (R-Prop) will also be explored. These methods refine the trajectory of learning by addressing issues like slow convergence or instability often encountered with basic gradient descent.

Strategies like the Cascade Correlation algorithm, Optimal Brain Damage (OBD) technique, and Genetic Propagation (GProp) introduce novel ways of updating weights or restructuring networks. These techniques incorporate higher-level optimization principles and heuristic-based approaches, extending beyond traditional gradient-based updates to achieve better performance and adaptability in complex learning scenarios.

In summary, supervised learning forms the backbone of many neural network applications and its success depends on the efficient computation of error gradients, robust weight update rules, and innovative algorithms. The subsequent subsections will delve into these aspects, building on the foundational understanding of supervised learning to provide a comprehensive exploration of its methodologies.

#### 2.3.1.1 Previous Epoch Error Derivative

The previous epoch error derivative is a technique in supervised learning where the gradient information from the previous training iteration is incorporated into the current weight update process. By leveraging past gradient information, the learning algorithm can achieve smoother convergence and address issues such as oscillations or stagnation in the weight update trajectory. This approach forms the basis for momentum-based optimization techniques, which build upon the idea of utilizing gradient history.

In a standard gradient descent algorithm, the weights of the network are updated at each epoch by computing the gradient of the loss function with respect to the weights for the current training data. Mathematically, the update rule is expressed as:

$$\Delta w^{(t)} = -\eta \frac{\partial J^{(t)}}{\partial w^{(t)}} \tag{2.40}$$

Here, $\Delta w^{(t)}$ represents the weight change at epoch $t$, $\eta$ is the learning rate, $J^{(t)}$ is the loss function, and $\frac{\partial J^{(t)}}{\partial w^{(t)}}$ is the gradient of the loss function with respect to the weights at the current epoch. While this approach works well for many cases, it often struggles in scenarios where the error surface is rugged or contains local minima, leading to unstable updates.

By incorporating the gradient information from the previous epoch, the weight update rule is modified to include a weighted sum of the current and prior gradients:

$$\Delta w^{(t)} = -\eta \left( \beta \frac{\partial J^{(t-1)}}{\partial w^{(t-1)}} + (1-\beta) \frac{\partial J^{(t)}}{\partial w^{(t)}} \right) \tag{2.41}$$

In this equation, $\beta$ is a weighting factor that controls the contribution of the previous epoch's gradient to the current update. A higher value of $\beta$ emphasizes the gradient from the previous epoch, creating a momentum-like effect. The inclusion of the historical gradient prevents abrupt changes in weight updates caused by sharp variations in the error surface.

This approach can be particularly beneficial in addressing oscillations in narrow ravines of the error landscape, where steep gradients in one direction and shallow gradients in another can slow convergence. By smoothing out the trajectory of weight updates, the previous epoch error derivative ensures a more stable and efficient search for the global minimum.

One practical consideration is the choice of the weighting factor $\beta$. A value of $\beta$ that is too small diminishes the impact of past gradients, making the approach equivalent to standard gradient descent. On the other hand, an excessively large $\beta$ may overemphasize outdated gradient information, slowing the network's responsiveness to current data. Empirically, a moderate value of $\beta$ (e.g., 0.5) is often chosen to balance these trade-offs.

By utilizing information from the previous epoch's error derivative, this technique paves the way for a more controlled learning process and lays the foundation for advanced gradient optimization strategies. These strategies, such as QuickProp, R-Prop, and momentum-based methods, build upon the principles introduced here and offer more sophisticated mechanisms for efficient weight updates.

### 2.3.1.2 Present Epoch Error Derivative

The present epoch error derivative represents the instantaneous response of the neural network to the error incurred during the current training epoch. It is a direct measure of how the weights in the network should be adjusted to minimize the error. This method, fundamental to many gradient-based optimization techniques, emphasizes the real-time calculation of gradients without relying on past error data, ensuring that the weight updates are aligned with the latest state of the network.

The relationship between the loss function and the weights is expressed as in eqn. (2.40). The process for computing the derivative involves the backpropagation algorithm, which calculates the gradient layer by layer. For a given weight $w_{ij}$ connecting neuron $i$ and neuron $j$, the gradient is determined as:

$$\frac{\partial J^{(t)}}{\partial w_{ij}} = \delta_j^{(t)} a_i^{(t)} \tag{2.42}$$

where

- $\delta_j^{(t)}$ is the error signal for neuron $j$ at the current epoch $t$, derived from the difference between the predicted and target outputs, and
- $a_i^{(t)}$ is the activation of neuron $i$ at the input to the weight $w_{ij}$.

The present epoch error derivative approach ensures that the weight updates reflect the most current error signal, making it highly responsive to changes in the input data or network behaviour. This direct responsiveness can sometimes lead to challenges, such as:

- Gradient oscillations: On rugged error surfaces, abrupt changes in the gradient may cause oscillations during training.
- Sensitive to noise: If the training data contains noise, the weight updates can become erratic due to highly variable gradients.

To address these issues, many algorithms introduce modifications that balance the immediate response of the present epoch derivative with the smoothing effect of additional terms, such as momentum or adaptive learning rates. These methods are elaborated in subsequent sections.

Despite its limitations, the present epoch error derivative remains the cornerstone of many supervised learning methods. Its simplicity and directness make it an invaluable tool for training neural networks, forming the foundation upon which more advanced techniques are built.

#### 2.3.1.3 Quick-prop Update Rule

The Quick-Prop update rule is an advanced optimization technique that builds upon the limitations of standard gradient descent methods by incorporating second-order information to accelerate convergence. Unlike basic gradient-based methods, which rely solely on the first derivative of the loss function, Quick-Prop uses an approximation of the second derivative to adaptively adjust weight updates. This approach is particularly useful in scenarios where steep or flat regions in the error surface slow the learning process.

In the standard gradient descent, weight updates are calculated using eqn. 2.40. However, this approach can struggle with convergence speed when encountering plateaus or sharp curvatures in the error surface. Quick-Prop aims to address these issues by leveraging a parabolic approximation of the error function. It assumes that the error function $J(w)$ with respect to a particular weight www can be represented locally as:

$$J(w) \approx aw^2 + bw + c \tag{2.43}$$

where $a$, $b$, and $c$ are coefficients and $a > 0$ ensures convexity in the neighbourhood of $w$. From this quadratic approximation, the second derivative of the error is derived:

$$\frac{\partial^2 J}{\partial w^2} = 2a \tag{2.44}$$

Using this approximation, the Quick-Prop update for the weight is calculated as:

$$\Delta w^{(t)} = -\frac{\dfrac{\partial J^{(t)}}{\partial w^{(t)}}}{\dfrac{\partial^2 J}{\partial w^2}} \tag{2.45}$$

This update rule allows for larger steps in regions where the curvature (second derivative) is small, accelerating convergence. Conversely, in areas of high curvature, the step size is automatically reduced to maintain stability.

One of the key advantages of Quick-Prop is its ability to escape regions of slow convergence, such as plateaus, by dynamically adjusting the step size. But then, it requires additional considerations to ensure robust learning:

- Handling non-convexity: In non-convex regions, where the second derivative is not positive, Quick-Prop may lead to instability. Practical implementations often include safeguards, such as capping the maximum step size, to address this issue.
- Initialization and early training: The method performs best after a few initial epochs of training with a standard gradient descent algorithm to establish reliable gradient estimates.

Quick-Prop finds widespread application in training neural networks for tasks where fast convergence is critical. However, its dependency on the quality of the second derivative approximation can limit its effectiveness in highly complex or noisy error surfaces. But still, it remains a foundational technique in optimization, bridging the gap between traditional gradient descent methods and more advanced second-order techniques like Newton's method.

#### 2.3.1.4 R-(Resilience) Prop Update Rule

The R-(Resilience) Propagation (R-Prop) algorithm was introduced as an enhancement to gradient-based optimization techniques for neural network training. Standard gradient descent methods depend heavily on the magnitude of the gradient to compute weight updates. This dependency can lead to problems such as vanishing step sizes in flat regions of the error surface or erratic updates in steep regions. R-Prop addresses these limitations by utilizing only the sign of the gradient to determine the direction of weight updates and independently adapting step sizes for each weight during training [11].

The key innovation in R-Prop is its ability to decouple the magnitude of the weight updates from the magnitude of the gradient, resulting in more robust optimization. For each weight $w_{ij}$, the sign of the gradient guides the direction of change:

$$\Delta w_{ij}^{(t)} = \begin{cases} -\Delta_{ij}^{(t)} & \text{if } \dfrac{\partial J^{(t)}}{\partial w_{ij}} > 0, \\ \\ +\Delta_{ij}^{(t)} & \text{if } \dfrac{\partial J^{(t)}}{\partial w_{ij}} < 0. \end{cases} \tag{2.46}$$

Here, $\Delta_{ij}^{(t)}$ denotes the step size for the weight update. The algorithm dynamically adjusts these step sizes using the behaviour of the gradient sign across successive iterations:

$$\Delta_{ij}^{(t+1)} = \begin{cases} \eta^{+}\Delta_{ij}^{(t)} & \text{if } \dfrac{\partial J^{(t)}}{\partial w_{ij}} \cdot \dfrac{\partial J^{(t-1)}}{\partial w_{ij}} > 0, \\ \\ \eta^{-}\Delta_{ij}^{(t)} & \text{if } \dfrac{\partial J^{(t)}}{\partial w_{ij}} \cdot \dfrac{\partial J^{(t-1)}}{\partial w_{ij}} < 0, \\ \\ \Delta_{ij}^{(t)} & \text{otherwise.} \end{cases} \tag{2.47}$$

In this formulation, $\eta^{+} > 1$ (typically 1.2) increases the step size when successive gradients point in the same direction, signalling smooth regions of the error surface, whereas $\eta^{-} < 1$ (typically 0.5) reduces the step size when the gradient direction reverses, preventing oscillations. This dynamic adjustment provides a balance between rapid convergence and numerical stability.

#### 2.3.1.5 R-prop Weight Updating

R-Prop weight updating implements the principles established in the Resilience Propagation rule to achieve efficient and reliable optimization. It relies on the dynamically adjusted step sizes $\Delta_{ij}^{(t)}$ for weight updates, ensuring that the weight adjustment magnitude is independent of the gradient's magnitude. The updated weight at epoch $t+1$ is computed as:

$$w_{ij}^{(t+1)} = w_{ij}^{(t)} + \Delta w_{ij}^{(t)} \tag{2.48}$$

Here, $\Delta w_{ij}^{(t)}$ is derived as shown in Equation (2.46), with the step size $\Delta_{ij}^{(t)}$ governed by Equation (2.47). R-Prop weight updating seamlessly integrates the gradient's directional information with step size adjustments. By treating each weight independently, it mitigates the problems associated with uniform step sizes, such as overshooting or being stuck in regions of small gradients.

The step size $\Delta_{ij}^{(t)}$ for each weight evolves dynamically over the course of training, driven by gradient trends, as defined in the R-(Resilience) Propagation rule. These adaptations ensure faster convergence, reduced sensitivity to initial learning rates and resilience against noisy gradients or high curvature regions of the error surface.

Through its design, the R-Prop weight updating mechanism highlights the practical application of the R-Prop methodology, making it a robust and effective algorithm for training feedforward neural networks and beyond.

#### 2.3.1.6 Cascade Correlation Algorithm

The Cascade Correlation algorithm represents a unique paradigm in neural network training, diverging from traditional static architecture designs. Unlike conventional models that require predetermined network structures, this algorithm incrementally builds the network during the training process. By introducing new units one at a time, it addresses the challenge of optimizing both network topology and connection weights simultaneously. This is equivalent to simultaneous updating of model order and parameter estimates' in system identification of time-series models, like ARMA models.

The fundamental principle of the Cascade Correlation algorithm is to start with a minimal network and expand it dynamically based on the performance requirements. Initially, the network consists of only the input and output layers. During training, if the current configuration fails to achieve the desired accuracy, new hidden units are added sequentially. These newly introduced units are trained to maximize their correlation with the network's residual error, effectively "cascading" into the structure as functional blocks [12].

At each step, the new unit receives input from all previous network units, including the inputs and any pre-existing hidden units. This arrangement allows the new unit to act as a feature extractor, capturing patterns not represented by the existing network structure. The correlation between the residual error and the candidate unit's output determines its effectiveness. Maximizing this correlation involves optimizing the weights of the new unit using gradient-based methods:

$$E = \sum_{n=1}^{N} \left(y_n - \hat{y}_n\right)^2 \tag{2.49}$$

where $y_n$ is the target output, $\hat{y}_n$ is the current network output, and $N$ is the total number of training samples. The residual error drives the creation of a new unit, ensuring that subsequent additions refine the model's predictive capability.

Once a new unit is incorporated into the network, its weights are frozen. This freezing prevents interference with previous network components and ensures stability during further training. The algorithm then proceeds to train the remaining network weights while monitoring the overall error. Additional units are introduced as required until the desired error threshold is met.

This incremental construction approach offers several advantages. It eliminates the need for pre-specifying the number of hidden units, a task often subject to trial-and-error. Additionally, the hierarchical structure formed by cascading units enables the network to represent complex, non-linear functions efficiently. By adapting its architecture to the problem at hand, the Cascade Correlation algorithm achieves a balance between network complexity and performance, often resulting in smaller networks with enhanced generalization capabilities.

The Cascade Correlation algorithm is particularly well-suited for applications requiring dynamic adaptability, such as time-series prediction and pattern classification. Its modular growth mechanism provides insights into the importance of individual network components and fosters an interpretable structure.

This innovative method represents a significant step forward in the development of adaptive neural network training strategies, offering a powerful alternative to traditional fixed-architecture approaches. Its success underscores the importance of coupling architectural flexibility with robust learning paradigms to address diverse computational challenges.

#### 2.3.1.7 Optimal Brain Damage (OBD) Technique

The Optimal Brain Damage (OBD) technique is a method for pruning neural networks by strategically eliminating redundant weights, thereby simplifying the network structure without significantly degrading its performance. The central idea of OBD is that not all connections in a trained neural network are equally important. Some connections contribute minimally to the network's output and can be removed to achieve a more efficient architecture.

After a neural network is trained, the OBD technique evaluates the contribution of each weight to the error function. This evaluation is based on the Taylor series expansion of the error function, which approximates the change in error when a weight is removed. Using a second-order approximation, the change in error due to pruning a weight $w_i$ is expressed as:

$$\Delta J \approx \frac{1}{2} H_{ii} w_i^2 \tag{2.50}$$

where $J$ is the error function, $H_{ii}$ is the diagonal element of the Hessian matrix, and $w_i$ is the weight being evaluated. The diagonal element $H_{ii}$ represents the sensitivity of the error function to changes in $w_i$. A smaller $H_{ii} w_i^2$ indicates that pruning $w_i$ will have a negligible impact on the network's performance.

The process of OBD involves the following steps:

1. Training the network: The neural network is initially trained using standard methods like gradient descent to minimize the error function.
2. Evaluating weights: After training, each weight is analyzed using the Hessian matrix to calculate its contribution to the error function.
3. Pruning: Weights with the smallest contributions (lowest values of ΔJ\Delta JΔJ) are pruned systematically. This is similar to pruning of Gaussian sum components in the Gaussian sum approximation EKF method used for state and parameter estimation.
4. Fine-tuning: The network is retrained after pruning to recover any performance loss caused by the removed weights.

By removing redundant weights, the OBD technique offers several advantages:

- Reduced complexity: The pruned network is smaller, leading to faster computation and reduced storage requirements.
- Improved generalization: Simplifying the network decreases the risk of overfitting, thereby enhancing its ability to generalize to unseen data.
- Enhanced interpretability: A reduced number of parameters makes the network's functionality easier to analyze and understand.

While the OBD technique is computationally expensive due to the requirement of computing the Hessian matrix, it is highly effective for applications requiring lightweight, deployable neural networks. This makes OBD a pioneering method for optimizing neural network architectures.

#### 2.3.1.8 Genetic Learning Rule – GProp

The Genetic Learning Rule (GProp) adopts principles of evolutionary algorithms to train neural networks. Unlike gradient-based methods that rely on derivative information, GProp simulates evolutionary processes such as selection, mutation, and crossover to optimize network weights. This biologically inspired approach excels in discovering global optima, especially for problems with non-linear or discontinuous error surfaces.

In GProp, the neural network's weights are encoded as genes within individuals of a population. The algorithm begins with a population of randomly initialized individuals, where each represents a unique set of network weights. The fitness of each individual is determined by evaluating the network's performance on a given task [13].

The key operations in GProp include:

- Selection: High-performing individuals are selected to create the next generation, ensuring that advantageous traits are carried forward.
- Crossover: Selected individuals exchange genetic information (network weights) to produce offspring. This process encourages exploration of new weight configurations by combining features from different solutions.
- Mutation: Random alterations to genes introduce diversity into the population, preventing premature convergence to suboptimal solutions.
- Survivor replacement: The least fit individuals are replaced by the new offspring, ensuring that the overall fitness of the population improves over successive generations.

This iterative process continues until a termination criterion is met, such as reaching a specific error threshold or completing a predefined number of generations.

One of GProp's strengths is its independence from gradient information, making it suitable for scenarios where gradient computation is infeasible or unreliable. Furthermore, its global search mechanism enables it to overcome local minima and identify optimal weight configurations.

However, GProp's reliance on population-based evaluation introduces computational overhead, making it slower than traditional gradient-based methods for large-scale problems. Nevertheless, its flexibility and robustness make it an excellent choice for complex optimization challenges where standard methods fall short.

GProp has been successfully applied to various domains, including function approximation, pattern recognition, and reinforcement learning, showcasing its potential as a versatile tool for training neural networks. By mimicking the evolutionary processes observed in nature, GProp highlights the efficacy of biologically inspired algorithms in tackling computational problems.

### 2.3.2 Unsupervised Learning

Unsupervised learning is a paradigm of machine learning (ML) where a model is trained on unlabelled data to identify patterns, groupings, or inherent structures. Unlike supervised learning, which relies on predefined labels, unsupervised learning enables the discovery of hidden insights within the data without explicit external guidance. This is especially beneficial in scenarios where labelled data is scarce, expensive to obtain, or unavailable.

In neural networks, unsupervised learning methods are used for tasks such as clustering, density estimation, and feature extraction. The algorithms employed rely on metrics like distance, probabilistic functions, or adaptive techniques that learn patterns in input data. Key approaches include the calculation of Euclidean distance, the application of Gaussian functions and Learning Vector Quantization (LVQ), all of which play a significant role in unsupervised neural network methodologies.

#### 2.3.2.1 Euclidean Distance

The Euclidean distance is a foundational metric in unsupervised learning, used to quantify the dissimilarity between two points in an non-dimensional space. For two points represented as $\mathbf{x} = \left[x_1, x_2, \ldots, x_n\right]$ and $\mathbf{y} = \left[y_1, y_2, \ldots, y_n\right]$, the Euclidean distance $d(\boldsymbol{x}, \boldsymbol{y})$ is calculated as:

$$d(\mathbf{x},\mathbf{y}) = \sqrt{\sum_{i=1}^{N}\left(x_i - y_i\right)^2} \tag{2.51}$$

This metric underpins various clustering algorithms and is a cornerstone of techniques such as $k$-means clustering. In neural networks, it is used extensively in self-organizing maps (SOMs) and competitive learning frameworks. By grouping data points based on the Euclidean distance, models can discern patterns and classify input features into cohesive clusters.

#### 2.3.2.2 Gaussian Function

The Gaussian function is widely employed in neural networks for modelling data density and as an activation function in techniques like RBFNNs. The general form of the Gaussian function for one-dimensional data is:

$$G(x) = \frac{1}{\sqrt{2\pi\sigma^2}} \exp\left(-\frac{\left(x-\mu\right)^2}{2\sigma^2}\right) \tag{2.52}$$

where $\mu$ is the mean, $\sigma^2$ is the variance, and $x$ represents the data input. For multi-dimensional data, the Gaussian function is extended as:

$$G(\mathbf{x}) = \frac{1}{\sqrt{\left(2\pi\right)^k \left|\Sigma\right|}} \exp\left(-\frac{1}{2}\left(\mathbf{x}-\boldsymbol{\mu}\right)^T \Sigma^{-1}\left(\mathbf{x}-\boldsymbol{\mu}\right)\right) \tag{2.53}$$

where $\mu$ is the mean vector, $\Sigma$ is the covariance matrix, and $k$ represents the dimensionality of the data. Gaussian functions facilitate clustering, density estimation, and classification. In RBFNNs, the parameters $\mu$ and $\Sigma$ define the center and shape of the activation, with the response of a neuron determined by the proximity of input data to the center. This localized response ensures fine-grained modelling of input features.

#### 2.3.2.3 Learning Vector Quantization

Learning Vector Quantization (LVQ) is a prototype-based algorithm designed to classify and organize data in feature space. While it is often associated with supervised methods due to the use of labelled examples during training, its underlying mechanisms of adaptive prototype placement make it highly relevant for unsupervised clustering [14].

The training involves initializing prototype vectors to represent distinct clusters. The adaptation of these prototypes is governed by the following update rules:

For a correct classification:

$$\mathbf{w}_p \leftarrow \mathbf{w}_p + \eta\left(\mathbf{x} - \mathbf{w}_\mathrm{p}\right) \tag{2.54}$$

where $w_p$ is the prototype vector, $\mathbf{x}$ is the input vector, and $\eta$ is the learning rate.

For an incorrect classification:

$$\mathbf{w}_p \leftarrow \mathbf{w}_p - \eta\left(\mathbf{x} - \mathbf{w}_\mathrm{p}\right) \tag{2.55}$$

Through iterative updates, LVQ optimizes the prototypes to align with their corresponding clusters, effectively partitioning the feature space. This method combines the geometric intuition of distance metrics with adaptive learning principles, making it particularly suitable for problems involving high-dimensional data.

By integrating these techniques – Euclidean distance, Gaussian functions, and LVQ – unsupervised learning provides robust frameworks for neural networks to analyze and interpret complex, unlabelled datasets, paving the way for innovative applications in pattern recognition, data mining, and feature extraction. While unsupervised learning focuses on identifying patterns and structures in data, network trajectory learning algorithms concentrate on optimizing the training process itself, ensuring more efficient and effective traversal through the network's weight space.

### 2.3.3 Network Trajectory Learning Algorithms

Network trajectory learning algorithms represent a specialized class of optimization techniques in neural networks designed to improve the training process by analyzing and optimizing the trajectory of weight updates. Instead of focusing solely on the convergence to a minima in the error surface, these algorithms aim to ensure the learning process is efficient, stable, and resistant to issues like oscillations or premature convergence.

The core idea of trajectory learning lies in tracking and refining how a neural network traverses its multidimensional error surface. The trajectory, a path followed by the weights of a network during training is influenced by the choice of learning rate, the optimization method employed, and the properties of the loss function. These algorithms optimize not just the final result but also the dynamic path taken to achieve convergence, which can significantly affect both training time and the generalization capability of the model.

To conceptualize the approach, consider a multidimensional weight space where the cost function forms a highly non-convex surface. The trajectory of learning is a path from an initial random weight configuration to an optimal weight configuration that minimizes the cost. If the path is smooth, convergence is quicker and computationally less intensive. But, if the path oscillates or diverges due to steep or flat regions on the surface, the process becomes inefficient and error-prone.

Key methodologies employed within trajectory learning algorithms include momentum-based techniques, dynamic learning rate adjustments, and second-order methods such as Newton's method and conjugate gradient descent. These algorithms ensure that the learning trajectory remains efficient and effective by adapting the weight updates dynamically to the curvature and topology of the error surface.

A fundamental advantage of trajectory learning algorithms is their ability to mitigate common pitfalls in neural network training. For example, these methods prevent overshooting in regions of sharp gradients by dynamically reducing the step size, while simultaneously accelerating convergence in flat regions by increasing the step size appropriately. Furthermore, they help navigate saddle points, areas of the cost surface with zero gradients but non-optimal solutions, by adjusting the trajectory to escape these plateaus effectively.

Mathematically, trajectory learning can be enhanced by incorporating terms that capture the direction and magnitude of updates in weight space. For instance, momentum-based updates can be expressed as:

$$\Delta \mathbf{w}_t = \eta \nabla E\left(\mathbf{w}_t\right) + \alpha \Delta \mathbf{w}_{t-1} \tag{2.56}$$

where $\mathbf{w}_t$ represents the weights at iteration $t$, $\eta$ is the learning rate, $\nabla E\left(w_t\right)$ is the gradient of the error function, and $\alpha$ is the momentum factor that incorporates previous weight changes.

Additionally, second-order methods like Newton's method employ curvature information through the Hessian matrix $\mathbf{H}$:

$$\Delta \mathbf{w} = -\mathbf{H}^{-1} \nabla E\left(\mathbf{w}\right) \tag{2.57}$$

enabling precise trajectory adjustments, especially in highly nonlinear error surfaces.

Network trajectory learning algorithms are particularly relevant in large-scale, deep neural networks, where traditional gradient-based optimization can become prohibitively slow or unreliable. By focusing on the pathway of learning, these algorithms provide a nuanced approach to training, ensuring that neural networks achieve better generalization and faster convergence. Their applications range from deep learning in image and speech recognition to reinforcement learning in dynamic systems, underscoring their versatility and importance in modern AI systems.

#### 2.3.3.1 Trajectory Matching Algorithms

A crucial refinement in trajectory-based learning is the Trajectory Matching Algorithm (TMA), which originates from state-space modelling and recurrent neural network

(RNN) parameter estimation [15]. This algorithm formulates weight updates as a trajectory-matching problem, where the neural network parameters are optimized to follow a predefined desired trajectory. By modelling the learning process in terms of trajectory evolution, this approach provides a structured method to ensure smooth and controlled convergence.

Four distinct RNN variants have emerged as powerful tools for trajectory matching, each offering unique perspectives on trajectory learning.

a) For HNN or RRN-S Model

The first of these is the Hopfield Neural Network (HNN), also known as RNN-S, where the sigmoid nonlinearity operates directly on the states. The network dynamics unfold according to the equation:

$$\dot{U}(t) = -U(t)R^{-1} + W\{f(U(t))\} + b \tag{2.58}$$

Here, $U$ represents the internal states of the neurons and $f(U(t))$ gives us the output state. The weight matrix $W$ and input vector $b$ work together to shape the network's behaviour, with $R$ representing the neuron impedance. This fundamental architecture serves as a foundation for understanding more complex variations.

The trajectory matching in HNN consists of a three-step process. First, the cost function measures the discrepancy between desired and actual trajectories:

$$E = \frac{1}{2}\int_{t_0}^{t}\left[\sum_{i=0}^{q}\left(d_i(t) - U_i(t)\right)^2\right]dt \quad \text{for} \quad t \to \infty \tag{2.59}$$

where $d(t)$ is the desired trajectory. This squared error formulation provides a smooth, differentiable surface for optimization. The algorithm then computes the neural state derivative gradient:

$$\frac{\partial \dot{U}_i}{\partial w_{kj}} = -\frac{\partial U_l}{\partial w_{kj}} + \delta_{ki} f(U_l) + \sum_{l=1}^{q} w_{il} f'(U_l)\frac{\partial U_l}{\partial w_{kj}} \tag{2.60}$$

Starting from zero initial conditions, this gradient guides the network toward better trajectory matches. $\delta$ is the Kronecker delta.

b) For RNN-FI Model

The RNN-FI variant represents the next evolution, where 'FI' indicates that the nonlinearity operates on the forcing input. Think of this as applying the nonlinear transformation to the combined effect of weighted states and network input. The dynamics follow:

$$\dot{U}(t) = -U(t)R^{-1} + f(WU(t) + b) \tag{2.61}$$

This formulation reveals an interesting relationship with HNN through an affine transformation, provided the weight matrix W is invertible. The trajectory matching process mirrors that of HNN, but with a modified neural state derivative gradient that reflects the forcing input structure:

$$\frac{\partial \dot{U}_i}{\partial w_{kj}} = -\frac{\partial U_l}{\partial w_{kj}} + f'(FI)\delta_{ki}U_j + \sum_{l=1}^{q} w_{il}\frac{\partial U_l}{\partial w_{kj}} \tag{2.62}$$

c) For RNN-WS Model

The third variation, RNN-WS, takes yet another approach by applying nonlinearity to the weighted states. Its dynamic equation:

$$\dot{U}(t) = -U(t)R^{-1} + f(WU(t)) + b \tag{2.63}$$

shows how the network processes the weighted combinations of states before applying the nonlinear transformation. This network maintains a linear relationship with HNN, again requiring an invertible weight matrix. The neural state derivative gradient is given as:

$$\frac{\partial \dot{U}_i}{\partial w_{kj}} = -\frac{\partial U_l}{\partial w_{kj}} + f'(s)\left(\delta_{ki}U_j + \sum_{l=1}^{q} w_{il}\frac{\partial U_l}{\partial w_{kj}}\right) \tag{2.64}$$

where $s = \sum_{j=1}^{q} w_{ij}U_j$

d) For RNN-E Model

The most distinctive approach comes from RNN-E, which applies nonlinearity directly to the residual error. This network's dynamics follow:

$$\dot{U}(t) = -U(t)R^{-1} + WU(t) + b \tag{2.65}$$

What makes RNN-E particularly interesting is its use of a convex cost function:

$$E = \frac{1}{2}\int_{t_0}^{t}\left[\frac{p}{\lambda}\sum_{i=1}^{q} \ln cosh\left(\lambda\left(d_i(t) - U_i(t)\right)\right)\right] \tag{2.66}$$

The neural state derivatives and cost function gradient take on forms that reflect this error-centric approach:

$$\frac{\partial \dot{U}_i}{\partial w_{kj}} = -\frac{\partial U_l}{\partial w_{kj}} + \delta_{ki}U_j + \sum_{l=1}^{q} w_{il}\frac{\partial U_l}{\partial w_{kj}}$$

$$\frac{\partial E}{\partial w_{kj}} = -\sum_{i=1}^{q}\left[\int_{t_0}^{t} \rho tanh\left(\lambda e_i(t)\right)\frac{\partial U_i}{\partial w_{kj}}dt\right] \tag{2.67}$$

In practical implementations, these algorithms require careful attention to several key parameters. The sigmoid nonlinearity parameters λ and μ significantly influence the network's learning behaviour. The neuron impedance R typically takes on high values to ensure stable dynamics. Initial conditions for neural state derivatives start at zero, allowing the network to build up its response from a neutral state.

The beauty of these trajectory matching algorithms lies in their systematic approach to learning dynamic behaviours. Whether through direct state manipulation (HNN), forcing input processing (RNN-FI), weighted state transformation (RNN-WS), or error-based learning (RNN-E), each variant offers unique advantages for different types of dynamic systems. The mathematical relationships between these variants – particularly the affine and linear transformations linking them to HNN – reveal the underlying unity in their diverse approaches to trajectory matching.

## EXERCISES

1. Explain the primary challenges faced by neural networks during training on highly non-convex error surfaces. How do trajectory learning algorithms address these issues? Discuss using momentum-based and second-order methods as examples.
2. Derive the weight update rule for a neural network using Newton's method. Discuss the role of the Hessian matrix in this algorithm and its limitations in terms of computational cost and application to deep learning.
3. Discuss the differences between traditional gradient-based optimization and trajectory learning algorithms in neural networks. In what scenarios would trajectory learning algorithms significantly outperform simple gradient descent?

## REFERENCES


1. Eberhart, R.C., and Dobbins, R.W., *Neural Network PC Tools – A Practical Guide*, Academic Press, Inc., New York, USA, 1993.
2. Irwin, G.W., Warwick, K., and Hunt, K.J. (Eds.), *Neural Network Applications in Control, IEE Control Engineering Series 53*, The IEE, London, U.K., 1995.
3. Widrow, B., and Lehr, M.A., "Thirty Years of Adaptive Neural Networks: Perceptron, Madaline and Back Propagation", *Proceeding of the IEEE,* Vol. 78, No. 9, pp. 1415–1442, Sept. 1990.
4. Linse, D.J., and Stengel, R.F., "Identification of Aerodynamic Coefficients Using Computational Neural Networks", *Journal of Guidance, Control and Dynamics*, Vol. 16, No. 6, pp. 1018–1025, Nov-Dec. 1993.
5. Narendra, K.S., and Parthasarthy, K., "Identification and Control of Dynamical Systems Using Neural Networks", *IEEE Transactions On Neural Networks*, Vol. 1, No. 1, pp. 4–27, March 1990.

6. Raol, J.R., "Feed Forward Neural Networks for Aerodynamic Modelling and Sensor Failure Detection", *Journal of Aeronautical Society of India*, Vol. 47, No. 4, pp. 193–199, Nov. 1995.
7. Raisinghani, S.C, Ghosh, A.K., and Kalra, P.K, "Two New Techniques for Aircraft Parameter Estimation Using Neural Networks", *Aeronautical Journal*, Vol. 102, No. 1011, pp. 25–29, 1998.
8. Werbos, P.J., "Back Propagation Through Time: What It Does and How to Do It", *Proceeding of the IEEE*, Vol. 78, No. 10, pp. 1550–1560, Oct. 1990.
9. Scalero, R.S., and Tepedelenliogh, N., "A Fast New Algorithm for Training Feed Forward Neural Networks", *IEEE Transactions on Signal Processing*, Vol. 40, No. 1, pp. 202–210, Jan. 1992.
10. Raol, J.R., "Parameter Estimation of State-Space Models by Recurrent Neural Networks", *IEE Proceeding Control Theory and Applications (U.K.)*, Vol. 142, No. 2, pp. 114–118, March 1995.
11. Riedmiller, M., and Braun, H., "A Direct Adaptive Method for Faster Backpropagation Learning: The RPROP Algorithm", *IEEE International Conference on Neural Networks*, San Francisco, CA, USA, pp. 586–591, 1993.
12. Fahlman, S.E., and Lebiere, C., "The Cascade-Correlation Learning Architecture", In D.S. Touretzky (Ed.), *Advances in Neural Information Processing Systems 2 (NIPS 1989)*, Morgan Kaufmann Publishers, San Francisco, CA, USA, pp. 524–532, 1990.
13. Angeline, P.J., Saunders, G.M., and Pollack, J.B., "An Evolutionary Algorithm that Constructs Recurrent Neural Networks", *IEEE Transactions on Neural Networks*, Vol. 5, No. 1, pp. 54–65, Jan. 1994.
14. Kohonen, T., "Learning Vector Quantization", *Neural Networks*, Vol. 19881, No. 1, pp. 303, 1988.
15. Raol, J.R., "Parameter Estimation of State Space Models by Recurrent Neural Networks", *IEE Proceedings-Control Theory and Applications*, Vol. 142, No. 2, pp. 114–118, March 1995.

# 3 System Identification

A study of dynamic systems can be performed via analytical or experimental means. An analytical approach yields a mathematical model based on several assumptions (because the real- life dynamic systems are mostly/often nonlinear and distributed in nature) and application of theory and concepts of modelling. In an experimental approach one has to design an appropriate experiment and generate data for analysis.

## 3.1 CLASSICAL IDENTIFICATION METHODS

The analysis is mainly carried out by using signal processing, system identification, and parameter estimation techniques, which is highly a mature field, and includes the modern and classical methods. Mathematical modelling of dynamic systems is very essential for analysis, design, and simulation of control systems and the prediction of the behaviour of these systems. A mathematical model characterizes the static and dynamic behaviour of a system. Some practical examples are: use of mathematical models in flight simulation and design of control systems for aerospace vehicles (like aircraft, rotorcraft, missile, unmanned aerial vehicles, micro air vehicles, and spacecraft), weather forecasting, control of robotic systems, use in optimization of chemical processes in industry, and analysis of communications and biomedical signals. In many problems in science and engineering, one finds a crucial need to represent a dynamic system by its mathematical model. Some simple and classical methods are briefly discussed next.

### 3.1.1 Transfer Function Models

As many of us might be aware, the studies of electric/electronic circuits, basic linear control systems, mechanical systems, and related analyses lend to the so-called transfer function (TF) analysis. A TF is defined, for single-input single-output (SISO) linear systems, as the ratio of the Laplace transform (LT) of the output and the Laplace transform of the input. It is often very important to quickly determine the frequencies and damping ratios of the dynamic systems; these parameters can be determined from the TFs.

We consider the TF analysis in which the coefficients of numerator and denominator polynomials (in discrete time or continuous time; 'z' or 's' – complex frequency) are estimated using some least squares method from the real data of a dynamic

DOI: 10.1201/9781003616580-3

system. This estimation can be performed to obtain either discrete-time series models (to obtain pulse transfer function – in ‘z’ domain) or directly the continuous-time models (by using appropriate methods in ‘s’ domain). Additionally one can obtain the Bode diagram, the amplitude ratio, and phase of the TF as a function of frequency.

#### 3.1.1.1 Continuous Time Model

Let a system be described by linear constant coefficient ordinary n-order differential equation:

$$d^n y(t)/dt^n + a_{n-1}\, d^{n-1} y(t)/dt^{n-1} + \ldots + a_0\, y(t) = \tau_p\, d^p\, u(t)/dt^p + \ldots + \tau_0\, u(t) \quad (3.1)$$

Here, *u(t)* is system input and *y(t)* its output. Taking the LT and assuming that the initial conditions are zero we obtain the continuous-time TF (CTTF):

$$y(s) = G(s)u(s) = \frac{\tau_p s^p + \tau_{p-1} s^{p-1} + \ldots + \tau_1 s + \tau_0}{s^n + a_{n-1} s^{n-1} + \ldots + a_1 s + a_0} u(s) \quad (3.2)$$

Here, *s* is a complex (frequency) variable. This is one type of the mathematical model of a SISO system. The linear theory of analysis can be applied to this model and the system’s properties can be analytically derived from this model itself. The natural dynamic behaviour of the system, in fact now of the model, can be evaluated by computing the roots of the denominator of G(s). These roots describe the various natural modes of the system, assuming the model is nearly a good representation of the system, and dictate the amplitude and the phase of the response. If the value of n is equal to 2 then this transfer function is called second order TF and its denominator has two roots (real/complex).

**Example 3.1**: Obtain the Bode-diagrams for the TF of an electrical circuit of Figure 3.1 for the time constant of 0.1 and 0.5 sec. The resistor is R Ohm and the capacitance is C microfarad. The time constant is given as RC= $\tau$ (sec.).

$$G(s) = \frac{V_o}{V_i} = \frac{1/\tau}{s + 1/\tau} \quad (3.3)$$

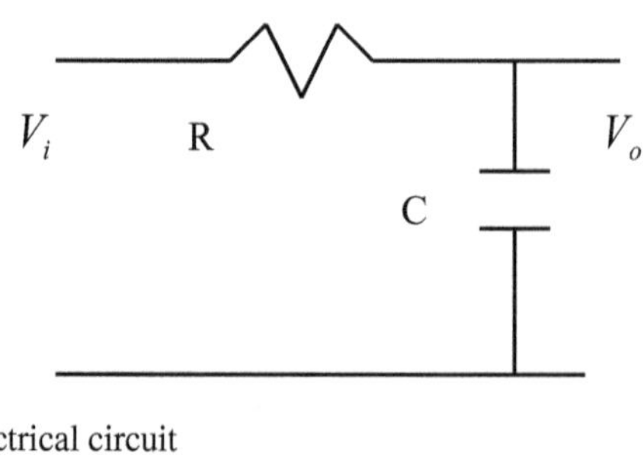


**FIGURE 3.1** RC-Electrical circuit

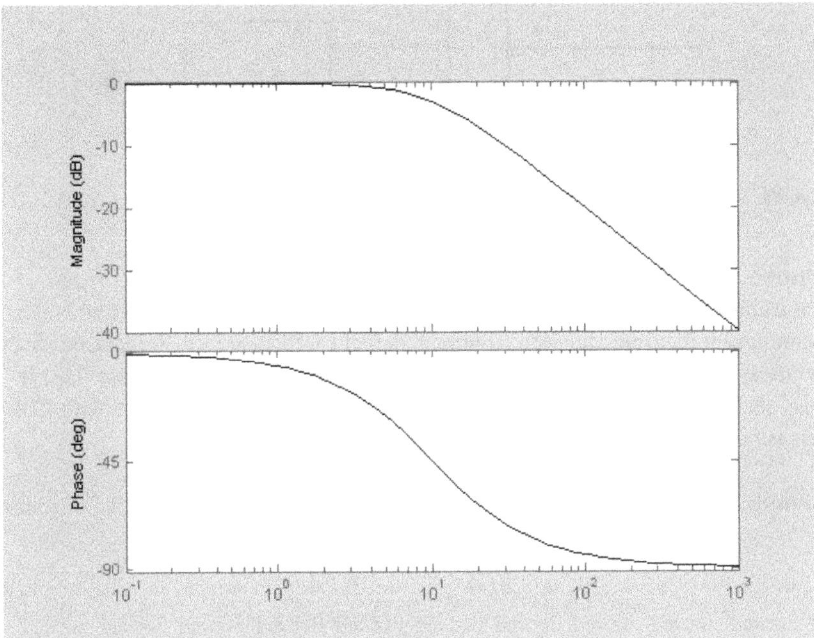


**FIGURE 3.2a** Bode plot of the first order continuous-time system with $\tau = 0.1$ s

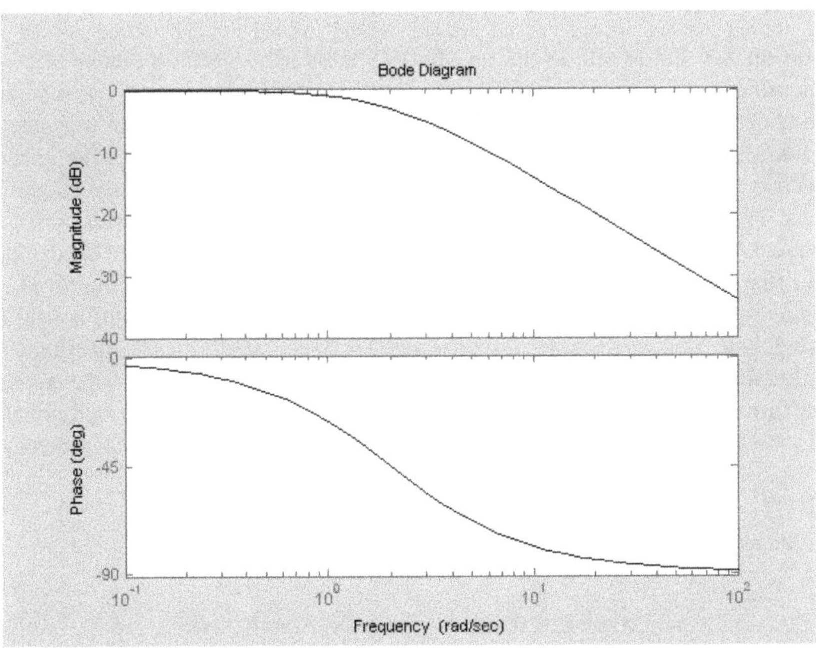


**FIGURE 3.2b** Bode plot of the first order continuous-time system with $\tau = 0.5$ s

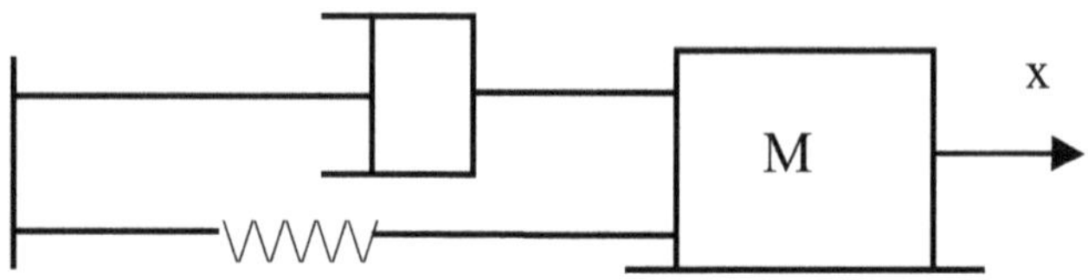


**FIGURE 3.3** Mechanical spring mass second order system

**Solution 3.1:** Specify numerator and denominator of the TF as num=[10], den=[1 10] (in the command window of the MATLAB). Then use sys=tf(num,den) and bode(sys) to obtain/plot the Bode diagram (Figure 3.2a and b). Repeat with 10 replaced by 2. At the cutoff frequencies (Cof, cutoff frequency=1/time constant) the phase (lag) is –45 degs., the property of the first order TF. The bandwidth of the system with Cof=10 rad/s is broader than that of with Cof = 2 rad/s.

**Example 3.2:** The TF of a second order mechanical system of Figure 3.3 is given as

$$G(s)=\frac{X(s)}{f(s)}=\frac{K/M}{s^2+(D/M)s+K/M} \tag{3.4}$$

Here, M is the mass, D is the damping constant, and K is the spring stiffness/constant. Use suitable values for M, D, and K and obtain the bode diagrams. Study the improvement in the damping property by increasing the value of D by a factor of 2.

**Solution 3.2:** Let M=10, D=10, and K=100, arbitrarily. Obtain the numerator and denominator of the TF as num=[10], den=[1 10/10 100/10]. Then use sys=tf(num,den) and bode(sys) to obtain/plot the Bode diagram. We see that the natural frequency is sqrt(K/M)=3.1623 rad/s and the damping ratio = 0.1581 (of the second order system), which is not the same as the damping constant. Due to low damping the magnitude of the TF has a peaking value, see Figure 3.4a. Repeat the solution with D=20, see Figure 3.4b for this case. The increased damping constant has increased the damping ratio (the magnitude of the peak is reduced). The step responses are improved, see Figure 3.4c. The step response is generated by step(sys). The first plot is held (by typing 'hold') and then the step response with D=20 gets automatically superimposed. We see that still the time response is not very well damped. With D=40 the response (with arrow) is very well damped and acceptable. The time constant is approximately the same but the transients are suppressed very well when the value of D is increased.

#### 3.1.1.2 Discrete-time Model

Let the system be described by the difference equation as

$$\begin{aligned} y(k)+a_1\, y(k-1)+\ldots+a_n y(k-n) &= b_0\, u(k) \\ &+ b_1 u(k-1)+\ldots+b_m\, u(k-m) \end{aligned} \tag{3.5}$$

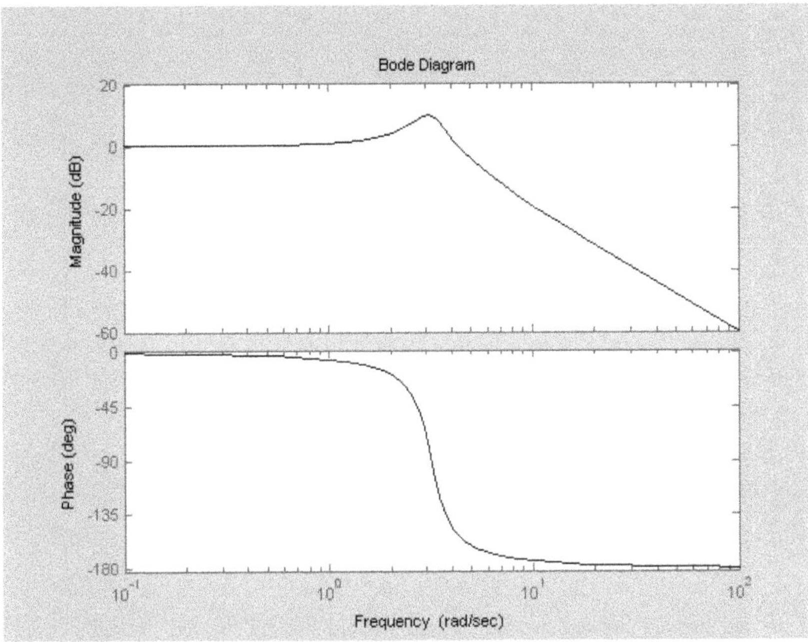


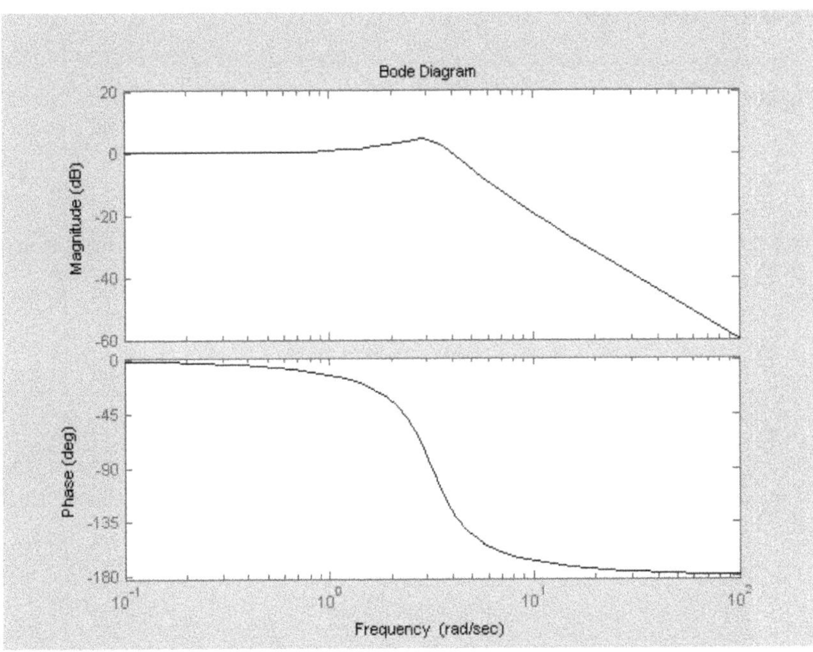


**FIGURE 3.4** (a) Bode plot of the second order continuous-time system with D=10. (b) Bode plot of the second order continuous-time system with D=20. (c) Step responses of the second order system with D=10, 20, 40

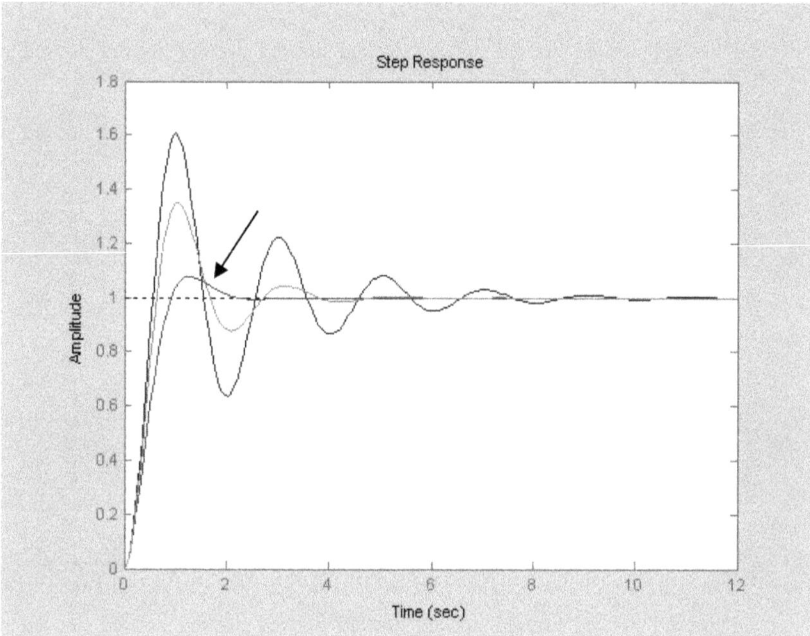


**FIGURE 3.4** (Continued)

Using the time shift operator we obtain

$$\mathbf{q}^{-1}y(k) = y(k-1) \tag{3.6}$$

From eqn. (3.5) the corresponding polynomials, on the left and right sides are obtained as

$$A\,(q^{-1}) = 1 + a_1\ q^{-1} + \ldots + a_n\ q^{-n}$$

$$B(q^{-1}) = b_0 + b_1\ q^{-1} + \ldots + b_m\ q^{-m} \tag{3.7}$$

Thus, we have A( $\mathbf{q}^{-1}$ ) y(k) = B( $\mathbf{q}^{-1}$ ) u(k) and applying the Z-transformation we get the so-called discrete transfer function (DTF/pulse transfer function or shift form TF) for the discrete time linear dynamic system:

$$G(z^{-1}) = \frac{b_o + b_1\ z^{-1} + \ldots + b_m\ z^{-m}}{1 + a_1\ z^{-1} + \ldots + a_n\ z^{-n}} \tag{3.8}$$

Here, z is a complex variable, m,n (n>m) are the orders of the respective polynomials. The discrete Bode diagram can also be obtained. The roots of the denominator polynomial will give modes of the system in terms of frequency and damping ratio that are important parameters to be ascertained from real data as first cut results. But, these are in the z domain and not in the continuous-time domain. For proper and formal interpretation, one should transform these into s-domain parameters. One major limitation of this approach is that there is no apparent correspondence between the numerical values of the coefficients of the DTF/SFTF and the corresponding CTTF of the real system.

### 3.1.2 Time-Series Models

The time series methods have gained considerable acceptance in system identification literature in view of their inherent simplicity and flexibility. These techniques provide external descriptions of systems under study and lead to parsimonious, minimum parameterization representation of the process. The accurate determination of the dynamic order of the time series models is a necessary first step in system identification. The coefficients of time-series and TF models are the parameters, which can be estimated by using the basic least squares.

The time-series modelling is one of the specialized aspects of system identification/parameter estimation study. It addresses the problem of determining coefficients of a differential or difference equation, which can be fitted to the empirical data, or obtaining coefficients of a transfer function model of a system from its input-output data. One of the main aims of time-series modelling is the use of the model for prediction of the future behaviour of the system or phenomena. One of the major applications of this approach is to understand various natural phenomena, e.g., rainfall-runoff prediction. In general, time-series are a result of stochastic (random) input to some system or some inaccessible random like influence on some phenomenon, e.g., the temperature variation at some point in a room at a certain time. Hence, a time-series can be considered as a stochastic phenomenon. The modelling and prediction of the seasonal time-series are equally important and can be handled using extended estimation procedures. Often, assumption of ergodicity is made in dealing with time-series modelling aspects. We will generally deal with discrete-time systems. Although many phenomena occurring in nature are of continuous type and can be described by continuous-time models, the theory of the discrete-time modelling is very handy and the estimation algorithms can be easily implemented using a digital computer. In addition, the discrete-time noise processes can be easily handled and represented by simple models. However, continuous-time phenomena can also be represented by a variety of (similar) time-series models. A general linear stochastic discrete time system/model is described here with the usual meaning for the variables:

$$\begin{aligned} x(k+1) &= \Phi_k x(k) + Bu(k) + w(k) \\ z(k) &= Hx(k) + Du(k) + v(k) \end{aligned} \tag{3.9}$$

However, for time-series modelling a canonical form known as Astrom's model is used:

$$A(q^{-1})z(k) = B(q^{-1})u(k) + C(q^{-1})e(k) \tag{3.10}$$

Here, $A$, $B$, and $C$ are polynomials in $q^{-1}$ which is a shift operator defined as:

$$q^{-n}z(k) = z(k-n) \tag{3.11}$$

For a SISO system, we have the expanded form as:

$$\begin{aligned} z(k)+a_1z(k-1)+\ldots+a_nz(k-n) &= b_0u(k)+b_1u(k-1)+\ldots+b_mu(k-m) \\ &\quad + e(k)+c_1e(k-1)+\ldots+c_pe(k-p) \end{aligned} \tag{3.12}$$

Here $z$ is the discrete measurement sequence, $u$ is the input sequence, and $e$ is the random noise/error sequence. We have the following equivalence:

$$\begin{aligned} A(q^{-1}) &= 1+a_1q^{-1}+\ldots+a_nq^{-n} \\ B(q^{-1}) &= b_0+b_1q^{-1}+\ldots+b_nq^{-m} \\ C(q^{-1}) &= 1+c_1q^{-1}+\ldots+c_nq^{-p} \end{aligned} \tag{3.13}$$

Here, $a_i, b_i$ and $c_i$ are the coefficients to be estimated. We also assume here that the noise processes w and v are uncorrelated and white. In addition, we assume that the time series we deal with are stationary in the sense that first and second order (and higher) statistics are not dependant on time '$t$' explicitly. There are certain special forms of these time-series models. We assume here that inputs are such that they excite the modes of the system. This means that the input contains sufficient frequencies to excite the dynamic modes of the system. This will in turn assure that in the output, there is sufficient effect of the modes and hence the information so that from input/output time-series data, one can accurately estimate the characteristics of the process.

a) **Astrom's model**: This is the most general linear time-series analysis model, with the full form of error/noise model. Given input ($u$)/output ($z$) data, the parameters can be estimated by some iterative process, e.g., ML method. The transfer function form is given by:

$$z = \frac{B(q^{-1})}{A(q^{-1})}u + \frac{C(q^{-1})}{A(q^{-1})}e \tag{3.14}$$

This model can be used to fit time-series data, which can be considered to be arising out of some system phenomenon with a controlled input $u$ and a random excitation.

b) **Autoregressive (AR) model**: By assigning $b_i = 0$ and $c_i = 0$ in the Astrom's model, we get:

$$z(k) = -a_1z(k-1)-\ldots-a_nz(k-n)+e(k) \tag{3.15}$$

The transfer function form can be easily obtained as:

$$z = \frac{1}{A(q^{-1})} e \tag{3.16}$$

Here, the output process $z(k)$ depends on its previous values (and hence the name autoregressive) and it is excited by the random signal $e$. It is assumed that the parameters $a_i$ are constants such that the process z is stationary.

We can consider that $1/A(q^{-1})$ is an operator, which transforms the process $e$ into the process $z$. The polynomial $A$ determines the characteristics of the output signal $z$ and the model is called an "all poles" model. This is because the roots of $A(q^{-1}) = 0$ are the poles of the transfer function model. The input process $e$ is inaccessible and immeasurable. The parameters of $A$ can be estimated by using the least squares method. In addition, this model is very useful for determining the spectrum of the signal $z$, if input process $e$ is considered as a white process noise, since the parameters of $A$ are estimated and hence known. This method of estimation of spectrum of a signal contrasts with the one using the Fourier Transform. However, both the methods are supposed to give similar spectra. It is most likely that the autoregressive spectrum will be smoother compared to the Fourier spectrum.

c) **Moving average (MA) model**: If we put $a_i = 0$ and $b_i = 0$ in the Astrom's model, we get:

$$z(k) = e(k) + c_1 e(k-1) + \ldots + c_p e(k-p) \tag{3.17}$$

The process $z$ is now a linear combination of the past and present values of the inaccessible random input process $e$.

The roots of $C(q^{-1}) = 0$ are the zeros of the model. The process z is called the MA process and is always stationary since $A(q^{-1}) = 1$. In this form, the output signal does not regress over its past values.

d) **Autoregressive moving average (ARMA) model**

Letting $b_i = 0$ in the Astrom's model, we obtain an ARMA model, since it contains both AR and MA parts. We emphasize here that the control input $u$ is absent:

$$z(k) + a_1 z(k-1) + \ldots + a_n z(k-n) = e(k) + c_1 e(k-1) + \ldots + c_p e(k-p) \tag{3.18}$$

$$z = \frac{C(q^{-1})}{A(q^{-1})} e \tag{3.19}$$

So this model is a zero/pole type model and has the structure of the output/input model. More complex time-series can be accurately modelled using this model.

e) **Least squares model**
By letting $c_i = 0$ in the Astrom's model, we get:

$$\begin{aligned} z(k)+a_1z(k-1)+\ldots+a_nz(k-n) &= b_0u(k) \\ &+b_1u(k-1)+\ldots+b_mu(k-m)+e(k) \end{aligned} \tag{3.20}$$

Here, control input $u$ is present. The model is so called since its parameters can be easily estimated by the LS method. The transfer function form is:

$$z = \frac{B(q^{-1})}{A(q^{-1})}u + \frac{1}{A(q^{-1})}e \tag{3.21}$$

It has an AR model for the noise part and the output/input model for the signal part. Determination of $B(q^{-1})/A(q^{-1})$ gives the transfer function model of the system. One can obtain a discrete Bode diagram of the system from this pulse transfer function and then convert it to continuous time domain to interpret the dynamic behaviour of the system. One can use complex curve fitting technique or bilinear/pade method.

### 3.1.3 Least Squares Method

The estimation of parameters of MA and ARMA can be done using the ML approach, since the unknown parameters appear in MA part, which represents itself as unknown time series '$e$'. However, parameters of AR and LS models can be estimated using the LS method. Assumption of the identifiability of the coefficients of the postulated models is pre-supposed.

From the LS model, eqn. (3.21), we define the equation error as:

$$e(k) = A(q^{-1})z(k) - B(q^{-1})u(k); r(k) = \hat{A}(q^{-1})z(k) - \hat{B}(q^{-1})u(k) \tag{3.22}$$

Eqn. (3.22) can be put in the form:

$$z = H\beta + e; \text{ with } z = \{z(n+1), z(n+2), \ldots z(n+N)\}^T \tag{3.23}$$

Here,

$$H = \begin{bmatrix} -z(n) & -z(n-1) & -z(1) & u(n) & u(n-1) & u(1) \\ -z(n+1) & -z(n) & -z(2) & u(n+1) & u(n) & u(2) \\ \vdots & & \vdots & \vdots & & \\ -z(N+n-1) & \cdots & -z(n) & u(N+n-1) & \cdots & u(N) \end{bmatrix} \tag{3.24}$$

$N$ = number of total data used ; $m$=$n$ and $b_0 = 0$. For example, let $n$=2 and $m$= 1, then

$$e(k) = z(k) + a_1 z(k-1) + a_2 z(k-2) - b_0 u(k) - b_1 u(k-1) \tag{3.25}$$

$$z(k) = \begin{bmatrix} -z(k-1) & -z(k-2) \end{bmatrix} \begin{bmatrix} a_1 \\ a_2 \end{bmatrix} + \begin{bmatrix} u(k) & u(k-1) \end{bmatrix} \begin{bmatrix} b_0 \\ b_2 \end{bmatrix} + e(k) \tag{3.26}$$

$$z(k+1) = \begin{bmatrix} -z(k) & -z(k-1) \end{bmatrix} \begin{bmatrix} a_1 \\ a_2 \end{bmatrix} + \begin{bmatrix} u(k+1) & u(k) \end{bmatrix} \begin{bmatrix} b_0 \\ b_2 \end{bmatrix} + e(k+1) \tag{3.27}$$

The above leads to:

$$z = H\beta + e \tag{3.28}$$

Using the LS method, we get:

$$\hat{\beta} = \left\{ \hat{a}_1, \hat{a}_2, \ldots, \hat{a}_n \ : \ \hat{b}_1, \hat{b}_2, \ldots, \hat{b}_m \right\} = (H^T H)^{-1} H^T z \tag{3.29}$$

Since, we have all the information on H and z, we can easily determine the coefficients of the time series/LS model; also, the parameters/coefficients of time-series/LS models can be estimated using the System identification Toolbox of MATLAB. The crucial aspect of time-series modelling is that of the selection of the model structure (AR, MA, ARMA or LS) and the number of coefficients for fitting this model to the time-series data [1].

## 3.2 PARALLEL AND SERIES IDENTIFICATION METHODS

Two principal approaches in system identification are parallel identification and series identification, which differ in their implementation strategies and applicability.

### 3.2.1 Parallel Identification

Parallel identification involves running an identification model simultaneously with the actual system, allowing for real-time parameter updates. The identification model takes the same inputs as the system and attempts to approximate the system output by adjusting its internal parameters adaptively [2]. A parallel identification model is defined by:

$$\hat{y}(k+1) = \sum_{i=0}^{n-1} a_i(k)\hat{y}(k+i) + \sum_{j=0}^{m-1} b_j(k)u(k+j) \tag{3.30}$$

where $\hat{y}(k+1)$ is the estimated system output, $a_i(k)$ and $b_j(k)$ are adjustable parameters, and $u(k)$ is the input signal. The error function used to optimize this model is:

$$e(t) = y(t) - \hat{y}(t) \tag{3.31}$$

where $y(t)$ is the actual system output. The parameters of the model are updated using dynamic backpropagation, a method that extends standard backpropagation to time-dependent systems.

A common gradient-based update rule is:

$$\theta(k+1) = \theta(k) + \eta \nabla_{\theta} e(t) \tag{3.32}$$

where $\theta(k)$ represents the parameter vector (combining coefficients $a_i$ and $b_i$) and $\eta$ is the learning rate. Properties of parallel identification are: i) real-time adaptation – parameters update continuously to track time-varying systems; ii) error minimization – updates minimize $e(t)$ using gradient descent or other adaptive techniques; and iii) application in adaptive control – useful for neural network-based adaptive controllers, where real-time adjustments improve system performance. On the other hand, a limitation of parallel identification is that it does not guarantee stability, particularly when feedback exists within the system model [2].

### 3.2.2 Series-parallel Identification

To overcome stability issues, series-parallel identification is often used. Unlike parallel identification, this approach feeds the actual system output back into the model, reducing error accumulation and improving the system's stability. The series-parallel identification model is represented as:

$$\hat{y}(k+1) = \sum_{i=0}^{n-1} a_i(k) y(k-i) + \sum_{j=0}^{m-1} b_j(k) u(k-j) \tag{3.33}$$

where the key difference from parallel identification is that the actual system output $y(k)$, rather than the estimated output $\hat{y}(k)$, is used in the equation. The error function remains the same as eqn. (3.31). Since the identification model does not include a direct feedback loop, its parameters can be updated using static backpropagation, reducing computational overheads. The gradient-based update rule also remains the same as in eqn. (3.32).

Properties of Series-Parallel Identification are: i) improved stability – reduces instability issues seen in parallel identification; ii) lower computational complexity – uses static backpropagation, simplifying parameter updates; and iii) application in neural network-based system identification – effective in training neural networks to model complex system behaviours. Because of these advantages, the series-parallel model is preferred in many system identification applications.

**TABLE 3.1**
**Comparison-parallel and series parallel identification**

| Feature | Parallel identification | Series-parallel identification |
|---|---|---|
| **Real-time adaptation** | Yes (Continuous Learning) | No (Batch Learning) |
| **System feedback** | Uses estimated output $y(k)$ | Uses actual system output y(k) |
| **Stability** | May be unstable | More stable |
| **Computational complexity** | Higher | Lower |
| **Training method** | Dynamic Backpropagation | Static Backpropagation |

Table 3.1 summarizes the differences between parallel and series-parallel identification:

## 3.3 IDENTIFICATION WITH NEURAL NETWORKS

System identification is the process of developing mathematical models of dynamic systems based on observed input-output data. Traditional approaches such as transfer functions and state-space models have been widely used, but they often rely on linearity assumptions and require detailed prior knowledge of system dynamics. Neural networks, particularly feedforward (FFNN) and recurrent (RNN) architectures, provide a powerful alternative by approximating complex nonlinear relationships directly from data.

Neural network-based system identification involves training a network to minimize the error between actual and predicted system outputs. The network learns an implicit mapping between inputs and outputs, making it particularly useful for modelling systems with unknown or highly nonlinear behaviour.

### 3.3.1 Using Feedforward Neural Networks (FFNNs) for System Identification

Feedforward Neural Networks (FFNNs) are widely employed in system identification due to their universal approximation capability. These networks consist of an input layer, one or more hidden layers with activation functions, and an output layer [3]. Unlike recurrent networks, FFNNs do not incorporate feedback loops, making them suitable for static system identification tasks where past states do not influence future predictions.

#### 3.3.1.1 Mathematical Representation

A single-layer FFNN can be represented as:

$$h_j = f\left(\sum_{i=1}^{n} w_{ij} x_i + b_j\right) \tag{3.34}$$

where $x_i$ represents the input variables, $w_{ij}$ are the weights, $b_j$ are biases, and $f(\cdot)$ is a nonlinear activation function such as sigmoid, hyperbolic tangent, or rectified linear unit (ReLU). The final network output is computed as:

$$y_k = g\left(\sum_{j=1}^{m} v_{jk} h_j + c_k\right) \tag{3.35}$$

where $v_{jk}$ are the weights connecting hidden neurons to the output layer, $c_k$ are output biases, and $g(\cdot)$ is the output activation function.

#### 3.3.1.2 Training of Feedforward Networks

The objective of training an FFNN for system identification is to minimize the prediction error. The most commonly used loss function is the Mean Squared Error (MSE), given by:

$$E = \frac{1}{N}\sum_{k=1}^{N}\left(d_k - y_k\right)^2 \tag{3.36}$$

where $d_k$ is the desired output and $y_k$ is the predicted output. The weight update follows the gradient descent rule:

$$w_{new} = w_{old} - \eta \frac{\partial E}{\partial w} \tag{3.37}$$

where $\eta$ is the learning rate.

#### 3.3.1.3 Application in System Identification

FFNNs are widely applied in system identification, particularly for approximating static nonlinear mappings in control systems, signal processing, and robotics. They are effective in cases where dynamic memory is not required, such as modelling steady-state system behaviour or capturing nonlinear relationships in empirical data.

### 3.3.2 Using Recurrent Neural Networks (RNNs) for System Identification

FFNNs are limited to static system identification because they do not retain information about past states. But, many real-world systems exhibit dynamic behaviour where future states depend on past inputs. RNNs overcome this limitation by incorporating memory through feedback connections, making them ideal for dynamic system identification [3].

#### 3.3.2.1 Mathematical Formulation

An RNN models sequential dependencies using a hidden state that evolves over time. The general form of an RNN is:

$$h_t = f\left(W_h h_{t-1} + W_x x_t + b\right) \tag{3.38}$$

where $h_t$ is the hidden state at time step $t$, $W_h$ and $W_x$ are weight matrices, $x_t$ is the current input, and $b$ is a bias term. The output is given by:

$$y_t = g\left(W_y h_t + c\right) \tag{3.39}$$

where $W_y$ and $c$ are the output parameters.

#### 3.3.2.2 Training of Recurrent Neural Networks

Training RNNs for system identification requires minimizing sequence prediction error using Backpropagation through Time (BPTT). The gradient update rule remains as:

$$w_{new} = w_{old} - \eta \sum_t \frac{\partial E}{\partial w_t} \tag{3.40}$$

Standard RNNs suffer from vanishing and exploding gradients, making long-term dependencies difficult to capture. This issue is mitigated using specialized architectures such as Long Short-Term Memory (LSTM) networks and Gated Recurrent Units (GRUs).

#### 3.3.2.3 LSTM for System Identification

The LSTM networks are a special type of RNNs designed to overcome the vanishing gradient problem, which hinders standard RNNs from learning long-range dependencies. LSTMs achieve this by incorporating memory cells and gating mechanisms that regulate the flow of information, allowing the network to selectively retain or discard information across different time steps [4]. Each LSTM cell consists of three key gates:

i) **Forget gate**: Determines how much past information should be discarded from the memory cell. The forget gate output is computed as:

$$f_t = \sigma\left(W_f x_t + U_f h_{t-1} + b_f\right) \tag{3.41}$$

where $f_t$ is the forget gate activation at time step $t$, $x_t$ is the input at time $t$, $h_{t-1}$ is the hidden state from the previous time step, $W_f$ and $U_f$ are weight matrices, $b_f$ i s the bias term, and $\sigma$ is the sigmoid activation function that outputs values in the range $[0,1]$, controlling the amount of information retained or discarded.

ii) **Input gate**: Controls how much new information should be added to the memory cell. The input gate operation is given by:

$$i_t = \sigma\left(W_i x_t + U_i h_{t-1} + b_i\right) \tag{3.42}$$

where $i_t$ represents the input gate activation, and $W_i, U_i$, and $b_i$ are the weight matrices and bias specific to the input gate. The candidate cell state update, which represents new information, is computed as:

$$\tilde{c}_t = tanh\left(W_c x_t + U_c h_{t-1} + b_c\right) \tag{3.43}$$

where $\tilde{c}_t$ is the candidate memory content before being scaled by $i_t$.

iii) **Memory cell update**: The memory cell updates its state using both forget and input gate contributions:

$$c_t = f_t \circ c_{t-1} + i_t \circ \tilde{c}_t \tag{3.44}$$

Where $c_t$ is the new cell state, and $\circ$ represents element-wise multiplication.

iv) **Output gate**: Determines how much memory content should be exposed as output. It is computed as:

$$o_t = \sigma\left(W_o x_t + U_o h_{t-1} + b_o\right) \tag{3.45}$$

The final hidden state (output of the LSTM cell) is then obtained by applying the output gate to the memory content:

$$h_t = o_t \circ \tanh\left(c_t\right) \tag{3.46}$$

where $h_t$ represents the updated hidden state that is propagated to the next time step.

LSTMs provide a robust mechanism for capturing long-term dependencies in sequential data, making them particularly effective for system identification tasks involving temporal dynamics.

#### 3.3.2.4 Gated Recurrent Units (GRUs)

These are a variant of LSTMs that offer a simplified architecture with fewer parameters while maintaining similar performance. GRUs eliminate the explicit memory cell and instead use a single hidden state to retain long-term dependencies [4].

A GRU consists of two primary gates:

i) **Update Gate**: Controls how much past information should be carried forward. It is computed as:

$$z_t = \sigma\left(W_z x_t + U_z h_{t-1} + b_z\right) \tag{3.47}$$

where $z_t$ is the update gate activation, and $W_z$, $U_z$, and $b_z$ are the corresponding weight matrices and bias. The update gate blends new input information with past hidden states, enabling efficient information flow across time steps.

ii) **Reset gate**: Determines how much past information should be ignored. It is computed as:

$$r_t = \sigma\left(W_r x_t + U_r h_{t-1} + b_r\right) \tag{3.48}$$

where $r_t$ allows selective forgetting of past information when necessary.

iii) **State update**: The hidden state is updated using a candidate state that blends past information (modulated by $r_t$) and new input information:

$$\tilde{h}_t = tanh\left(W_h x_t + U_h (r_t \circ h_{t-1}) + b_h\right) \tag{3.49}$$

The final hidden state is then computed as a linear interpolation between the previous hidden state and the candidate state:

$$h_t = \left(1 - z_t\right) \circ h_{t-1} + z_t \circ \tilde{h}_t \tag{3.50}$$

where $h_t$ is the new hidden state, $\left(1 - z_t\right)$ determines how much of the previous state is retained, and $z_t$ determines how much of the new candidate state contributes to the updated hidden state. GRUs are computationally more efficient than LSTMs due to the reduced number of gates, making them preferable in real-time applications.

#### 3.3.2.5 Applications of LSTM and GRU in System Identification

LSTM and GRU architectures are widely applied in system identification scenarios where memory and sequential dependencies are crucial. Some key applications include: i) aerospace and robotics: modelling of dynamics of flight systems and adaptive robotic controllers; ii) biomedical engineering: identifying physiological signal patterns (e.g., ECG, EEG); iii) control systems: real-time adaptive control in industrial automation; and iv) economic forecasting: predicting stock market trends using historical data.

## 3.4 GENERALIZED ADALINE NEURAL NETWORK

The Adaptive Linear Neuron (ADALINE) is a fundamental neural network model introduced by Bernard Widrow and Marcian Hoff in the 1960s. Unlike the perceptron, ADALINE uses a linear activation function and employs the Least Mean Squares (LMS) learning rule for weight updates, making it more suitable for function approximation and adaptive filtering. The core concept of ADALINE is to adjust its weights iteratively based on the error between the actual and desired outputs, ensuring convergence to an optimal solution in linearly separable problems. The basic structure of an ADALINE unit consists of a set of inputs ($x_1$, $x_2$, ..., $x_0$), associated weights ($w_1$, $w_2$, ..., $w_0$), a bias input (typically set to 1) with its weight ($w_0$), a summation function that computes the weighted sum of inputs and a linear activation function. Mathematically, ADALINE operates as a single-layer neural network where the output is given by:

$$y = w_0 + w_1 x_1 + w_2 x_2 + \ldots + w_n x_n = w_0 + \sum_{i=1}^{n} w_i x_i \tag{3.51}$$

The weight update follows the LMS rule:

$$w(k+1) = w(k) + \eta e(k) x(k) \tag{3.52}$$

where $e(k) = d(k) - y(k)$ is the error signal, $d(k)$ is the desired output, and $\eta$ is the learning rate. ADALINE has been widely used in signal processing, system identification, and control applications due to its simplicity and ability to adapt to changing environments. While ADALINE represented a significant advancement in neural network development, it has several limitations like slow convergence speed, making it less suitable for online applications, limited effectiveness for identifying time-varying systems, and potential instability with larger learning rates, causing zigzag patterns during convergence.

The Generalized ADALINE (GADALINE) extends the ADALINE framework by incorporating adaptive learning mechanisms to improve convergence speed and tracking performance in dynamic environments. GADALINE extends the original ADALINE in two significant ways:

i) Enhanced input structure: GADALINE incorporates Tapped Delay Lines (TDLs) for both system input signals and output feedback, creating a recurrent neural network structure; and
ii) Modified learning algorithm: GADALINE adds a momentum term to the weight adjustment process during the convergence period, which is then turned off once the error falls below a specified threshold.

### 3.4.1 GADALINE LEARNING ALGORITHM

The GADALINE learning algorithm builds upon the LMS rule by incorporating a momentum term in the weight adjustment phase. This enhancement significantly improves convergence speed and reduces oscillations during training. The traditional LMS algorithm minimizes the error function:

$$E(w) = \frac{1}{2}e^2(k) \tag{3.53}$$

The weight adjustment in standard ADALINE is:

$$\Delta w(k) = \eta e(k)x(k) \tag{3.54}$$

GADALINE modifies this update rule by adding a momentum term:

$$\Delta w(k) = \eta e(k)x(k) + \alpha \Delta w(k-1) \tag{3.55}$$

where $\alpha$ is a momentum coefficient ($0 < \alpha < 1$). The impact of this momentum term can be understood by expanding the equation:

$$\begin{aligned}\Delta w(k) &= \eta e(k)x(k) + \alpha \Delta w(k-1)\\ &= \eta e(k)x(k) + \alpha\left[\eta e(k-1)x(k-1) + \alpha \Delta w(k-2)\right]\\ &= \eta e(k)x(k) + \alpha \eta e(k-1)x(k-1) + \alpha^2 \eta e(k-2)x(k-2) + \ldots\end{aligned} \tag{3.56}$$

This expansion reveals that the weight adjustment becomes an exponentially weighted time series of previous gradient terms. The momentum term provides two key benefits: i) when consecutive gradients have opposite signs (indicating oscillation around a minimum), the momentum term reduces the magnitude of the weight adjustment, dampening zigzag patterns; and ii) when consecutive gradients have the same sign (indicating a consistent direction), the momentum term increases the magnitude of the weight adjustment, accelerating descent in steady downhill regions. To prevent the weights from continuing to adjust after convergence, GADALINE turns off the momentum term once the error magnitude $|e|$ falls below a small threshold $\varepsilon$. This allows the weights to stabilize after convergence while still enabling quick adaptation when system parameters change.

### 3.4.2 System Identification Using GADALINE

For linear systems, this typically involves estimating the coefficients of a difference equation that relates the current output to past inputs and outputs. A discrete-time linear SISO system can be represented by the following difference equation:

$$y(k)+a_1y(k-1)+a_2y(k-2)+\ldots+a_ny(k-n)$$
$$=b_1u(k-1)+b_2u(k-2)+\ldots+b_mu(k-m) \tag{3.57}$$

where $y(k)$ is the system output at time $k$, $u(k)$ is the system input at time $k$, $n$ is the system order, $m \le n$ represents the number of input terms, and $a_i$ and $b_i$ are system parameters. Using the delay operator $z^{-1}$ (where $z^{-1}y(k)=y(k-1)$), this can be written in polynomial form:

$$A(z^{-1})y(k)=B(z^{-1})u(k) \tag{3.58}$$

where:

$$A(z^{-1})=1+a_1z^{-1}+a_2z^{-2}+\ldots+a_nz^{-n} \text{ and } B(z^{-1})=b_1z^{-1}+b_2z^{-2}+\ldots+b_mz^{-m}$$

Adding an error term to account for noise and modelling inaccuracies gives the AutoRegressive with eXogenous input (ARX) model:

$$A(z^{-1})y(k)=B(z^{-1})u(k)+e(k) \tag{3.59}$$

The system output can be expressed as:

$$y(k)=B(z^{-1})u(k)+\left(1-A(z^{-1})\right)y(k) \tag{3.60}$$

This formulation shows that the current output depends on past inputs and outputs, making it suitable for modelling with GADALINE's recurrent structure.

### 3.4.3 GADALINE Configuration for System Identification

GADALINE can be configured in two ways for system identification: i) parallel model: the system input is passed through a TDL, and the GADALINE's own output is fed back through another TDL; and ii) series-parallel model: the system input is passed through a TDL, and the actual measured system output is fed back through another TDL. The series-parallel model is generally preferred for online identification as it avoids potential stability issues that can arise in the parallel model. In the series-parallel configuration, the input vector to the GADALINE is constructed as:

$$u(k)=\left[u_s(k-1),u_s(k-2),\ldots,u_s(k-m),-y_s(k-1),\right.$$
$$\left.-y_s(k-2),\ldots,-y_s(k-n)\right] \tag{3.61}$$

where $u_s(k)$ is the system input and $y_s(k)$ is the measured system output. The corresponding weight vector:

$$w(k)=\left[w_1(k),w_2(k),\ldots,w_m(k),w_{m+1}(k),\ldots,w_{m+n}(k)\right] \tag{3.62}$$

This maps to an estimate of the system parameters:

$$\theta=\left[b_1,b_2,\ldots,b_m,a_1,a_2\ldots,a_n\right] \tag{3.63}$$

### 3.4.4 System Identification Procedure

The process of identifying a system using GADALINE involves the following steps:

- Experiment design: Select an input signal that is "persistently exciting" enough to activate all modes of the system. Common choices include: Random signals, Multi-sine signals, Random Binary Sequence (RBS) signals that switch between two levels with a given probability.
- Data collection: Measure the input-output data pairs $[u_s(k),y_s(k)]$ and construct the input vector for GADALINE.
- Parameter estimation: Apply the GADALINE learning algorithm to estimate the system parameters $\theta(k)=w(k)$.

GADALINE is particularly well-suited for identifying time-varying systems where the parameters $a_i$ and $b_j$ change over time. These changes might occur as slow continuous variations or sudden switches between different parameter values. One of the key advantages of GADALINE for time-varying system identification is its faster convergence, as the momentum term accelerates the learning process, allowing the network to quickly adapt to parameter changes. Additionally, GADALINE produces smoother parameter trajectories by reducing zigzag oscillations during convergence, ensuring more stable and accurate parameter estimates. Its improved tracking capability further enhances its effectiveness in identifying time-varying parameters more efficiently than traditional ADALINE. Also, the relative simplicity of the GADALINE algorithm results in low computational complexity, making it particularly suitable for real-time applications and online system identification.

While GADALINE offers significant improvements over traditional ADALINE for system identification, certain limitations should be considered. One of these challenges is the curse of dimensionality, which, like most neural networks, can cause performance degradation when dealing with high-order systems with numerous parameters. Another critical aspect is the selection of the momentum parameter; if the momentum coefficient $\alpha$ is too small, it provides little benefit, while a value that is too large can cause overshooting and instability. The error threshold $\varepsilon$ for deactivating the momentum term also influences the stability of parameter estimates after convergence, requiring careful tuning. Furthermore, GADALINE assumes a linear ARX model structure, which may limit its identification performance for systems that do not conform to this assumption.

Despite these limitations, GADALINE represents a significant advancement in neural network-based system identification, balancing computational simplicity with enhanced performance for adaptive modelling of dynamic systems. It extends the traditional ADALINE neural network by incorporating tapped delay lines for input and output feedback, along with a momentum term in the learning algorithm. These enhancements significantly improve convergence speed and parameter tracking capability, making GADALINE particularly well-suited for online identification of time-varying systems. The key innovations of GADALINE include a recurrent neural network structure that captures system dynamics through tapped delay lines, an enhanced learning algorithm with momentum that accelerates convergence while reducing oscillations, and an adaptive momentum control mechanism that ensures stability after convergence. For system identification applications, GADALINE provides a computationally efficient method for estimating and tracking parameters of linear dynamic systems, even in the presence of parameter variations and measurement noise.

## 3.5 IDENTIFICATION OF HELICOPTER DYNAMICS USING RECURRENT NEURAL NETWORKS

Helicopter dynamics are inherently complex, arising from the strong interactions among subsystems such as the rotor, fuselage, power plant, transmission, empennage, and tail rotor. Traditional methods for system identification, which depend on predefined model structures, often struggle to capture these intricate relationships and the nonlinear behaviour that is characteristic of helicopters. RNNs present a promising alternative, as they can model temporal dependencies and nonlinear dynamics without needing an explicit model structure [5]. This approach has been explored using flight data, focusing on three types of RNN architectures: i) nonlinear autoregressive exogenous input (NARX) models, ii) memory neuron networks (MNN), and iii) recurrent multi-layer perceptron (RMLP) networks.

A helicopter's dynamics can be modelled as a multiple-input multiple-output (MIMO) system, with significant interaction between inputs and outputs. This system can be represented discretely by a set of equations where the state, input, and output vectors evolve over time, with the system's dynamics captured by nonlinear functions.

$$x(k+1) = F\left(x(k), u(k), k\right)$$

$$y(k) = G\left(x(k), u(k), k\right) \tag{3.64}$$

where $x(k)$ represents the state vector, $u(k)$ represents the input vector, $y(k)$ is the output vector, and $F(\cdot)$ and $G(\cdot)$ are nonlinear functions describing the system dynamics. The goal of system identification is to develop a model $\hat{y}(k)$ that accurately predicts the output of the helicopter system, minimizing the error between predicted and actual outputs within a specified threshold $\epsilon$ :

$$\|y(k) - \hat{y}(k)\| \leq \epsilon \qquad \forall k > 0 \tag{3.65}$$

### 3.5.1 Nonlinear Auto-regressive Exogenous Input (NARX)

The NARX model is a widely used RNN architecture for system identification. It employs tapped-delay lines to introduce dynamics into the network, with the model relating the current output to past inputs and outputs. The model is described by:

$$\hat{y}(k+1) = N_F\left(u(k).u(k-1),\ldots,u\left(k-n_u+1\right),y(k),y(k-1),\ldots,y\left(k-n_y+1\right)\right) \tag{3.66}$$

Here $N_F(\cdot)$ is the neural network approximation of the system dynamics and $n_u$ and $n_y$ are the number of past inputs and outputs respectively. The network is trained using the Backpropagation algorithm, minimizing the error between predicted and actual outputs. The NARX model has shown good performance in capturing both longitudinal and lateral dynamics, with low training and generalization errors.

### 3.5.2 Memory Neuron Network

The Memory Neuron Network is a type of RNN that incorporates internal memory neurons, which store past values of the network's activations. Unlike traditional feed-forward neural networks, the MNN includes feedback connections that allow it to retain information about previous states. This feature makes the MNN well-suited for modelling dynamic systems, such as helicopter dynamics, where the behaviour of the system is influenced by its history. The architecture of the MNN consists of network neurons and memory neurons, with each memory neuron connected to a network neuron and responsible for storing a weighted combination of the current and past outputs. This feedback mechanism enables the MNN to capture temporal dependencies without the need for tapped-delay lines, as in the NARX model. Training the MNN involves using the Backpropagation Through Time (BPTT) algorithm, which is commonly used for RNNs. During training, the memory neurons are initialized using the first few samples of the training data to provide a reasonable estimate of the system's initial state. The weights of the network and memory neurons are then updated iteratively using gradient descent to minimize the error between the predicted and actual outputs. Errors are computed at each time step, and the gradients are propagated backward through time to update the weights.

One of the main advantages of the MNN is its flexibility, as it does not require prior knowledge of the system's order or delay. This flexibility allows the MNN to capture complex temporal dependencies. A significant limitation is that the MNN tends to overfit the training data, especially if not properly regularized. This overfitting reduces its generalization capabilities, which can be problematic for tasks requiring the model to perform well on unseen data. Additionally, training the MNN can be computationally expensive, particularly for systems with many states or long time dependencies.

In the context of helicopter dynamics, the MNN was used to model both longitudinal and lateral dynamics. While the model showed low training errors, it struggled to generalize to new data, especially in the presence of off-axis corrections made by the pilot. This indicates that while the MNN can effectively capture the dynamics of the system, it may not be the best choice for applications where generalization to unseen data is essential.

### 3.5.3 Recurrent Multi-layer Perceptron (RMLP)

The RMLP network is a fully recurrent network where the output of hidden neurons is fed back to the input layer after a delay of one time step. The network dynamics are described by:

$$z(k)=\begin{cases} u_l(k) & if\ l\in I \\ y_l(k) & if\ l\in U \end{cases} \tag{3.67}$$

Here $I$ and $U$ are the sets of input and hidden states, respectively. The network is trained using the Real-Time Recurrent Learning algorithm, which updates weights in real time based on the error at each time step. Although the RMLP performs well for longitudinal dynamics, it demands more computational resources due to its fully recurrent structure. The training and evaluation of RNN models for helicopter system identification are critical steps in ensuring that the models accurately capture the system's dynamics and generalize well to new data. The process begins with the collection of flight data, which typically includes measurements of state variables such as pitch, roll, and yaw rates, as well as control inputs like longitudinal cyclic, lateral cyclic, collective, and pedal inputs. This data is divided into training and testing sets, with the training set used to adjust the model's weights and the testing set reserved for evaluating the model's performance on unseen data. The choice of model architecture depends on the specific requirements of the application. The performance of the trained models is evaluated using several metrics to assess their accuracy and generalization capabilities. One of the primary metrics is the Training Error (TE), which measures how well the model fits the training data. It is computed as the Root Mean Square Error (RMSE) between the predicted and actual outputs on the training set:

$$TE=\sqrt{\frac{1}{N_1}\sum_{k=1}^{N_1}\left(y(k)-\hat{y}(k)\right)^2} \tag{3.68}$$

Here $y(k)$ is the actual output, $\hat{y}(k)$ is the predicted output, and $N_1$ is the number of data points in the training set. A low TE indicates that the model has learned the training data well, but it does not guarantee good performance on unseen data.

To evaluate the model's ability to generalize, the Generalization Error (GE) is computed using the testing data. The GE is also calculated as the RMSE between the predicted and actual outputs, but on the testing set:

$$GE = \sqrt{\frac{1}{N_2}\sum_{k=1}^{N_2}\left(y(k)-\hat{y}(k)\right)^2} \tag{3.69}$$

Here $N_2$ is the number of data points in the testing set. A low GE indicates that the model generalizes well to new data, which is essential for real-world applications.

In addition to TE and GE, the Coefficient of Determination ( $R^2$ ) is often used to assess the quality of the model fit. The $R^2$ statistic measures how well the model explains the variability in the data and is computed as:

$$R^2 = 1 - \frac{SSR}{SST} \tag{3.70}$$

Here $SSR$ is the sum of squares of residuals (the difference between the actual and predicted outputs), and $SST$ is the total sum of squares (the variance in the actual outputs). The $R^2$ value ranges from 0 to 1, with values closer to 1 indicating a better fit. For example, an $R^2$ value of 0.95 means that 95% of the variability in the data is explained by the model.

The application of RNNs in helicopter system identification has significant implications for flight control system design, fault detection, and autonomous rotorcraft navigation. By capturing the intricate dependencies between control inputs and system states, RNNs facilitate the development of adaptive control strategies that enhance stability and performance under varying flight conditions. Their ability to model the nonlinearities and time delays inherent in rotorcraft dynamics makes them a valuable tool for real-time parameter estimation and system monitoring. The robustness of RNN-based models also enables their application in adverse flight conditions where conventional models may fail. The integration of RNN-based models with conventional physics-based approaches, such as Kalman filtering and nonlinear observers, can further improve accuracy and interpretability, enabling robust and efficient solutions for helicopter system identification.

## 3.6 DIRECT IDENTIFICATION OF STRUCTURAL PARAMETERS

Direct identification of structural parameters represents an important advancement in structural health monitoring and damage assessment of engineering structures. As infrastructure continues to age in many developed countries, there is a growing need for efficient and accurate methods to evaluate structural integrity. Traditional structural identification methods often rely on complex eigenvalue analyses and optimization processes, which can be computationally intensive and sometimes numerically unstable, especially for large-scale infrastructures with many degrees of freedom. Artificial intelligence approaches, particularly neural networks, have emerged as powerful tools for direct parameter identification. These methods can learn complex relationships between structural responses and parameters without requiring explicit mathematical formulations of these relationships. For a structural system under dynamic excitation, the equation of motion can be expressed as:

$$[M]\{\ddot{x}(t)\}+[C]\{\dot{x}(t)\}+[K]\{x(t)\}=\{u(t)\} \tag{3.71}$$

where $[M]$,$[C]$ and $[K]$ are the mass, damping, and stiffness matrices; $\{\ddot{x}(t)\}$,$\{\dot{x}(t)\}$ and $\{x(t)\}$ are the acceleration, velocity, and displacement response vectors; and $\{u(t)\}$ is the input vector representing external excitation. Typically, a Rayleigh damping matrix is considered for structures:

$$[C]=\alpha[M]+\beta[K] \tag{3.72}$$

Here $\alpha$ and $\beta$ are coefficients that define the damping characteristics. When comparing an object structure to a reference structure, the stiffness and damping coefficients can be expressed as:

$$K_i = s_i K_{ri}$$

$$\alpha = d_\alpha \alpha_r$$

$$\beta = d_\beta \beta_r \tag{3.73}$$

Here $K_i$ and $K_{ri}$ are the interval stiffness coefficients, $s_i$ is the stiffness scaling factor (SSF), and $d_\alpha$ and $d_\beta$ are the damping coefficient scaling factors (DCSFs). Several

methodologies have been developed for direct identification of structural parameters. These include time-domain methods, frequency-domain methods, modal-based identification, and artificial intelligence approaches. While each method has its merits, artificial intelligence techniques, particularly neural networks, have demonstrated exceptional capabilities in handling complex, nonlinear relationships between structural responses and parameters.

Neural networks offer a powerful framework for direct identification of structural parameters by learning the complex relationships between dynamic responses and structural properties. The approach eliminates the need for eigenvalue analysis and complex optimization algorithms typically required in traditional methods. Also, to enhance the accuracy and robustness of direct parameter identification, adaptations of FFNNs and RNNs have been explored. Unlike general neural network applications, here the networks are specifically trained to recognize patterns in dynamic structural responses and estimate the underlying parameters.

While both FFNN and RNN approaches provide effective frameworks for parameter identification, they face challenges when dealing with complex structures having many degrees of freedom or when the relationship between responses and parameters becomes highly nonlinear. To address these limitations, a specialized dual neural network strategy has been developed that divides the identification problem into manageable sub-tasks [6]. The dual neural network strategy leverages the strengths of neural networks while mitigating their limitations by decomposing the complex nonlinear mapping into two distinct stages. This approach employs two specialized networks: an emulator neural network and a parametric evaluation neural network.

The emulator neural network serves as a nonparametric model of a reference structure. Its purpose is to predict the dynamic responses of the reference structure at the next time step based on current responses and excitation. Mathematically, the emulator network maps:

$$\left\{x_r(t_k), \dot{x}_r(t_k), u(t_k)\right\} \rightarrow \left\{x_r(t_{k+1}), \dot{x}_r(t_{k+1})\right\} \tag{3.74}$$

Here $x_r(t_k)$ and $\dot{x}_r(t_k)$ represent displacement and velocity responses of the reference structure at time $t_k$, and $u(t_k)$ represents the excitation. The emulator neural network typically consists of a three-layer architecture: i) input layer: current displacements, velocities, and excitations; ii) hidden layer: typically triple the input neurons with sigmoid activations; and iii) output layer: next-step displacements and velocities. The network is trained using the backpropagation algorithm to minimize the mean squared error between predicted and actual responses. The parametric evaluation neural network identifies structural parameters based on differences between predicted and actual responses. It maps:

$$\{d\} \rightarrow \left\{s_1, s_2, \ldots, s_n, d_\alpha, d_\beta\right\} \tag{3.75}$$

Here $\{d\}$ is the root mean square (RMS) difference vector between predicted and actual responses, and $\left\{s_1, s_2, \ldots, s_n, d_\alpha, d_\beta\right\}$ are the structural parameters to be identified. The RMS difference vector is defined as:

$$\{d\} \rightarrow \left\{d_{v1}, d_{v2}, \ldots, d_{vn}, d_{d1}, d_{d2}, \ldots, d_{dn}\right\} \tag{3.76}$$

$$d_{vi} = \sqrt{\frac{1}{N}\sum_{k=1}^{N_2}\left(\dot{x}_i(t_k) - \hat{\dot{x}}_i(t_k)\right)^2}$$

$$d_{di} = \sqrt{\frac{1}{N}\sum_{k=1}^{N_2}\left(x_i(t_k) - \hat{x}_i(t_k)\right)^2} \tag{3.77}$$

$\dot{x}(t_k)$ and $x_i(t_k)$ are the actual velocity and displacement responses at the $i^{th}$ degree of freedom, and $\hat{\dot{x}}_i(t_k)$ and $\hat{x}_i(t_k)$ are the corresponding predicted responses from the emulator network. The parametric evaluation neural network also has a three-layer architecture which includes the: i) input layer representing components of the RMS difference vector; ii) hidden layer with sigmoid activation functions; and iii) ouput layer representing stiffness scaling factors and damping coefficient scaling factors.

The training of the dual neural network system proceeds in three main steps: i) training the emulator network using reference structure data; ii) creating associated structures with varied parameters and calculating their responses; and iii) training the parametric evaluation network to map response differences to parameter variations.

Once both neural networks are trained, the parameter identification process for an unknown object structure proceeds as follows: i) measure the dynamic responses of the object structure under the same excitation used for the reference structure; ii) use the emulator neural network to predict the responses of the reference structure; iii) calculate the RMS difference vector $\{d\}$ between the predicted responses and the measured responses of the object structure; iv) input the RMS difference vector $\{d\}$ into the parametric evaluation neural network to obtain the identified structural parameters $\{s_1, s_2, \ldots, s_n, d_\alpha, d_\beta\}$; and v) calculate the actual structural parameters of the object structure using the identified scaling factors: $K_i = s_i K_{ri}, \alpha = d_\alpha \alpha_r, \beta = d_\beta \beta_r$.

This approach eliminates the eigenvalue analysis and optimization algorithms required in traditional methods while maintaining accuracy and computational efficiency, even with measurement noise up to 7% of maximum response signals.

The neural network approach to direct identification of structural parameters has been successfully applied to various structural systems. For multi-storey buildings, this approach has demonstrated the ability to identify storey stiffness and damping coefficients with high accuracy. The identification results have been used for post-earthquake damage assessment, retrofit planning and evaluation, and long-term health monitoring. The application of neural networks in bridge structures has focused on identifying cable tensions in cable-stayed bridges, bearing stiffness, girder flexural rigidity, and foundation support conditions.

For large-scale structures such as dams and offshore platforms, the neural network approach has been adapted to handle the spatial distribution of stiffness parameters, soil-structure interaction effects, and hydrodynamic interaction effects. The ability of neural networks to efficiently and accurately identify both stiffness and damping coefficients makes them a valuable tool for evaluating the integrity of existing structures. Their robustness to noise and computational efficiency suggest strong potential for practical applications in real-time structural health monitoring systems.

As computational capabilities continue to advance, neural network methods for structural parameter identification will likely become more refined, enabling more accurate, efficient, and comprehensive structural assessment. The integration of these methods with other artificial intelligence techniques promises to enhance our ability to maintain and protect aging infrastructure, contributing to improved safety.

## 3.7 LINEAR RECURRENT NEURAL NETWORK FOR MULTIVARIABLE SYSTEM IDENTIFICATION

The identification of multivariable dynamic systems presents significant challenges due to the presence of strong interdependencies between state variables, nonlinearities, and uncertainties in real-world conditions. Traditional system identification techniques, such as least squares estimation and subspace methods, often struggle with capturing temporal dependencies effectively, especially when dealing with highly coupled multivariable systems. RNNs, particularly their linear variants, provide a structured and computationally efficient approach to multivariable system identification by modelling the underlying dynamics in a recursive manner, ensuring the preservation of historical state dependencies. A general multivariable system can be described in the discrete-time state-space representation as follows:

$$x(k+1) = Ax(k) + Bu(k) + w(k)$$

$$y(k) = Cx(k) + Du(k) + v(k) \tag{3.78}$$

where $x(k)$ is the state vector, $u(k)$ is the input vector, $y(k)$ is the output vector, and $w(k)$ and $v(k)$ represent process and measurement noise, respectively. The goal of system identification is to estimate the system matrices $A, B, C$ , and $D$ from observed input-output data. Linear RNNs, also known as Echo State Networks (ESNs) or Linear Dynamical Systems (LDS) with recurrent structures, modify the standard RNN framework by constraining the activation functions to linear mappings, ensuring that the learned model adheres to the principles of linear time-invariant (LTI) systems. The governing equation for an RNN-based multivariable system identification model can be written as:

$$x(k+1) = \phi x(k) + \Gamma u(k)$$

$$y(k) = \psi x(k) + \Omega u(k) \tag{3.79}$$

where $\phi, \Gamma, \psi$, and $\Omega$ are the weight matrices corresponding to the recurrent connections, input contributions, output transformation, and direct feedthrough terms, respectively. These matrices serve as the learned parameters of the neural network model and are updated iteratively using optimization techniques such as gradient descent or recursive least squares.

One of the primary advantages of using a linear RNN for multivariable system identification is its ability to capture temporal correlations across multiple inputs and outputs while maintaining computational efficiency. The recursive structure allows the model to process long sequences of data without excessive memory requirements, making it suitable for real-time applications such as flight dynamics modelling, industrial process control, and adaptive filtering in signal processing.

The training process for an RNN-based system identification model involves defining an appropriate loss function that minimizes the prediction error between the observed and estimated outputs. The cost function is typically formulated as:

$$J(\theta) = \sum_{k=1}^{N} \left\| y(k) - \hat{y}(k) \right\|^2 \tag{3.80}$$

where $\theta$ represents the set of learnable parameters, and $\hat{y}(k)$ is the predicted output at time step $k$. Optimization algorithms such as Adam or RMSprop are used to update the network parameters iteratively. A compelling practical application of linear RNN-based multivariable system identification is in complex industrial processes with interconnected dynamic variables. Consider a chemical processing plant where multiple input parameters – such as temperature, pressure, reactant concentrations, and flow rates – simultaneously influence several output variables including reaction yield, product composition, and energy efficiency. In such a multivariable system, traditional linear modelling approaches struggle to capture the intricate, nonlinear interactions between different process parameters. A linear RNN can effectively model these complex interdependencies by learning the recursive relationships across input and output variables, enabling precise prediction of system behaviour under varying operational conditions.

## 3.8 FUNCTIONAL APPROXIMATION AND SYSTEM IDENTIFICATION

Functional approximation is a fundamental concept in mathematical modelling and computational intelligence, enabling the representation of complex functions through simpler mathematical structures. In the context of system identification, functional approximation seeks to determine an unknown function that maps inputs to outputs using a finite set of observed data. The approximation function should generalize well to new inputs while preserving essential characteristics of the underlying system.

Given an unknown function $f(x)$ defined over a domain $\mathbb{R}^n$, the goal is to approximate $f(x)$ with a function $\hat{f}(x)$ from a chosen function class such that:

$$\hat{f}(x) \approx f(x), \qquad \forall x \in \mathbb{R}^n \tag{3.81}$$

The error of approximation is measured using a norm, commonly the squared error norm:

$$E = \int \left| f(x) - \hat{f}(x) \right|^2 dx \tag{3.82}$$

The choice of function class determines the accuracy and computational efficiency of the approximation. Several function approximation techniques have been developed, including polynomial approximation, radial basis functions, wavelet transforms, and kernel methods.

a) **Polynomial approximation:** Polynomial interpolation and regression techniques approximate a function using polynomials of varying degrees:

$$P_n(x) = a_0 + a_1 x + a_2 x^2 + \ldots + a_n x^n \tag{3.83}$$

The coefficients $a_i$ are estimated using least squares minimization to best fit the given data.

b) **Fourier series expansion:** Periodic functions are approximated using a sum of sinusoidal basis functions:

$$f(x) = a_0 + \sum_{n=1}^{N} \left( a_n \cos(n\omega x) + b_n \sin(n\omega x) \right) \tag{3.84}$$

This method is particularly useful for approximating functions with periodic components.

c) **Radial Basis Function (RBF) approximation:** RBFs approximate functions using a weighted sum of basis functions centred at selected points:

$$f(x) = \sum_{i=1}^{N} w_i \phi\left( \left\| x - c_i \right\| \right) \tag{3.85}$$

Common choices for $\phi$ include Gaussian and thin-plate splines.

d) **Wavelet transforms:** Wavelet-based approximation decomposes functions into localized wavelet coefficients, allowing multi-resolution analysis:

$$f(x) = \sum_{j,k} c_{j,k} \psi_{j,k}(x) \tag{3.86}$$

where $\psi_{j,k}(x)$ are wavelet basis functions.

Neural networks provide a flexible and powerful approach to function approximation, capable of learning arbitrary nonlinear mappings from data. The Universal Approximation Theorem states that a sufficiently large neural network with a single hidden layer can approximate any continuous function to an arbitrary degree of accuracy. Several neural architectures are commonly used for function approximation.

e) **Feedforward Neural Networks (FFNNs)**
FFNNs approximate functions using a multilayer structure with nonlinear activation functions. The function approximation using a single-layer FFNN with $M$ hidden neurons is given by:

$$\hat{f}(x)=\sum_{j=1}^{M}w_j\sigma\left(w_j^T x+b_j\right) \tag{3.87}$$

where $\sigma(\cdot)$ is the activation function (e.g., sigmoid, ReLU), $w_j$ are weights, and $b_j$ are biases. Training the network involves adjusting these parameters using optimization algorithms such as gradient descent to minimize the approximation error. Consider approximating a nonlinear function such as $f(x)=\sin(x)+0.5x$. An FFNN with a single hidden layer is trained on a dataset of $\{x_i, f(x_i)\}$ pairs. The network learns to approximate the function by minimizing the error using a loss function such as Mean Squared Error. After training, the FFNN generalizes to unseen values of $x$, demonstrating its effectiveness in functional approximation and system identification.

f) **Feedforward Neural Networks (FFNNs)**
RBFNs employ radial basis functions in the hidden layer to approximate localized features of a function:

$$\hat{f}(x)=\sum_{i=1}^{M}w_i e^{-\frac{\|x-c_i\|^2}{2\sigma^2}} \tag{3.88}$$

where $c_i$ are the centers of the RBF neurons and $\sigma$ determines their spread. RBFNs are particularly effective for function approximation in system identification problems where local variations in the function need to be captured accurately.

g) **Deep Neural Networks (DNNs) and Physics-Informed Neural Networks (PINNs)**
Deep networks extend function approximation by stacking multiple layers, enabling hierarchical feature extraction:

$$\hat{f}(x)=W_L\sigma\left(W_{L-1}\cdots\sigma\left(W_1x+b_1\right)+b_{L-1}\right)+b_L \tag{3.89}$$

where $L$ represents the number of layers. DNNs excel at modelling complex, high-dimensional functions and are widely used in time-series forecasting and dynamic system modelling.

PINNs incorporate known physical laws into the neural network training process by embedding differential equation constraints into the loss function. This ensures that the approximated function adheres to underlying physical principles, enhancing interpretability and generalization.

Functional approximation is widely applied in diverse engineering and scientific domains. In aerospace engineering, aerodynamic modelling relies on neural networks to approximate the nonlinear forces and moments acting on an aircraft. Flight control system identification utilizes data-driven models to design adaptive autopilot systems that respond to changing flight conditions. Engine health monitoring employs radial basis functions to predict engine performance based on real-time sensor data. Biomedical signal processing applications use neural networks to approximate ECG and EEG signals, enabling accurate disease classification and anomaly detection. Power systems rely on functional approximation techniques for load forecasting and fault detection in power grids. Structural health monitoring uses RBFNs and DNNs to detect structural degradation and predict failures in bridges, buildings, and mechanical components. These applications highlight the versatility of functional approximation in solving complex system identification problems. The integration of classical mathematical techniques with modern neural network-based methods ensures robust and accurate function estimation across various domains.

Functional approximation is a cornerstone of system identification, enabling the representation of complex functions through mathematical models. While classical methods such as polynomial regression and Fourier analysis remain relevant, neural networks have emerged as the dominant approach due to their flexibility, robustness, and capability to handle high-dimensional nonlinear mappings. The integration of neural networks with physics-informed constraints further enhances their reliability, making them indispensable tools for modern system modelling and identification.

## 3.9 MULTI-LAYERED NEURAL NETWORK USING HEURISTIC SELF-ORGANIZATION

The process of system identification using neural networks has gained significant traction due to its ability to approximate complex nonlinear relationships and adapt to varying system dynamics. Among various architectures, multi-layered neural networks have been particularly effective in learning intricate mappings between input and output data. But, conventional training methods often suffer from limitations such as local minima convergence, slow learning rates, and sensitivity to initialization. To address these issues, heuristic self-organization strategies have been developed, allowing the network to dynamically adjust its structure and learning parameters based on data characteristics.

A standard multi-layered neural network consists of an input layer, multiple hidden layers, and an output layer. The depth and width of the hidden layers dictate the network's ability to approximate complex functions. The function approximation capability of such networks is based on the universal approximation theorem, which states that a sufficiently large feedforward neural network with at least one hidden layer can approximate any continuous function to an arbitrary degree of accuracy. Consider a dynamical system represented as:

$$y(t) = f\big(u(t), u(t-1), \ldots, u(t-n)\big) + e(t) \tag{3.90}$$

where $y(t)$ is the system output, $u(t)$ is the input, $f(\cdot)$ is an unknown nonlinear mapping, and $e(t)$ represents noise. The goal of system identification is to construct a neural network model $\hat{f}$ such that:

$$\hat{y}(t) = \hat{f}\left(u(t), u(t-1), \ldots, u(t-n)\right) \tag{3.91}$$

approximates $y(t)$ as accurately as possible. A multi-layered neural network achieves this approximation by using weight matrices $W^{(l)}$ and bias vectors $b^{(l)}$ across multiple layers. The forward propagation in such a network is given by:

$$a^{(l)} = \sigma\left(W^{(l)} a^{(l-1)} + b^{(l)}\right) \tag{3.92}$$

where $a^{(0)} = u$ is the input, $\sigma(\cdot)$ is the activation function, and $a^{(L)}$ is the predicted output $\hat{y}$, where $L$ denotes the final layer of the network. The network learns by minimizing the error function:

$$J = \sum_{t=1}^{N} \left(y(t) - \hat{y}(t)\right)^2 \tag{3.93}$$

Unlike traditional backpropagation, which relies on a fixed network structure, heuristic self-organization dynamically modifies the network topology and learning parameters based on input data characteristics. Instead of predefining the number of hidden neurons, the network grows adaptively. Neurons are added when the current model exhibits high error or fails to capture the system dynamics effectively. The criteria for adding neurons are typically based on:

$$\frac{\partial J}{\partial W} > \theta_{add} \tag{3.94}$$

where $\theta_{add}$ is a predefined threshold indicating the significance of weight updates.

Excessive neurons can lead to overfitting and increased computational costs. The pruning process removes neurons with minimal contribution, determined by weight magnitudes:

$$\left\| W^{(l)} \right\| < \theta_{prune} \tag{3.95}$$

where $\theta_{prune}$ is a small threshold defining negligible weight contributions. Instead of using a fixed learning rate, heuristic self-organization adjusts the learning rate η\eta based on error trends:

$$\eta_{t+1} < \eta_t + \alpha\left(\left|J_t - J_{t-1}\right|\right) \tag{3.96}$$

where $\alpha$ is an adaptation coefficient ensuring stable convergence. To optimize weight updates, evolutionary techniques such as genetic algorithms or particle swarm optimization are employed. These methods iteratively refine the network parameters based on fitness functions defined by prediction accuracy. The weight update rule in heuristic self-organization follows an adaptive gradient descent approach:

$$W_{t+1}^{(l)} = W_t^{(l)} - \eta \frac{\partial J}{\partial W^{(l)}} + \gamma W_t^{(l)} \tag{3.97}$$

where $\gamma$ is a momentum term enhancing convergence speed. The neuron growth rule is given by:

$$N_{hidden} = N_{hidden} + 1 \quad \text{if } \max\left|e(t)\right| > \theta_{add} \tag{3.98}$$

and the pruning rule:

$$N_{hidden} = N_{hidden} - 1 \quad \text{if } \left\|W^{(l)}\right\| < \theta_{prune} \tag{3.99}$$

A practical example of heuristic self-organizing neural networks can be observed in the identification of an aircraft flight dynamics model. The aircraft dynamics are governed by nonlinear differential equations, and traditional methods often require precise prior knowledge of aerodynamic coefficients. By employing an adaptive multi-layered neural network, the system can iteratively refine its structure to capture the complex dependencies between control surface deflections, airspeed, and angular velocities.

Multi-layered neural networks using heuristic self-organization provide a powerful framework for system identification, dynamically adapting their structure to optimize performance. By integrating adaptive network growth, pruning, dynamic learning rates, and evolutionary parameter tuning, these networks overcome limitations associated with traditional backpropagation-based training. Their application in complex dynamical systems, such as aircraft modelling, demonstrates their potential in capturing highly nonlinear behaviours with improved accuracy and robustness. The incorporation of self-organization principles ensures that the network remains efficient, avoiding unnecessary computational complexity while maintaining high predictive fidelity.

## 3.10 PARAMETER OPTIMIZATION METHOD FOR RADIAL BASIS TYPE MODELS

Radial Basis Function Neural Networks (RBFNNs) are widely used for system identification, classification, and function approximation. Their performance depends critically on optimizing key parameters such as the selection of radial basis function centres, the determination of spread values, and the adjustment of output layer weights. Effective optimization methods significantly enhance the generalization ability, stability, and computational efficiency of RBFNNs in practical applications. Unlike other neural network architectures, RBFNNs consist of localized activation functions centred around specific points in the input space. This localized nature makes them particularly sensitive to parameter tuning, requiring systematic methods to determine optimal values. The primary optimization challenges in RBFNNs include: i) selecting an appropriate number and placement of basis function centres; ii) determining the spread of the radial basis functions to ensure smooth interpolation; and iii) optimizing output layer weights for minimal error in predictions.

### 3.10.1 Optimization of Radial Basis Function Centres

The placement of radial basis function centres plays a crucial role in defining the network's approximation capability [7]. Several techniques are used for selecting centres, each with its own advantages and trade-offs: i) **Data-Driven Clustering Approaches:** A common approach to selecting centres is clustering input data points into representative groups. Popular methods include: a) **K-Means Clustering:** The dataset is partitioned into mmm clusters, and the cluster centroids serve as the radial basis function centres. The objective function minimized by K-Means is:

$$J = \sum_{i-1}^{N} \sum_{j=1}^{m} \left\| x_{i-} c_j \right\|^2 \tag{3.100}$$

where $c_j$ is the center of the $j^{th}$ cluster; and b) **Fuzzy C-Means Clustering:** This method assigns a degree of membership to each data point, allowing for a more flexible allocation of centres. The update rule for centres is given by:

$$c_j = \frac{\sum_{i=1}^{N} u_{ij}^m x_i}{\sum_{i=1}^{N} u_{ij}^m} \tag{3.101}$$

where $u_{ij}$ represents the membership of $x_i$ in cluster $j$; and ii) **Orthogonal Least Squares (OLS) Selection:** Instead of selecting centres arbitrarily, OLS incrementally chooses centres that contribute most to reducing the overall network error. The process involves: i) Computing the error reduction ratio for each candidate centre; ii) Selecting the centre that minimizes the residual error; and iii) Iteratively adding new centres until a predefined threshold is met. This method reduces redundancy in centre placement, ensuring that selected centres maximize variance reduction in the dataset; and iii) **Randomized Selection with Refinement:** In some cases, centres are chosen randomly from the training dataset and later refined using gradient-based methods. A centre $c_j$ is updated as:

$$c_j^{new} = c_j^{old} - \eta \frac{\partial E}{\partial c_j} \tag{3.102}$$

where $E$ is the network error, and $\eta$ is the learning rate.

### 3.10.2 Optimization of Spread Parameters

The spread $\sigma_j$ of a radial basis function determines the width of influence of each center. Selecting an appropriate spread is crucial to balancing model complexity and generalization: i) **Fixed spread based on data distribution:** A simple heuristic is to set $\sigma_j$ proportional to the average distance between selected centers:

$$\sigma_j = \beta \frac{1}{m} \sum_{i=1}^{m} \left\| c_{j-} c_i \right\| \tag{3.103}$$

where $\beta$ is a scaling factor that controls the degree of function overlap; ii) **Adaptive spread based on neighbouring centres:** A more dynamic approach adapts $\sigma_j$ based on the distance to the nearest neighboring center:

$$\sigma_j = \alpha \min_{i \neq j} \sum_{i=1}^{m} \left\| c_{j-} c_i \right\| \tag{3.104}$$

where $\alpha$ is an empirical parameter ensuring that each function covers a meaningful region without excessive overlap; and iii) **Gradient descent-based spread tuning:** Instead of heuristics, $\sigma_j$ can be optimized using gradient descent by minimizing the cost function:

$$E = \sum_{k=1}^{N} \left( y_k - \hat{y}_k \right)^2 \tag{3.105}$$

where $y_k$ is the actual output and $\hat{y}_k$ is the predicted output. The update rule for $\sigma_j$ follows:

$$\sigma_j^{new} = \sigma_j^{old} - \eta \frac{\partial E}{\partial \sigma_j} \tag{3.106}$$

This approach allows dynamic adaptation of spread values to minimize network error.

### 3.10.3 Optimization of Output Weights

Once centres and spreads are determined, the final step is optimizing the output layer weights $w_{jk}$. The optimization methods include: i) **Least squares regression:** A direct method for computing optimal weights is solving the least squares problem:

$$W = H^{+} Y \tag{3.107}$$

where $H^+$ is the Moore-Penrose pseudoinverse of the hidden layer activation matrix $H$, and $Y$ is the target output. This provides a closed-form solution but may be computationally expensive for large datasets; ii) **Regularized least squares (ridge regression):** To improve robustness, ridge regression introduces a penalty term:

$$W = \left(H^T H + \lambda I\right)^{-1} H^T Y$$

where $\lambda$ prevents overfitting by controlling weight magnitudes; and iii) **Iterative gradient descent optimization:** For large-scale applications, weights can be optimized using:

$$w_{jk}^{new} = w_{jk}^{old} - \eta \frac{\partial E}{\partial w_{jk}} \tag{3.108}$$

where $\eta$ is the learning rate. This method is suitable for online learning applications where data arrives sequentially.

### 3.10.4 Practical Considerations for RBFNN Optimization

To ensure efficient learning and generalization, the following practical strategies are commonly used: i) Regularization techniques: Adding an $L_2$ norm penalty to the cost function prevents excessive weight magnitudes, reducing overfitting; ii) Cross-validation: Splitting data into training and validation sets helps assess generalization performance and fine-tune parameters; and iii) Incremental learning: Instead of fixing the network structure, dynamically adding or removing neurons based on error thresholds improves adaptability.

### 3.10.5 Applications of Optimized RBFNNs

Optimized RBFNNs find applications in various domains, including: i) System identification: Modelling nonlinear dynamic systems in robotics and control; ii) Pattern recognition: Improving classification accuracy in image and speech processing; iii) Medical diagnosis: Enhancing disease detection from physiological signals; and iv) Financial forecasting: Predicting stock prices and economic trends.

Parameter optimization is crucial for achieving high accuracy and efficiency in RBFNNs. The selection of centres using clustering or OLS methods, adaptive tuning of spread values, and efficient weight optimization ensures robust performance. By systematically applying these techniques, RBFNNs can effectively model complex relationships in data, making them a powerful tool in system identification and function approximation.

## EXERCISES

1. Explain the significance of functional approximation in system identification. How do classical methods like polynomial approximation and Fourier series compare to neural network-based approaches?
2. Consider a system described by the equation:

$$y(t) = f\left(u(t), u(t-1), \ldots, u(t-n)\right) + e(t)$$

where $y(t)$ is the system output, $u(t)$ is the input, and $e(t)$ represents noise.
(a) How does a Multi-Layered Neural Network approximate such a system?
(b) What is the significance of heuristic self-organization in neural network-based system identification?
3. What are Physics-Informed Neural Networks (PINNs), and how do they differ from standard deep neural networks in system identification? Provide an example of a PINN application.
4. Describe the Parameter Optimization Methods for Radial Basis Function Neural Networks (RBFNNs). What challenges arise in optimizing RBFNN parameters?
5. Explain the Dual Neural Network Strategy for Direct Structural Parameter Identification. How does it improve upon traditional structural identification methods?

## REFERENCES

1. Raol, J.R., Girija, G., and Singh, J., *Modelling and Parameter Estimation of Dynamic Systems. IEE/ITE Control Series* Vol. 65, Institution of Electrical Engineers (IEE), London, UK, 2004.
2. Narendra, K.S., and Anuradha, M., *Annaswamy, Stable Adaptive Systems*, Courier Corporation, New York, 2012.
3. Narendra, K.S., and Parthasarathy, K., "Identification and Control of Dynamical Systems Using Neural Networks", *IEEE Transactions on Neural Networks*, Vol. 1, No. 1, pp. 4–27, March. 1990.
4. Chung, J., Gulcehre, C., Cho, K., and Bengio, Y., "Empirical Evaluation of Gated Recurrent Neural Networks on Sequence Modeling", In *NIPS 2014 Workshop on Deep Learning*, Dec. 2014.
5. Kumar, M.V., Omkar, S.N., Ganguli, R., Sampath, P., and Suresh, S., "Identification of Helicopter Dynamics Using Recurrent Neural Networks and Flight Data", *Journal of the American Helicopter Society*, Vol. 51, No. 2, pp. 164–174, 2006.
6. Xu, B., Wu, Z., Chen, G., and Yokoyama, K., "Direct Identification of Structural Parameters from Dynamic Responses with Neural Networks", *Engineering Applications of Artificial Intelligence*, Vol. 17, No. 8, pp. 931–943, 2004.
7. Lowe, D., and Broomhead, D., "Multivariable Functional Interpolation and Adaptive Networks", *Complex Systems*, Vol. 2, No. 3, pp.321–355, 1988.

# 4 Filtering and Parameter Estimation

The filtering and parameter estimation can be considered as higher level of digital signal processing (DSP) which is the area of engineering that deals with signals from physical processes, and which has made definite strides in other areas in science, economics, business, and environmental data analyses. The continuous signals are converted to digital form and using algorithms and techniques which operate on these digital signals, some useful information is generated. DSP applications range from image processing and, speech processing to communication and control systems. With the advent of fast digital computers, the application of DSP has reached every aspect of engineering, commercial market, aerospace, and mobile telephones. DSP is an interdisciplinary area encompassing analogue and digital electronics, numerical analysis, statistics and probability, and decision and communication/information theory. Signals could be represented as a function of time, frequency, or distance depending on whether the analysis is carried out using time domain, frequency domain, or spatial domain methods. Mathematical models to understand and analyze the system characteristics often include parameters which describe the stability and control behaviour of the system. These parameters are estimated using the output responses generated by applying appropriate forcing functions (input signals) to the system and the input forcing function itself. A suitable method/algorithm based on minimization of some criterion (based on suitable signal or system norm) is used along with the I/O data to estimate the unknown parameters [1]-[4]. The most common basis is the minimization of output error (estimation error = actual signal – estimated/predicted signal).

## 4.1 FILTERING CONCEPTS AND METHODS

Filtering, in general, is used for either separating two signals that have been combined or to restore signals that are distorted in some way. Signal separation is required when interference by other signals or noise contaminates the desired/true signal as in signal transmission through a noisy channel/medium. Signal restoration is required when distortion of signals occurs as in de-blurring of an image acquired by a shaky camera. In analogue filtering applications, we encounter low pass, high pass, and band pass filters which are used in signal processing applications to remove unwanted frequencies above, below, or in a band of frequency of interest. Analog filters are designed using electronic circuits. In DSP, the equivalent digital filters are designed


DOI: 10.1201/9781003616580-4

by choosing appropriate impulse response functions which are also called filter kernels. While analogue filters are cheap, fast, and have a large dynamic range, digital filters have far superior levels of performance in terms of accuracy, range, and any criterion that one can specify.

### 4.1.1 Simple Recursive Filtering

Recursive filters use input value at the current instant and previously estimated output values to calculate the filtered output at the current instant. Given a system with input signal $u$ and output signal $y$, a recursive filter could be represented by

$$y(k) = a_0 u(k) + a_1 u(k-1) + a_2 u(k-2) + \ldots + b_1 y(k-1) + b_2 y(k-2) + \ldots \tag{4.1}$$

In eqn. (4.1) $u(k), u(k-1), \ldots$ represent input samples and $y(k-1), y(k-2), \ldots$ represent the output samples, and coefficients a(.) and b(.) are designed to achieve desired performance from the filter.

Recursive filters can be formulated using measurements of the dynamic process in such a manner that for state/parameter estimation at the current instant, past measurements need not be stored. The measurements are represented by

$$z(i) = x + v(i),\ i = 1, 2, \ldots, k \tag{4.2}$$

In eqn. (4.2) $v(i)$ is the measurement noise which is assumed to be white. It follows from LS principle that by averaging the measurements up to k-th instant, we obtain an unbiased, minimum variance estimate given by

$$\hat{x}(k) = \frac{1}{k} \sum_{i=1}^{k} z(i) \tag{4.3}$$

At (k+1) instant, if we acquire a new measurement, the estimate would be given by

$$\hat{x}(k+1) = \frac{1}{k+1} \sum_{i=1}^{k+1} z(i) \tag{4.4}$$

Eqn. (4.4) can be rewritten as

$$\hat{x}(k+1) = \frac{1}{k+1}\left(\frac{k}{k}\sum_{i=1}^{k} z(i)\right) + \frac{1}{k+1} z(k+1) = \frac{k}{k+1}\hat{x}(k) + \frac{1}{k+1} z(k+1) \tag{4.5}$$

Eqn. (4.5) gives the current estimate using the previous estimate which utilizes all the measurements up to that instant and the current measurement thereby eliminating the need to store all prior measurements. The recursive form of the estimate can also be written as

$$\hat{x}(k+1) = \hat{x}(k) + \frac{1}{k+1}\left(z(k+1) - \hat{x}(k)\right) \tag{4.6}$$

Eqn. (4.6) shows that the current estimate is obtained by adding a weighted difference of the current measurement and the prior estimate which is nothing but its expected value. It is ascertained here that the form of eqn. (4.6) is the fundamental structure of (almost) all the recursive filtering algorithms, differing only in the 'weight' that is computed using a variety of criteria, and this leads to several types of filtering algorithms.

### 4.1.2 Kalman Filtering

Real-life systems give rise to signals, which are invariably contaminated with so-called random noise. This noise could arise due to measurement errors from the sensors, instruments, data transmission channels, or human error. In most cases, the errors are random in nature and can be described best by a probabilistic model. A usual characteristic of such a random noise that affects the signal is Gaussian (normally distributed) noise with zero mean and a finite variance. This variance measures the power-strength of the noise and it is often compared to the power of the signal that is influenced by the random noise. This leads to a measure called signal to noise ratio (SNR). Often the noise is assumed a white process. The aim is then to maximize SNR by filtering out the noise from the signal/data of the dynamic system. We use the model-based approach and especially the one based on the state-space model of a dynamic system. Therefore, our major goal is to get a best estimate or prediction of the signal, which is buried in the random noise. This noise could be white or time-correlated (non-white). It could be coloured noise, i.e., output of a linear lumped parameter system excited by a white noise. We mainly study the problem of filtering and prediction using the Kalman filtering (KF) method.

For linear systems, it is an optimal state observer. Extended KF (EKF) can be used for nonlinear systems and the innovations computed from this approach are likely to be white Gaussian if we assure that the measurements are very frequent or dense [4].

A dynamic system is assumed as:

$$x(k+1) = \varphi x(k) + Bu(k) + Gw(k) \tag{4.7}$$

$$z(k) = Hx(k) + Du(k) + v(k) \tag{4.8}$$

Here, $x$ is $n$x1 state vector; $u$ is $p$x1 deterministic control input to the system; $z$ is $m$x1 measurement vector; $w$ is a WGN sequence with zero mean and covariance matrix $Q$. It is also called process noise with associated matrix $G$; $v$ is a WGN sequence with zero mean and covariance matrix $R$. It is also called measurement noise; $\varphi$ is the $n$x$n$ transition matrix that takes states from $k$ to $k$+1; $B$ is the input gain/magnitude vector/matrix; $H$ is the $m$x$n$ measurement model/sensor dynamics matrix; and $D$ is the $m$x$p$ feedforward/direct control input matrix. Often $D$ is dropped from the Kalman filter development.

We emphasize here that, although most dynamic systems are continuous-time, the KF is best discussed using the discrete-time model. We feel that understanding and implementing a discrete-time Kalman filter is easier. The problem of state estimation using KF can be formulated as: given the model of the dynamic system, statistics of the noise processes (Q,R) and the noisy measurement data, and the input, determine the best estimate of the state, $x$, of the system. Since, it is assumed that the dynamic system is known, it means that the (form and) numerical values of the elements of $\varphi$, $B$, and $H$ are accurately known. If some of these elements are not known, then these can be considered as additional unknown states and appended to the otherwise state vector $x$ yielding the extended state vector. In most circumstances this will lead to a nonlinear dynamic system for which EKF can be used. The process noise accounts for modelling errors as well as an artefact to do filter tuning for trajectory matching. Since, our aim is to obtain an estimate of the state of the dynamic system, we need to have measurements of the state; often these are available indirectly through a measurement model.

#### 4.1.2.1 Discrete Time Filtering Algorithm

We assume that the matrix $H$ and a priori covariance matrix $R$ are given or known.

Given: H, R and measurements z; assume: $\tilde{x}(k) \rightarrow$ a priori estimate of state at time k, i.e., before the measurement data is incorporated; $\hat{x}(k) \rightarrow$ updated estimate of state at time k, i.e., after the measurement data is incorporated; $\tilde{P} \rightarrow$ a priori covariance matrix of state estimation error. The structure of the filter has now been well defined:

*Current estimate = previous estimate + gain* x
*(error in measurement prediction)*

The term $[z(k) - H\,\tilde{x}(k)]$ is called the measurement prediction error or the residual of the measurement. It is also called innovation. The KF equations/algorithm is given as:

***State propagation/time evolution***

$$\text{State estimate:} \quad \tilde{x}(k+1) = \varphi\,\hat{x}(k) \tag{4.9}$$

$$\text{Covariance (a priori):} \quad (k+1) = \varphi\hat{P}(k)\varphi^T + GQG^T \tag{4.10}$$

***Measurement update***

$$\text{Residual:} \quad r(k+1) = z(k+1) - H\,\tilde{x}(k+1) \tag{4.11}$$

$$\text{Kalman Gain:} \quad K = \tilde{P}H^T(H\tilde{P}H^T + R)^{-1} \tag{4.12}$$

$$\text{Filtered estimate:} \quad \hat{x}(k+1) = \tilde{x}(k+1) + K\,r(k+1) \tag{4.13}$$

$$\text{Covariance (posteriori):} \quad \hat{P} = (I - KH)\tilde{P} \tag{4.14}$$

Although, $K$ and $P$ would vary as the filter is running, the time index is dropped for simplicity. However, $Q$ and $R$ are assumed to be pre-determined and constant.

We note here that $K = \tilde{P}H^T S^{-1}$ with $S = H\tilde{P}H^T + R$. This matrix $S$ is the covariance matrix of residuals or innovations.

### 4.1.2.2 Continuous Time KF

Although the discrete-time filtering algorithm is widely preferred for digital implementation, we give continuous-time filtering algorithm here. Let us define the continuous time model of the dynamical system as:

$$\dot{x}(t) = Ax(t) + w(t) \tag{4.15}$$

$$z(t) = Hx(t) + v(t) \tag{4.16}$$

We have the following assumptions:

i) The noise processes $w$(t) and $v$(t) are uncorrelated Gaussian random processes with spectral density matrices $Q$(t) and $R$(t), respectively.
ii) $E\{x(0)\} = \hat{x}_0$ ; $E\{(\hat{x}_0 - x(0))\,(\hat{x}_0 - x(0))^T\} = P_0$
iii) We have very accurate knowledge of $A$, $H$, $Q$ and $R$.

Then, continuous time KF is given as:

$$\dot{\hat{x}}(t) = A\hat{x}(t) + K(t)[z(t) - H\hat{x}(t)] \quad (\text{state evolution}) \tag{4.17}$$

$$\dot{P}(t) = AP(t) + P(t)A^T + Q(t) - KRK^T; \quad P(0) = P_0 \tag{4.18}$$

$$K = PH^T R^{-1} \quad (\text{Kalman gain}) \tag{4.19}$$

Eqn. (4.18) is called the matrix Riccati equation, which needs to be solved for obtaining $P$, which is used in computation of the Kalman gain.

## 4.2 PARAMETER ESTIMATION

To address the parameter estimation problem, we begin with the assumption that the real or experimental data are contaminated with random noise or measurement errors. We use these data in an identification-estimation procedure (IEP) to arrive at optimal estimates of the unknown parameters which best describe the behaviour of the data-system dynamics. This process of determining the unknown parameters of a postulated mathematical model from noisy IO data is termed as 'parameter estimation'. A closely related problem is that of 'state estimation' wherein the estimates of the so-called 'states' of the dynamic process/system (a power plant or an aircraft) are obtained by using the optimal linear or the non-linear filtering theory.

In parameter estimation, given a set of discrete-time measurements $Z = \{z(k), z(k-1), \ldots, z(1)\}$ which includes both I/O data (and measurements) of a system, it is required to estimate a parameter vector $\beta$ which is used to characterize the dynamic system. Thus, the problem reduces to determining the mapping of the measured data to the parameters in the mode $Z \to \hat{\beta}(t,Z)$ where $\hat{\beta}$ is the estimate of the parameters based on the information contained in Z. In batch procedures, all the data collected are used for processing and the mapping is described by $Z \to \hat{\beta}_N(N,Z)$. In recursive procedures, the estimate $\hat{\beta}(at\, t\, or\, k)$ is to be computed at each instant of time with the constraint that the required memory space and computation time should not increase with t; or increase only linearly with time in some cases, then such an estimation/filtering is called a finite dimensional filtering algorithm. The recursive algorithm is accomplished by formulating the current estimate to be a function of previous data, the current measurement, and an auxiliary memory of fixed size so that only the current parameter vector and the auxiliary variable are stored at each instant of time:

$$\hat{\beta}(t) = f(\hat{\beta}(t-1), P(t), z(t)) \tag{4.20}$$

$$(t) = h(\hat{\beta}(t-1), P(t-1), z(t)) \tag{4.21}$$

Here, f and h are known functions and the current estimate uses the previous estimate, the current data and the auxiliary variable P, that would often determine the variance/energy of the estimation errors. The parameters and the auxiliary variable are updated using an algorithm with a fixed number of steps that is independent of time t. The adequacy and validity of the estimated parameters has to be established using suitable theoretical criteria. Once the parameters of the model are estimated, they can be used to generate the system response to any input, and further for forecasting, simulation, and control applications.

The parameters of the dynamic model of the system are iteratively adjusted until the responses of the model match closely with the measured output/s of the system under investigation in the sense specified by the optimization criterion. Good response match is necessary but not sufficient for obtaining accurate estimates.

### 4.2.1 Equation Error Method

The equation error method (EEM) is a special case of the maximum likelihood estimator (MLE) for the data containing only process noise and no measurement noise. The output error method is a MLE for data containing only measurement noise and no process noise. The EEM has a cost function that is linear in parameters. It is simple and easy to implement.

EEM is based on the principle of least squares. It minimizes a quadratic cost function of the error in the (state) equations to estimate the parameters. It is assumed that states, their derivatives, and control inputs are available or accurately measured. The EEM is relatively fast and simple, and applicable to linear as well as linear-in-parameter systems. If the system is described by the state equation

$$\dot{x} = Ax + Bu \qquad \text{with} \qquad x(0) = x_0 \tag{4.22}$$

The equation error can be written as

$$e(k) = \dot{x}_m - Ax_m - Bu_m \tag{4.23}$$

Here, $x_m$ is the measured state, subscript $m$ denoting 'measured'. Parameter estimates are obtained by minimizing the equation error w.r.t. $\beta$. Eqn. (4.23) can be written as

$$e(k) = \dot{x}_m - A_a x_{am} \tag{4.24}$$

where $A_a = [A \quad B]$ and $x_{am} = \begin{bmatrix} x_m \\ u_m \end{bmatrix}$. In this case, the cost function is given by

$$J(\beta) = \frac{1}{2}\sum_{k=1}^{N}\left[\dot{x}_m(k) - A_a x_{am}(k)\right]^T \left[\dot{x}_m(k) - A_a x_{am}(k)\right] \tag{4.25}$$

The estimator is given as:

$$\hat{A}_a = \dot{x}_m (x_{am}^T)(x_{am} x_{am}^T)^{-1} \tag{4.26}$$

We illustrate the above formulation as follows:

$$\begin{bmatrix} \dot{x}_1 \\ \dot{x}_2 \end{bmatrix} = \begin{bmatrix} a_{11} & a_{12} \\ a_{21} & a_{22} \end{bmatrix}\begin{bmatrix} x_1 \\ x_2 \end{bmatrix} + \begin{bmatrix} b_1 \\ b_2 \end{bmatrix} u \tag{4.27}$$

Then, if there are say two measurements, we have:

$$x_{am} = \begin{bmatrix} x_{11m} & x_{12m} \\ x_{21m} & x_{22m} \\ u_{1m} & u_{2m} \end{bmatrix}_{3x2}; \quad u_m = \begin{bmatrix} u_{1m} & u_{2m} \end{bmatrix}; \quad \dot{x}_m = \begin{bmatrix} \dot{x}_{11m} & \dot{x}_{12m} \\ \dot{x}_{21m} & \dot{x}_{22m} \end{bmatrix} \tag{4.28}$$

Then, $\left[\hat{A}_a\right]_{2x3} = \left[\left[\hat{A}\right]_{2x2} \vdots \left[\hat{B}\right]_{2x1}\right] = \left[\dot{x}_m\right]_{2x2}\left[x_{am}^T\right]_{2x3}\left\{\left[x_{am}\right]_{3x2}\left[x_{am}^T\right]_{2x3}\right\}^{-1}$

(4.29)

Application of the EEM to parameter estimation requires accurate measurements of the states and their derivatives. In addition, it can be applied to unstable systems because it does not involve any numerical integration of the dynamic system that would otherwise cause divergence. Utilization of measured states and state-derivatives for estimation in the algorithm enables estimation of the parameters of even an unstable system directly. However, if the measurements are noisy, the method will give biased estimates. We would like to mention here that EE is amenable to be programmed in the structure of recurrent neural network as is shown later on.

### 4.2.2 Output Error Method (OEM)

The OEM is based on the assumption that only the observations contain measurement noise and there is no noise in the state equations [3]. The mathematical model of a linear system consists of the vector $x$ representing the system states, vector $y$ representing the computed system response (model output), vector $z$ representing the measured variables, and $u$ representing the control input vector. The matrices $A$, $B$, and $H$ contain the parameters to be estimated. The OEM assumes that the measurement vector $z$ is corrupted with noise which is zero mean and has a Gaussian distribution with covariance $R$, i.e., $v \sim N(0,R)$. The aim is to minimize the error between the measured and model outputs by adjusting the unknown parameters contained in matrices $A$, $B$, and $H$. Let the parameter vector to be estimated be represented by $\Theta$, where $\Theta = [\text{elements of } A, B, H, \text{ initial condition of } x]$. Then, an estimate of $\Theta$ is obtained by minimizing the cost function

$$J = \frac{1}{2}\sum_{k=1}^{N}[z(k)-y(k)]^T R^{-1}[z(k)-y(k)] + \frac{N}{2}\ln|R| \tag{4.30}$$

This cost function is like the weighted least squares (WLS) criterion with weighting matrix as $W$ and with one extra term. The estimate of $R$ can be obtained from:

$$\hat{R} = \frac{1}{N}\sum_{k=1}^{N}[z(k)-\hat{y}(k)][z(k)-\hat{y}(k)]^T \tag{4.31}$$

This matrix is obtained once the predicted measurements are computed. The estimate of $\Theta$ at the $(i+1)^{th}$ iteration is obtained as:

$$\Theta(i+1) = \Theta(i) + \left[\nabla_\Theta^2 J(\Theta)\right]^{-1}\left[\nabla_\Theta J(\Theta)\right] \tag{4.32}$$

In (4.32), the first and the second gradients are defined as:

$$\nabla_\Theta J(\Theta) = \sum_{k=1}^{N}\left[\frac{\partial y}{\partial \Theta}(k)\right]^T R^{-1}[z(k)-y(k)] \tag{4.33}$$

$$\nabla_\Theta^2 J(\Theta) = \sum_{k=1}^{N}\left[\frac{\partial y}{\partial \Theta}(k)\right]^T R^{-1}\left[\frac{\partial y}{\partial \Theta}(k)\right] \tag{4.34}$$

Starting with a set of suitable initial parameter values, the model response is computed with the input (that was used for obtaining measurement data). The estimated response and the measured response are compared and the response error is used to compute the cost function. The first and second gradients of the cost function are computed, and then eqn. (4.32) is used to update the model parameter values. The updated parameter values are once again used in the mathematical model to compute the new estimated response and the new response error. This iterative updating procedure continues until convergence is achieved.

The Gauss-Newton approximation for the second gradient in eqn. (4.34), also called the Fisher Information Matrix (FIM), provides a measure of relative accuracy of the estimated parameters. The diagonal elements of the inverse of the information matrix give the individual covariances, and the square root of these elements is a measure of the standard deviations called the Cramer-Rao bounds (*CRB*):

$$\text{Fisher Information Matrix} = \nabla^2_{\Theta} J(\Theta) \tag{4.35}$$

$$\begin{aligned} &\text{Standard deviation of estimated parameters} = CRB(\Theta) \\ &= \text{diag}\left[\sqrt{[\nabla^2_{\Theta} J(\Theta)]^{-1}}\right] \end{aligned} \tag{4.36}$$

The OEM also can be applied with equal ease to any nonlinear system, in principle:

$$\dot{x}(t) = f[x(t), u(t), \Theta] \quad \text{with initial } x(0) = x_0 \tag{4.37}$$

$$y(t) = h[x(t), u(t), \Theta] \tag{4.38}$$

$$z(k) = y(k) + v(k) \tag{4.39}$$

where $f$ and $h$ are general nonlinear functions, and the initial values $x_0$ of the state variables need to be estimated along with the parameter vector $\Theta$. It is evident that estimation of parameters with OEM would require computation of the state vector $x$, model output vector $y$ and sensitivity coefficients $\partial y/\partial \Theta$. The sensitivity coefficients for a linear system can be obtained analytically by partial differentiation of the system equations. However, for a nonlinear system, each time the model structure changes, partial differentiation of the system equations needs to be carried out to obtain $\partial y/\partial \Theta$. A better approach is to approximate the sensitivity coefficients by finite differences. In this procedure, the parameters in $\Theta$ are perturbed one at a time and the corresponding perturbed model response $y_p$ is computed. The sensitivity coefficient is then given by:

$$\partial y/\partial \Theta_j = (y_p - y)/\Delta \Theta_j \tag{4.40}$$

The use of finite differencing in calculating $\partial y/\partial \Theta$ results in a program code that is more flexible and user friendly.

## 4.3 PARAMETER ESTIMATION USING ANNS

We are aware now, how based on a priori knowledge of the system and the underlying physics, mathematical models are developed and subjected to parameter estimation using conventional techniques like EEM and OEM. The FFNNs and RNNs can also be used for explicit parameter estimation of the postulated math models [1].

### 4.3.1 Using FFNNs

The very fact that an FFNN provides for nonlinear mapping of the I/O data suggests that it should be possible to use it for system characterization, and hence math modelling. An FFNN however, works with a black-box model structure, which cannot be physically interpreted. The parameters of the network are the weights, which have no interpretation in terms of the actual system's parameters. The parameter estimation procedure using a FFNN has three steps: i) the network is given the measured data and is trained to reproduce the clean/predicted responses which are compared with the system responses in the sense of minimization of the output error, Figure 4.1; ii) these predicted responses are perturbed in turn for each parameter to be estimated and the changed predicted response is obtained; assume that $z = \beta x$ and the network is trained to produce clean $z$, the trained network is used to produce $z + \Delta z$ and $z - \Delta z$ when $x$ is changed to $x + \Delta x$ and $x - \Delta x$; and iii) then $\beta$ is obtained as

$$\beta = \frac{z^{+} - z^{-}}{x^{+} - x^{-}} \tag{4.41}$$

This method is called the Delta method. Since the variables are the signals, the parameter time histories are obtained and hence, the estimates are obtained by averaging these respective parameter time histories. This procedure is illustrated next.

**Example 4.1** Generate the simulated data using:

$$z = a + bx_1 + cx_2 \tag{4.42}$$

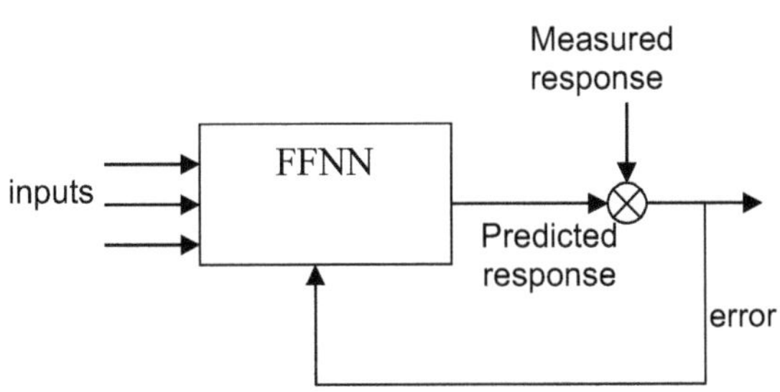


**FIGURE 4.1** Parameter estimation with FFNN

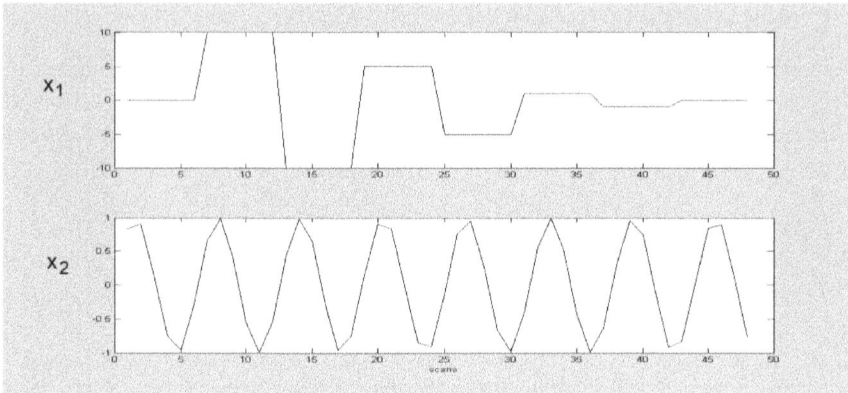


**FIGURE 4.2** Time histories of input signals; $x_1$, and $x_2$ (Example 4.1)

Here, the parameters are $a$=1, $b$=2, $c$=1 and $x_1$, $x_2$ are the inputs to the model and $z$ is the output of the model; i) train a neural network for the input variables $x_1$, $x_2$ and output variable $z$ using FFNN with backpropagation (FFNN-BPN) training algorithm, and ii) estimate $a$, $b$, and $c$ using the Delta method with the help of the trained network for various levels of noise added to input signals.

**Solution 4.1**

The data generation is carried out using eqn. (4.42) with constant value of parameters a, b, and c. The input signals $x_1$, and $x_2$ are shown in Figure 4.2. The input signal $x_2$ is generated using inbuilt MATLAB function 'sin($k$)' with $k$ varying from 1 to 48. The signal $x_1$ is generated as a periodic pulse with decreasing amplitude.

i) The simulated input and output signals were scaled and subsequently used to train FFNN using the backpropagation algorithm. The training constants/parameters were set to $\mu = 0.2, \Omega = 0.4$, in the FFNN with backpropagation algorithm with four neurons in the hidden layer. The sigmoid slope parameters $\lambda_1, \lambda_2$ for hidden and output layers were taken as 0.8 and 0.75, respectively. The training was stopped after 10,000 iterations and the percentage fit error (PFE) of the predicted data from network w.r.t. true data was found to be 0.1. Figure 4.3 shows the time history match of predicted signal $\hat{z}$ to the true signal $z$. The training is done using file '**trainffnn.m**' residing in folder **'ch4FFNNex1'**.
ii) After optimal training of the network, the Delta method was used for estimation of parameters a, b, and c. The estimated parameters and the parameter estimation error norm (PEEN) are given in Table 4.1. We see that with an increase in noise, the parameter estimation error norm increases, but still the estimates are just acceptable. The estimation is accomplished by using file '**peffnndm.m**' placed in folder **'Ch4FFNNex1**

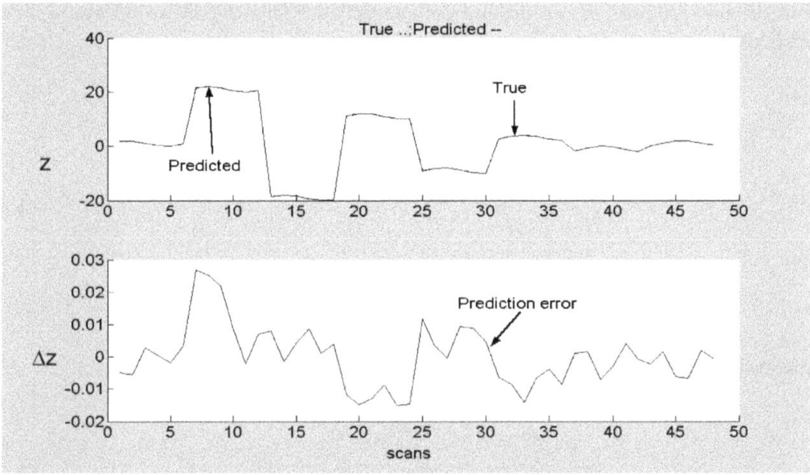


**FIGURE 4.3** Time history match using FFNN-BPN prediction phase (Example 4.1)

**TABLE 4.1**
**Estimated parameters with FFNN-BPN (Example 4.1)**

| Parameters | True values | Estimated values using delta method for different noise levels | | |
|---|---|---|---|---|
| | ---- | SNR= ∞ (no noise) | SNR=100 | SNR=10 |
| a | 1 | 0.9989 | 1.0272 | 1.1188 |
| b | 2 | 1.999 | 1.957 | 1.928 |
| c | 1 | 1.0004 | 0.9862 | 0.9441 |
| PEEN | ---- | 0.048 | 2.15 | 6.105 |

The Delta method when applied to aircraft flight test data can yield the aircraft stability and control derivatives. The method makes use of the basic definition of derivative which states that a derivative represents the change in the aerodynamic force or moment caused by a small change in the motion or control variable about its nominal position. For example, the derivative $C_{m_\alpha}$ can be defined as the change in the aircraft pitching moment $C_m$ due to a small change in the angle-of-attack $\alpha$ with all other motion and control variables held constant.

To estimate aircraft stability and control derivatives, the input layer of the neural network contains the motion and control variables, such as angle-of-attack, sideslip angle, rates, and control inputs. The output layer comprises the aerodynamic force and moment coefficients. In the following examples, the application of the Delta method and FFNN to estimate aircraft derivatives from simulated flight test data is demonstrated for better understanding.

**Example 4.2** Generate the simulated data using the following state and aerodynamic models of the aircraft dynamics:

$$\begin{aligned}
\dot{V} &= -\frac{\bar{q}S}{m}C_D - g\sin(\theta-\alpha)\\
\dot{\alpha} &= -\frac{\bar{q}S}{mV}C_L + \frac{g}{V}\cos(\theta-\alpha)+q\\
\dot{q} &= \frac{\bar{q}S\bar{c}}{I_y}C_m\\
\dot{\theta} &= q
\end{aligned} \tag{4.43}$$

The aerodynamic model is:

$$\begin{aligned}
C_D &= C_{D0} + C_{D\alpha}\alpha + C_{D\delta_e}\delta_e\\
C_L &= C_{L0} + C_{L\alpha}\alpha + C_{L\delta_e}\delta_e\\
C_m &= C_{m_0} + C_{m\alpha}\alpha + C_{m_q}\frac{q\bar{c}}{2V} + C_{m\delta_e}\delta_e
\end{aligned} \tag{4.44}$$

For a given set of parameter values (true values) do the following:

i) Generate the time histories of variables $\dot{V}, \dot{\alpha}, \dot{q}, \dot{\theta}, V, \alpha, q, \theta, \delta_e$, and coefficients $C_D, C_L$, and $C_m$ with sinusoidal input data.
ii) Train the FFNN for the variables $\dot{\alpha}, \dot{q}$ using
   - Network with backpropagation (FFNN-BPN, see Chapter 2) and
   - Network with backpropagation recursive least square filter algorithm with linear output layer (FFNN-BPNRLSFL, see Chapter 2).
iii) Train the FFNN for the aerodynamic coefficients $C_D, C_L$, and $C_m$ using
   - Network with backpropagation recursive least square filter algorithm with non-linear output layer (FFNN-BPNRLSFNL).
iv) Use the Delta method to estimate the aerodynamic derivatives appearing in eqn. (4.44), using the predicted time-histories of the aerodynamic coefficients obtained by training the neural network for each of the aerodynamic coefficients individually and with different noise levels added to the variables $V, \alpha, q, \theta$.

**Solution 4.2**

i) Time histories of variables $\dot{V}, \dot{\alpha}, \dot{q}, \dot{\theta}, V, \alpha, q, \theta, \delta_e$, and coefficients $C_D, C_L$, and $C_m$ are generated using eqns. (4.43) and (4.44) with sinusoidal input $\delta_e = A\sin(\theta)$; $A = 1$, $\theta = 0:\pi/8:n\pi$, and $n = 25$. For the simulation, true values of the aerodynamic coefficients given in Table 4.2 are used. The other parameters related to simulated aircraft are $\bar{c} = 10$ m, $S = 23.0$ m$^2$, $m = 7000$ Kg, $I_y = 50000$ Kg-m$^2$, $V = 100$ m/s, $\bar{q} = 5000$ Kg/m-s$^2$, and $g = 9.81$ m/s$^2$. The initial value of $\alpha, q$, and $\theta$ were taken as 0.1 rad, 0.0 rad/s, and 0.1 rad,

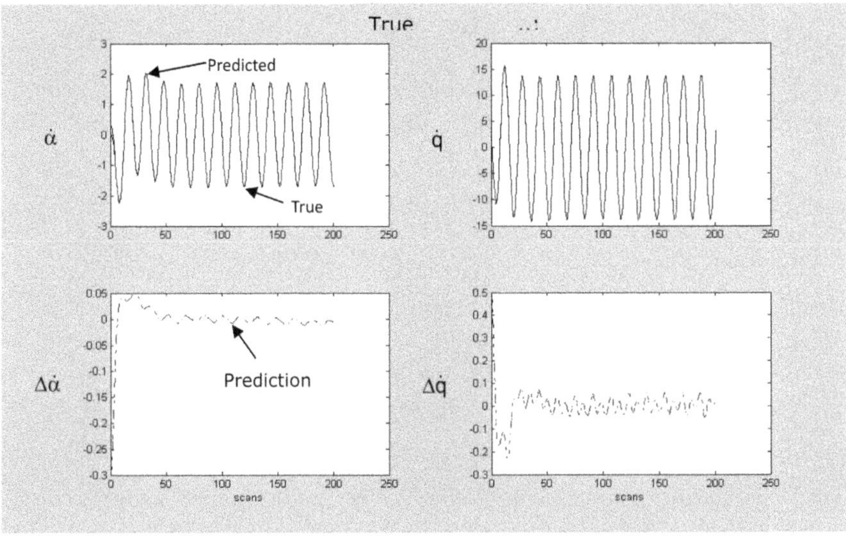


**FIGURE 4.4** Time history match and prediction error for $\dot{\alpha}$ and $\dot{q}$ using FFNN-BPN (Example 4.2)

respectively. A total number of 200 data samples are simulated for analysis. The programs for data simulation, training, prediction, and parameter estimation are contained in the folder **Ch4FFNNex2.**

ii) The following model is used for the purpose of training the network:

$$\dot{\alpha} = h_1(V, \alpha, q, \delta_e) \tag{4.45}$$

$$\dot{q} = h_2(V, \alpha, q, \delta_e) \tag{4.46}$$

Here, $h$ is a non-linear functional relationship. The signals $V, \alpha, q$ , and $\delta_e$ are presented to the network as inputs and signals $\dot{\alpha}$ and $\dot{q}$ as outputs. The network was trained using both FFNN-BPN and FFNN-BPNRLSFL algorithms. The tuning parameters used for training the algorithms for $\dot{\alpha}$ and $\dot{q}$ signals are given in Table 4.2. Figures 4.4 and 4.5 show the time history match for prediction phase using FFNN-BPN and FFNN-BPNRLSL algorithms, respectively, and we see that the latter gives somewhat better results.

iii) Next, the FFNN-BPNRLSNL algorithm was used for prediction of the aerodynamic coefficients (time histories) $C_D, C_L$, and $C_m$ as a function of $\alpha, q, V$ and $\delta_e$. The coefficient time histories are used as the outputs and $\alpha, q, V$ , $\delta_e$ as inputs to the network. The tuning parameters used for training are given in Table 4.2 and the time history match for the coefficients $C_D, C_L$, and $C_m$ are shown in Figure 4.6.

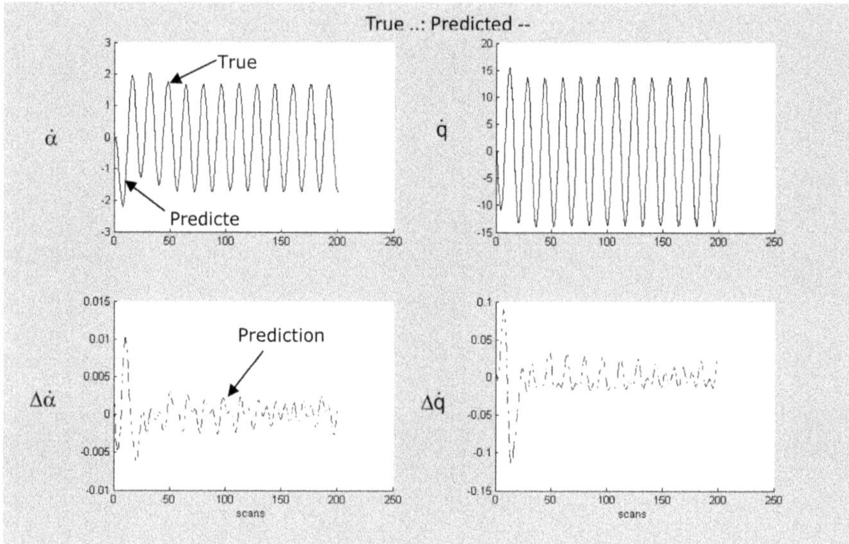


**FIGURE 4.5** Time history match and prediction error for $\dot{\alpha}$ and $\dot{q}$ using FFNN-BPNRLSFL (Example 4.2)

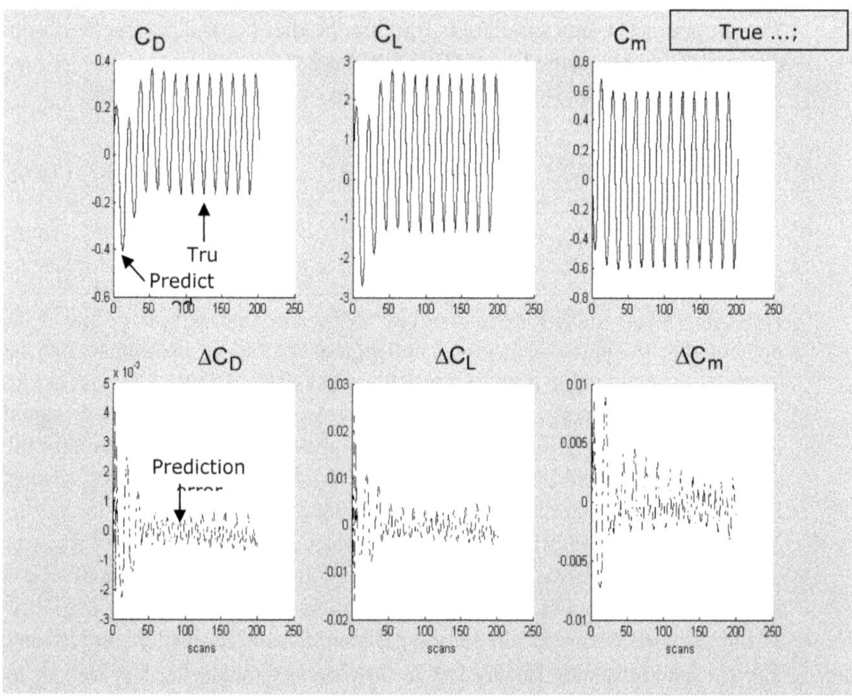


**FIGURE 4.6** Time history match and prediction error for $C_D$,$C_L$ and $C_m$ using FFNN-BPNRLSFNL (Example 4.2)

**TABLE 4.2**
**Tuning parameters used for feedforward neural network training for step (ii) and step (iii)**

| | $\dot{\alpha}, \dot{q}$ | | $C_D, C_L, C_m$ |
|---|---|---|---|
| **Tuning parameter** | **BPN** | **BPNRLSFL** | **BPNRLSFNL** |
| Function slope of hidden layer $\lambda_1$ | 0.8 | 0.8 | 0.8 |
| Function slope of hidden layer $\lambda_2$ | 0.75 | 0.75 | 0.75 |
| Number of hidden layers | 1 | 1 | 1 |
| Number of nodes in the hidden layer | 6 | 6 | 6 |
| Data scaling range | ± 0.1 | ±0.1 | ±0.1 |
| Learning rate parameter μ | 0.2 | 0.2 | 0.2 |
| Momentum parameter Ω | 0.4 | NA | NA |
| Training iterations | 10000 | 10000 | 2000 |

**TABLE 4.3**
**Tuning parameters used for FFNN network training for step (iv)**

| | $C_D$ | | $C_L$ | | $C_m$ | |
|---|---|---|---|---|---|---|
| **Tuning parameter** | **BPN** | **BPNRLSFNL** | **BPN** | **BPNRLSFNL** | **BPN** | **BPNRLSFNL** |
| Function slope of hidden layer $\lambda_1$ | 0.9 | 0.9 | 0.9 | 0.9 | 0.8 | 0.9 |
| Function slope of hidden layer $\lambda_2$ | 0.85 | 0.85 | 0.85 | 0.85 | 0.75 | 0.85 |
| Number of hidden layers | 1 | 1 | 1 | 1 | 1 | 1 |
| Number of nodes in the hidden layer | 6 | 6 | 6 | 6 | 6 | 6 |
| Data scaling range | ± 0.1 | ±0.1 | ±0.1 | ±0.1 | ±0.2 | ± 0.1 |
| Learning rate parameter μ | 0.2 | 0.2 | 0.2 | 0.2 | 0.2 | 0.2 |
| Momentum parameter Ω | 0.4 | NA | 0.4 | NA | 0.2 | NA |
| Training iterations | 10000 | 2000 | 10000 | 2000 | 50000 | 5000 |

iv) FFNN-BPN and FFNN-BPNRLSFNL algorithms are used to train the network and predict the coefficients time histories for $C_D, C_L$ and $C_m$ one at a time. The tuning parameters used for training the network are listed in Table 4.3. Once the network maps the input variables to output variables correctly, Delta method is used for estimation of derivatives $C_{D0}, C_{D\alpha}, C_{D\delta_e}, C_{L0}, C_{L\alpha}, C_{L\delta_e}, C_{m_0}, C_{m_\alpha}, C_{m_q}$, and $C_{m\delta_e}$. Having trained the network, any one variable in the input layer can be perturbed to cause a corresponding change in the output response. For

**TABLE 4.4**
**Parameter estimation with feedforward neural network (Example 4.2)**

| Parameters | True values | Estimated values using delta method for different noise levels: SNR = ∞ BPN | SNR = ∞ BPNRLSFNL | SNR = 100 BPN | SNR = 100 BPNRLSFNL | SNR = 10 BPN | SNR = 10 BPNRLSFNL |
|---|---|---|---|---|---|---|---|
| $C_{D0}$ | 0.046 | 0.0480 | 0.0465 | 0.0487 | 0.0472 | 0.0552 | 0.0534 |
| $C_{D\alpha}$ | 0.543 | 0.5406 | 0.5467 | 0.5392 | 0.5456 | 0.5069 | 0.5121 |
| $C_{D\delta_e}$ | 0.138 | 0.1383 | 0.1368 | 0.1284 | 0.1270 | 0.1160 | 0.1149 |
| PEEN | ---- | **0.565** | **0.696** | **1.893** | **2.024** | **7.688** | **6.981** |
| $C_{L0}$ | 0.403 | 0.4177 | 0.4030 | 0.4279 | 0.4138 | 0.5002 | 0.4859 |
| $C_{L\alpha}$ | 3.057 | 3.0475 | 3.0540 | 3.0708 | 3.0779 | 2.9733 | 2.9824 |
| $C_{L\delta_e}$ | 1.354 | 1.3542 | 1.3530 | 1.2703 | 1.2690 | 1.1818 | 1.1804 |
| PEEN | ---- | **0.520** | **0.094** | **2.625** | **2.619** | **6.375** | **6.127** |
| $C_{m_0}$ | 0.010 | -0.0175 | -0.0383 | -0.0170 | -0.0377 | -0.0132 | -0.0321 |
| $C_{m\alpha}$ | -0.119 | -0.1160 | -0.1219 | -0.1170 | -0.1226 | -0.1219 | -0.1272 |
| $C_{m_q}$ | -1.650 | -1.6385 | -1.6298 | -1.6560 | -1.6454 | -1.6191 | -1.6065 |
| $C_{m\delta_e}$ | -0.571 | -0.5696 | -0.5664 | -0.5274 | -0.5238 | -0.5162 | -0.5118 |
| PEEN | ---- | **1.715** | **3.007** | **2.956** | **3.852** | **3.837** | **4.859** |

example, with the weights in the network frozen after training, changing the value of $\alpha$ to $\alpha + \Delta\alpha$ at all points (other input variables remain unchanged) yields values of $C_D^+$ that are slightly different from $C_D$. Likewise, changing $\alpha$ to $\alpha - \Delta\alpha$ will yield the response $C_D^-$. Then $C_{D\alpha}$ derivative is given by $C_{D\alpha} = (C_D^+ - C_D^-)/2\Delta\alpha$. Following this procedure, the other derivatives can be determined. It is to be noted that the network produces as many estimates of the derivatives as the number of data points used to train the network. The final value of the derivative is obtained by taking the mean of these values for the corresponding derivative. After computing $C_{D\alpha}$, $C_{D\delta e}$ at all points, an estimate of $C_{D0}$ can be obtained as: $C_{D0} = C_D - [C_{D\alpha}\alpha + C_{D\delta_e}\delta_e]$. The results of estimation are given in Table 4.4.

We see that backpropagation recursive least squares filter algorithm with non-linear output layer gives somewhat better results compared to the backpropagation with the steepest descent method in certain cases, as can be seen from Table 4.4. Some improvement is surely possible.

**Example 4.3** Consider the aircraft aerodynamic model:

$$\begin{aligned} C_x &= C_{x0} + C_{x\alpha}\alpha + C_{x\alpha^2}\alpha^2 \\ C_z &= C_{z0} + C_{z\alpha}\alpha + C_{zq}q + C_{z\delta}\delta \\ C_m &= C_{m0} + C_{m\alpha}\alpha + C_{mq}q + C_{m\delta}\delta \end{aligned} \tag{4.47}$$

For a given set of (true) parameter values, simulation is carried out to generate time histories consisting of 250 data samples for the variables $\alpha, \alpha^2, q$, $\delta$ and coefficients $C_x, C_z$ and $C_m$. Using a FFNN in conjunction with Delta method, estimate the model parameters $C_{x0}, C_{x\alpha}, C_{x\alpha^2}, \ldots, C_{m\delta}$ appearing in eqn. (4.47). Apply the regression method (LS approach) to the simulated data and determine the parameter values. Compare the parameter values estimated using the Delta and regression methods with true values.

**Solution 4.3** The input layer of the network consists of the variables $\alpha, \alpha^2, q$ and $\delta$, and the output layer consists of the measured values of the non-dimensional force and moment coefficients $C_x, C_z$ and $C_m$. The (FFNN-BPN) network can be trained using one of the two options: i) considering all the three measurements in the output layer, or ii) considering only one coefficient at a time in the output layer. In the present example, we adopt the second approach of training the network to predict only one coefficient at a time. Following this procedure gives the user more freedom to come up with a suitable sets of tuning parameters that can lead to better prediction of $C_x, C_z$ and $C_m$. Once the network maps the input variables to output variables, the Delta method is used to estimate derivatives $C_{x\alpha}, C_{x\alpha^2}, C_{z\alpha}, C_{zq}, C_{z\delta}, C_{m\alpha}, C_{mq}$ and $C_{m\delta}$. After computing $C_{m\alpha}$, $C_{mq}$ and $C_{m\delta}$ at all points, an estimate of $C_{m0}$ can be obtained as:

$$C_{m0} = C_m - [C_{m\alpha}\alpha + C_{mq}q + C_{m\delta}\delta] \tag{4.48}$$

The values of the tuning parameters used for network training are listed in Table 4.5; as seen, the data scaling range selected for each of the coefficients for the FFNN training is different. For this example, it is observed that the choice of different scaling range for $C_x, C_z$ and $C_m$ leads to improved prediction of measured coefficients. The results of parameter estimation are provided in Table 4.6. Estimates obtained from applying the regression error method to the simulated data are also listed for comparison. It is concluded that if one can tune the feedforward neural network to yield good prediction of training data, one can expect to achieve satisfactory values of the parameter estimates using the Delta method. The training & estimation are accomplished by using file '**trainffnn.m**' placed in folder '**Ch4FFNNex3**'.

We see from Table 4.6 that the Delta method gives estimates slightly different from the true values compared to the regression method. It is surprising that despite very low values of percentage fit error, the parameter estimation error norms are a bit high. We see that the FFNN-based parameter estimation approach offers

## TABLE 4.5
## Tuning parameters used for feedforward neural network training (Example 4.3)

| Tuning parameter | Values of tuning parameters selected in FFNN to predict | | |
|---|---|---|---|
| | $C_x$ | $C_z$ | $C_m$ |
| Non-linear function slope of hidden layer $\lambda_1$ | 0.8 | 0.8 | 0.8 |
| Non-linear function slope of hidden layer $\lambda_2$ | 0.75 | 0.75 | 0.75 |
| Number of hidden layers | 1 | 1 | 1 |
| Number of nodes in the hidden layer | 6 | 6 | 6 |
| Data scaling range | -0.2 to 0.2 | -0.1 to 0.1 | -0.15 to 0.15 |
| Learning rate parameter $\mu$ | 0.2 | 0.2 | 0.2 |
| Momentum parameter $\Omega$ | 0.4 | 0.4 | 0.4 |

## TABLE 4.6
## Parameter estimation with FFNN (BPN/Example 4.3)

| Derivatives | True value[+] | Estimated values using Delta Method | Regression | Comments |
|---|---|---|---|---|
| $C_{x0}$ | -0.054 | -0.058 | -0.0539 | Fit error (PFE) after 10,000 iterations was 0.53 %. Thereafter, change in PFE was < 0.011% |
| $C_{x\alpha}$ | 0.233 | 0.279 | 0.2318 | |
| $C_{x\alpha^2}$ | 3.609 | 3.532 | 3.6129 | |
| PEEN | ---- | 2.475 | 0.11 | |
| $C_{z0}$ | -0.12 | -0.121 | -0.1188 | Fit error (PFE) after 10,000 iterations was 0.11 %. Thereafter, change in PFE was < 2.72e-6 % |
| $C_{z\alpha}$ | -5.68 | -5.679 | -5.6799 | |
| $C_{zq}$ | -4.32 | -4.406 | -4.1452 | |
| $C_{z\delta}$ | -0.407 | -0.407 | -0.3961 | |
| PEEN | ---- | 1.20 | 2.449 | |
| $C_{m0}$ | 0.055 | 0.056 | 0.055 | Training was stopped at 10,000 iterations and the PFE achieved was 0.95 %. Subsequent change in PFE was of the order 0.001% |
| $C_{m\alpha}$ | -0.729 | -0.733 | -0.729 | |
| $C_{mq}$ | -16.3 | -16.61 | -16.3 | |
| $C_{m\delta}$ | -1.94 | -1.956 | -1.94 | |
| PEEN | ---- | 1.887 | 0.00 | |

**+ Parameter values used to generate simulated data**

an alternative method and could be made more robust and accurate by choosing the training parameters automatically and optimally.

### 4.3.2 Using RNNs

In this section we present the parameter estimation approaches based on the RNNs. It must be emphasized that the coefficients of this network are not necessarily the parameters of the dynamic system under consideration. In fact, the states of the network are the parameters of the system of which the parameters are to be estimated. The formulation of the parameter estimation problem is based on the classical equation error method.

#### 4.3.2.1 Parameter Estimation with Hopfield Neural Networks

Consider the dynamic system

$$\dot{x} = Ax + Bu; x(0) = x_0 \tag{4.49}$$

For parameter estimation using Hopfield neural networks (HNN), $\beta = \{A, B\}$ represents the parameter vector to be estimated and $n$ is the number of parameters to be estimated. Based on the theory of HNNs, a suitable functional can be associated with it, which iterates to a stable parameter estimation solution. In this network, the neurons change their states $x_i$ according to eqn. (4.49). We can consider that the dynamics are affected by the nonlinear function *f*, i.e., $\beta_i = f(x_i)$. Let the cost function be given as

$$E(\beta) = 1/2 \sum_{k=1}^{N} e^T(k)\, e(k) = 1/2 \sum_{k=1}^{N} (\dot{x} - Ax - Bu)^T (\dot{x} - Ax - Bu) \tag{4.50}$$

Here, *e(k)* is the equation error:

$$e = \dot{x} - Ax - Bu \tag{4.51}$$

Using optimization theory we have from eqn. (4.50):

$$\frac{d\beta}{dt} = -\frac{\partial E(\beta)}{\partial \beta} = -\frac{1}{2} \frac{\partial \left\{ \sum_{k=1}^{N} e^T(k)\, e(k) \right\}}{\partial \beta} \tag{4.52}$$

Since, $\beta$ as a parameter vector contains the elements of A and B, we can obtain expressions $\partial E / \partial A$ and $\partial E / \partial B$ for A and B vectors, with $\sum(.) = \sum_{k=1}^{N}(.)$

$$\frac{\partial E}{\partial A} = \sum (\dot{x} - Ax - Bu)\,(-x^T) = A\sum x\,x^T + B\sum u\,x^T - \sum \dot{x}\,x^T$$
$$\frac{\partial E}{\partial B} = \sum (\dot{x} - Ax - Bu)\,(-u) = A\sum x\,u + B\sum u^2 - \sum \dot{x}\,u \tag{4.53}$$

Expanding we get, for $A(2,2)$ and $B(2,1)$, as a specific case:

$$\begin{bmatrix} \frac{\partial E}{\partial a_{11}} & \frac{\partial E}{\partial a_{12}} \\ \frac{\partial E}{\partial a_{21}} & \frac{\partial E}{\partial a_{22}} \end{bmatrix} = \begin{bmatrix} a_{11} & a_{12} \\ a_{21} & a_{22} \end{bmatrix} \begin{bmatrix} \Sigma x_1^2 & \Sigma x_1 x_2 \\ \Sigma x_2 x_1 & \Sigma x_2^2 \end{bmatrix} + \begin{bmatrix} b_1 \\ b_2 \end{bmatrix} \begin{bmatrix} \Sigma u x_1 & \Sigma u x_2 \end{bmatrix} - \begin{bmatrix} \Sigma \dot{x}_1 x_1 & \Sigma \dot{x}_1 x_2 \\ \Sigma \dot{x}_2 x_1 & \Sigma \dot{x}_2 x_2 \end{bmatrix} \tag{4.54}$$

Simplifying eqn. (4.54), we get:

$$\frac{\partial E}{\partial a_{11}} = a_{11} \Sigma x_1^2 + a_{12} \Sigma x_2 x_1 + b_1 \Sigma x_1 u - \Sigma \dot{x}_1 x_1$$
$$\frac{\partial E}{\partial a_{12}} = a_{11} \Sigma x_1 x_2 + a_{12} \Sigma x_2^2 + b_1 \Sigma u x_2 - \Sigma \dot{x}_1 x_2$$
$$\frac{\partial E}{\partial a_{21}} = a_{21} \Sigma x_1^2 + a_{22} \Sigma x_2 x_1 + b_2 \Sigma u x_1 - \Sigma \dot{x}_2 x_1 \tag{4.55}$$
$$\frac{\partial E}{\partial a_{22}} = a_{21} \Sigma x_1 x_2 + a_{22} \Sigma x_2^2 + b_2 \Sigma u x_2 - \Sigma \dot{x}_2 x_2$$

Additionally, we have:

$$\frac{\partial E}{\partial b_1} = a_{11} \sum x_1 u + a_{12} \sum x_2 u + b_1 \sum u^2 - \sum \dot{x}_1 u$$
$$\frac{\partial E}{\partial b_2} = a_{21} \sum x_1 u + a_{22} \sum x_2 u + b_2 \sum u^2 - \sum \dot{x}_2 u \tag{4.56}$$

Next, assuming that the impedance R is very high, we describe the dynamics of RNN-S as:

$$\dot{x}_i = \sum_{j=1}^{n} w_{ij} \beta_j + b_i \tag{4.57}$$

We also have $E = -1/2 \sum_i \sum_j W_{ij} \beta_i \beta_j - \sum_i b_i \beta_i$ as the energy landscape of the recurrent neural network. Then, we get

$$\frac{\partial E}{\partial \beta_i} = -\sum_{j=1}^{n} w_{ij}\beta_j - b_i \tag{4.58}$$

$$\text{or } \frac{\partial E}{\partial \beta_i} = -\left[\sum_{j=1}^{n} w_{ij}\beta_j + b_i\right] = -\dot{x}_i \text{ Or } \dot{x}_i = -\frac{\partial E}{\partial \beta_i} \tag{4.59}$$

$$\text{Since } \beta_i = f(x_i), \dot{x}_i = (f^{-1})'\,\dot{\beta}_i;\text{ Thus we have } (f^{-1})'\,\dot{\beta}_i = -\frac{\partial E}{\partial \beta_i} \tag{4.60}$$

Here $'$ denotes derivative w.r.t $\beta$. Hence,

$$\dot{\beta}_i = -\frac{1}{(f^{-1})'(\beta_i)}\frac{\partial E}{\partial \beta_i} = \frac{1}{(f^{-1})'(\beta_i)}\left[\sum_{j=1}^{n} w_{ij}\beta_j + b_i\right] \tag{4.61}$$

Next, we get expressions for the weight matrix $W$ and the bias vector $b$ as:

$$W = -\begin{bmatrix} \sum x_1^2 & \sum x_2x_1 & 0 & 0 & \sum ux_1 & 0 \\ \sum x_1x_2 & \sum x_2^2 & 0 & 0 & \sum ux_2 & 0 \\ 0 & 0 & \sum x_1^2 & \sum x_2x_1 & 0 & \sum ux_1 \\ 0 & 0 & \sum x_1x_2 & \sum x_2^2 & 0 & \sum ux_2 \\ \sum x_1u & \sum x_2u & 0 & 0 & \sum u^2 & 0 \\ 0 & 0 & \sum x_1u & \sum x_2u & 0 & \sum u^2 \end{bmatrix} \tag{4.62}$$

$$b = -\begin{bmatrix} \sum \dot{x}_1x_1 \\ \sum \dot{x}_1x_2 \\ \sum \dot{x}_2x_1 \\ \sum \dot{x}_2x_2 \\ \sum \dot{x}_1u \\ \sum \dot{x}_2u \end{bmatrix} \tag{4.63}$$

Thus, the algorithm for parameter estimation of a dynamic system using RNN-EE formulation can be given as: i) compute $W$-matrix, and b-vector, since the measurements of $x$, $\dot{x}$ and $u$ are available (equation error formulation) for a certain time interval T, using eqns. (4.62), and (4.63); ii) choose the initial values of $\beta_i$ randomly; and iii) then, solve the following differential equation: As $\beta_i = f(x_i)$ and since, the sigmoid non-linearity is a known function $f$, by differentiating and simplifying, we get

$$\frac{d\beta_i}{dt} = \frac{\lambda\,(\rho^2 - \beta_i^2)}{2\rho}\left[\sum_{j=1}^{n} w_{ij}\beta_j + b_i\right]; f(x_i) = \rho\left(\frac{1-e^{-\lambda x_i}}{1+e^{-\lambda x_i}}\right);$$

$$\text{here}, f(x_i) = \rho\left(\frac{1-e^{-\lambda x_i}}{1+e^{-\lambda x_i}}\right) \qquad 4.64)$$

Integration of eqn. (4.64) yields the solution to parameter estimation problem posed in the structure of HNN (RNN-S). For good convergence of the estimates to the true parameters, proper tuning of $\lambda$ and $\rho$ is very essential. Often $\lambda$ is chosen to be small, i.e., less than 1.0. The $\rho$ is chosen such that when $x_i$ (of RNN) approaches $\pm\infty$, the function $f$ approaches $\pm\rho$. The eqn. (4.64) can be discretized to obtain the estimates by recursion. Also, it is possible to use the inverse of the weighting matrix $W$ on the right-hand side of eqn. (4.64) to enhance the rate of convergence of the algorithm. The matrix $W$ can be regarded as the information matrix for the parameter estimator defined by eqn. (4.64). The foregoing scheme is termed as non-recursive, since the required computation of elements of W and $b$ is performed by considering all the data. The discrete form of eqn. (4.64) is given as:

$$\beta_i(k+1) = \beta_i(k) + \frac{\lambda(\rho^2 - \beta_i^2(k))}{2\rho}\left[\sum_{j=1}^{n} w_{ij}\beta_j(k) + b_j\right] \tag{4.65}$$

The $\Delta t$ is absorbed in the constants of the 2nd term of eqn. (4.65).

**Example 4.4**

Consider the second-order system described by

$$\dot{x} = \begin{bmatrix} -0.7531 & 1 \\ -1.3760 & -1.1183 \end{bmatrix} x + \begin{bmatrix} 0 \\ -2.49 \end{bmatrix} u \tag{4.66}$$

i) obtain the response of the system to a doublet input, ii) use $x, \dot{x},$ and u in the RNN-S algorithm to estimate all 6 parameters, and iii) comment on the accuracy of the results.

**Solution 4.4**

i) The 100 data samples are generated using a doublet input and initial state of the system $x(0) = [0.1 \;\; 0.01]$. The input signal and system response are shown in Figure 4.7.
ii) The equation error formulation is used in RNN-S (HNN) for parameter estimation. The estimation was carried out using noise-free data and data with additive noise. The tuning parameters $\lambda$ and $\rho$ were kept at 0.1 and 100, respectively. It was noted that RNN-S took around 350 iterations before the convergence of estimated parameters to true values. Figure 4.8 shows the estimated parameters for noisy data with SNR=10, and noise-free data. It can be concluded from the figure that convergence pattern for both cases are similar. Table 4.7 shows estimated parameters and PEENs for different SNRs. The system simulation and parameter estimation are accomplished by using file '**parestrnn1.m**' placed in folder **'Ch4RNNex4'**.

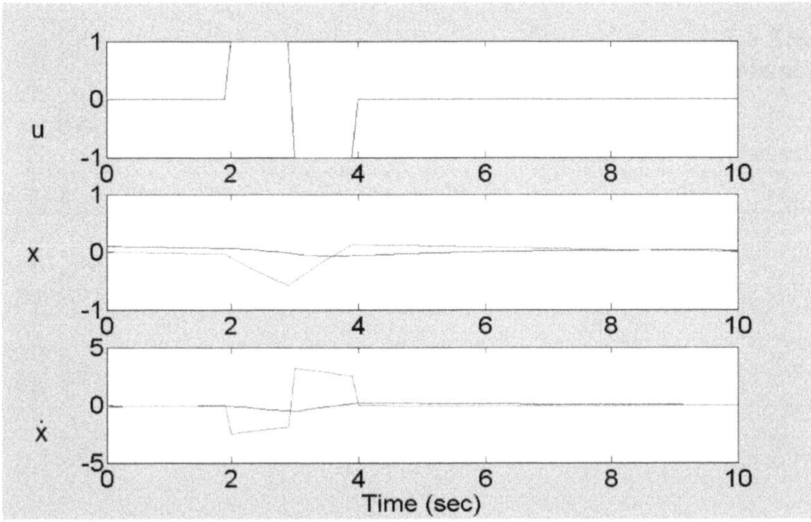


**FIGURE 4.7** Doublet input and system states (Example 4.4)

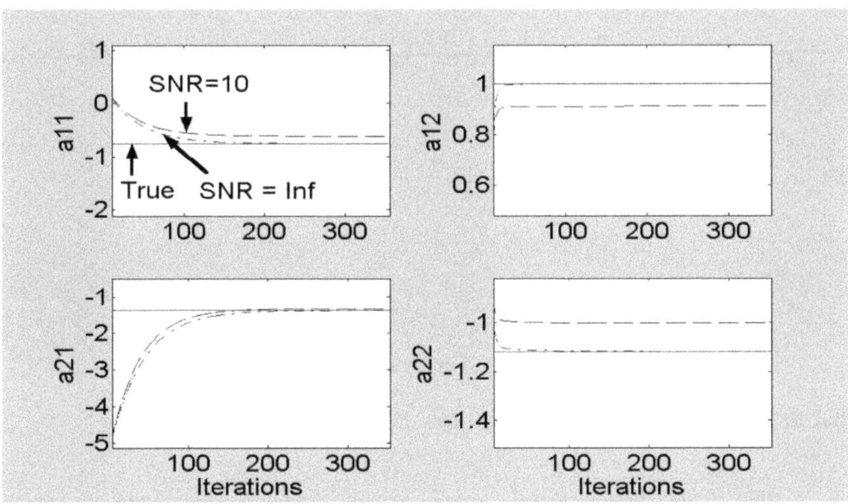


**FIGURE 4.8** Estimated parameters for different SNRs (Example 4.4)

**Example 4.5** Consider the second-order unstable system described by

$$\dot{x} = \begin{bmatrix} -1.1 & 0.8 \\ 0.12 & -0.05 \end{bmatrix} x + \begin{bmatrix} -0.12 \\ 0.8 \end{bmatrix} u \tag{4.67}$$

i) simulate the system of (4.67) with doublet input using a sampling interval $\Delta t = 0.1$ sec (number of data points=100), and ii) use $x, \dot{x},$ and $u$ in the

**TABLE 4.7**
**Parameter estimation with RNN-S (Example 4.4)**

| Parameters | True values | Estimated values using RNN-S (HNN) for different noise levels | | |
|---|---|---|---|---|
| | ---- | SNR = ∞ (no noise) | SNR = 100 | SNR = 10 |
| $a_{11}$ | -0.7531 | -0.7531 | -0.758 | -0.707 |
| $a_{12}$ | 1.0 | 1.0000 | 1.004 | 0.947 |
| $a_{21}$ | -1.376 | -1.3760 | -1.369 | -1.276 |
| $a_{22}$ | -1.1183 | -1.1183 | -1.108 | -1.017 |
| $b_{11}$ | 0.0 | -0.0000 | -0.002 | -0.011 |
| $b_{21}$ | -2.49 | -2.4900 | -2.485 | -2.477 |
| PEEN | ---- | 0.0 | 0.451 | 4.840 |

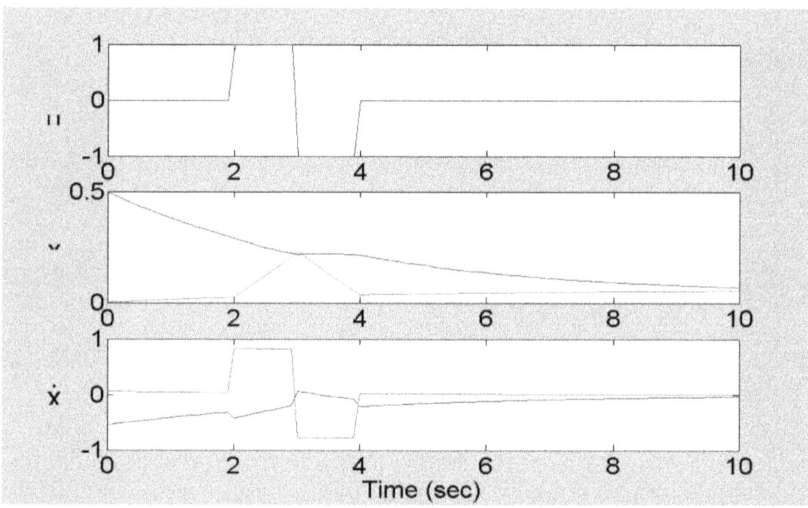


**FIGURE 4.9** Doublet input and system states (Example 4.5)

RNN-S algorithm to estimate the parameters and comment on the accuracy of the results.

**Solution 4.5**

The dynamic system described by eqn. (4.67) is unstable (its eigenvalues are $\lambda_1 = -1.18$ and $\lambda_1 = 0.03$, this one being +ve,), because one of the roots lies in the right half of the $s$-plane. The system response is obtained using doublet input with initial state of the system $x(0) = [0.5 \;\; 0.002]$. The input signal and system response are shown in Figure 4.9.

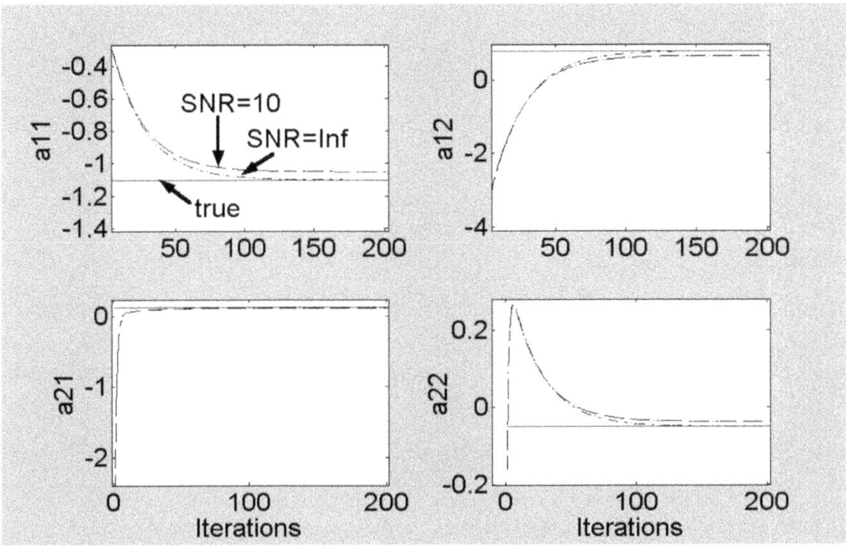


**FIGURE 4.10** Estimated parameters for different SNRs (Example 4.5)

i) The equation error formulation is used in RNN-S (HNN) for parameter estimation. The estimation was carried out using noise-free data and data with additive noise. The tuning parameters $\lambda$ and $\rho$ were kept at 0.1 and 100, respectively. It was noted that RNN-S took around 350 iterations before the convergence of estimated parameters to true values. Figure 4.10 shows the estimated parameters for noisy data with SNR=10, and noise-free data. It can be concluded from the figure that the convergence patterns for both cases are similar. Figure 4.11 shows the true and estimated system state ( $x_1$ and $\dot{x}_1$ ) for SNR=10. Table 4.8 shows the estimated parameters and PEENs for different SNRs. The system simulation and parameter estimation are accomplished by using file '**parestrnn2.m**' placed in folder **'Ch4RNNex5'**.

#### 4.3.2.2 Parameter Estimation of Unstable Dynamic System

Next, consider the following system

$$
\begin{aligned}
\dot{x}_1 &= \beta_1 x_4 \\
\dot{x}_2 &= \beta_2 x_5 \\
\dot{x}_3 &= \beta_3 x_6 \\
\dot{x}_4 &= \beta_4 u \\
\dot{x}_5 &= \beta_5 u \\
\dot{x}_6 &= \beta_6 u
\end{aligned}
\tag{4.68}
$$

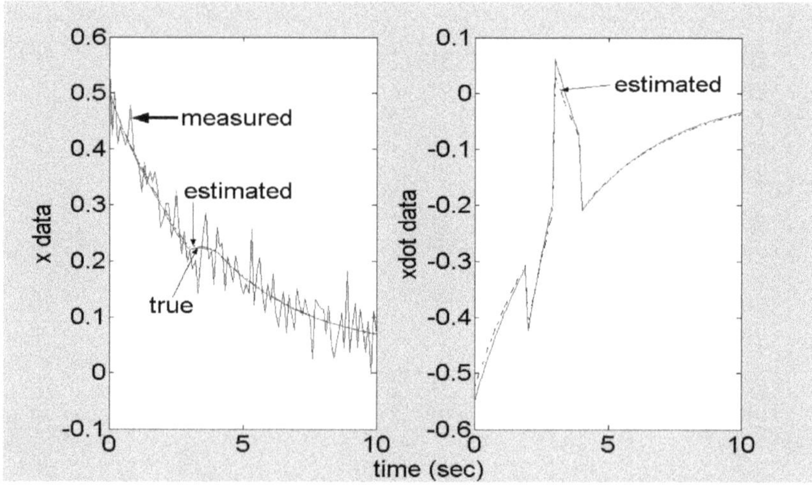


**FIGURE 4.11** True, measured, and estimated system states for SNR=10 (Example 4.5)

**TABLE 4.8**
**Parameter estimation with RNN-S (Example 4.5)**

| Parameters | True values | Estimated values using RNN-S (HNN) method for different noise levels | | |
|---|---|---|---|---|
| | ---- | SNR = ∞ (no noise) | SNR = 100 | SNR = 10 |
| $a_{11}$ | -1.1 | -1.1 | -1.10 | -1.070 |
| $a_{12}$ | 0.8 | 0.8 | 0.81 | 0.745 |
| $a_{21}$ | 0.12 | 0.12 | 0.12 | 0.117 |
| $a_{22}$ | -0.05 | -0.05 | -0.05 | -0.046 |
| $b_{11}$ | -0.12 | -0.12 | -0.12 | -0.121 |
| $b_{21}$ | 0.8 | 0.8 | 0.80 | 0.800 |
| PEEN | ---- | 0.0 | 0.710 | 4.067 |

Here, $\beta_1,\beta_2,\beta_3,\beta_4,\beta_5$, and $\beta_6$ are the parameters to be estimated using HNN and u is the input to the system. Cost function is defined as

$$J(\beta)=\frac{1}{2}\sum_{k=1}^{N}e^T(k)e(k)=\frac{1}{2}\sum_{k=1}^{N}(\dot{x}-f(x))^T(\dot{x}-f(x)) \tag{4.69}$$

Here, $\dot{x}=[\dot{x}_1 \quad \dot{x}_2 \quad \dot{x}_3 \quad \dot{x}_4 \quad \dot{x}_5 \quad \dot{x}_6]$, $f(x)=[\beta_1 x_4 \quad \beta_2 x_5 \quad \beta_3 x_6 \quad \beta_4 u \quad \beta_5 u \quad \beta_6 u]$

For the optimal estimation, we have:

$$\frac{d\beta}{dt} = -\frac{\partial J(\beta)}{\partial \beta} = -\frac{1}{2}\frac{\partial\left\{\left[\sum_{k=1}^{N} e^{T}(k)\ e(k)\right]\right\}}{\partial \beta} = \sum_{k=1}^{N}\frac{\partial f(x)}{\partial \beta}.e(k) \tag{4.70}$$

For simplification of expressions, let us assume $\sum(.) = \sum_{k=1}^{N}(.)$. Now putting the expression of *e (k)* in eqn. (4.70), we get,

$$\frac{\partial J(\beta)}{\partial \beta} = -\sum \begin{bmatrix} \dot{x}_1 - \beta_1 x_4 \\ \dot{x}_2 - \beta_2 x_5 \\ \dot{x}_3 - \beta_3 x_6 \\ \dot{x}_4 - \beta_4 u \\ \dot{x}_5 - \beta_5 u \\ \dot{x}_6 - \beta_6 u \end{bmatrix}^T \begin{bmatrix} x_4 & 0 & 0 & 0 & 0 & 0 \\ 0 & x_5 & 0 & 0 & 0 & 0 \\ 0 & 0 & x_6 & 0 & 0 & 0 \\ 0 & 0 & 0 & u & 0 & 0 \\ 0 & 0 & 0 & 0 & u & 0 \\ 0 & 0 & 0 & 0 & 0 & u \end{bmatrix} \tag{4.71}$$

The dynamics of RNN-S is described as:

$$\frac{\partial J(\beta)}{\partial \beta_i} = -\left[\sum_{j=1}^{n} w_{ij}\beta_j + b_i\right] \tag{4.72}$$

Here, $n$ is the total number of parameters to be estimated. From eqns. (4.71) and (4.72) we have; let us say $i$=1, and then expanding eqn. (4.72) we get,

$$\frac{\partial J(\beta)}{\partial \beta_1} = -w_{11}\beta_1 - w_{12}\beta_2 - w_{13}\beta_3 - w_{14}\beta_4 - w_{15}\beta_5 - w_{16}\beta_6 - b_1 \tag{4.73}$$

Similarly, by expanding eqn. (4.71) for $i$=1 we have:

$$\frac{\partial J(\beta)}{\partial \beta_1} = -\sum \dot{x}_1 x_4 + \beta_1 \sum x_4^2 \tag{4.74}$$

By comparing expressions from eqn. (4.73) and eqn.(4.74), we get expressions for 1st row elements of weight matrix W and bias vector $b$ as:

$$w_{11} = -\sum x_4^2, w_{12} = w_{13} = w_{14} = w_{15} = w_{16} = 0, \text{and } b_1 = \sum \dot{x}_1 x_4 \tag{4.75}$$

One can get full expressions of $W$ and $b$ for $i$=2,…,$n$. After complete evaluation, we get $W$ and $b$ as:

$$W = \begin{bmatrix} -\sum x_4^2 & 0 & 0 & 0 & 0 & 0 \\ 0 & -\sum x_5^2 & 0 & 0 & 0 & 0 \\ 0 & 0 & -\sum x_6^2 & 0 & 0 & 0 \\ 0 & 0 & 0 & -\sum u^2 & 0 & 0 \\ 0 & 0 & 0 & 0 & -\sum u^2 & 0 \\ 0 & 0 & 0 & 0 & 0 & -\sum u^2 \end{bmatrix}, \quad b = \begin{bmatrix} \sum \dot{x}_1 x_4 \\ \sum \dot{x}_2 x_5 \\ \sum \dot{x}_3 x_6 \\ \sum \dot{x}_4 u \\ \sum \dot{x}_5 u \\ \sum \dot{x}_6 u \end{bmatrix} \tag{4.76}$$

The values of *W* and *b* can be used for parameter estimation.

**Example 4.6** Consider the system with all eigenvalues at the origin:

$$\begin{aligned} \dot{x}_1 &= b_1 x_4 & \dot{x}_2 &= b_2 x_5 \\ \dot{x}_3 &= b_3 x_6 & \dot{x}_4 &= 0 \\ \dot{x}_5 &= 0 & \dot{x}_6 &= b_4 \end{aligned} \tag{4.77}$$

Here, true parameters are $b_1 = 1$, $b_2 = 1$, $b_3 = 1$, and $b_4 = -9.8$.

i) Simulate the system of eqn. (4.77) with a unit step input signal and a sampling interval $\Delta t = 0.1$ sec (number of data points=10).
ii) Use $x, \dot{x}$, and $u$ in the RNN-S algorithm to estimate the parameters $b_1, b_2, b_3, b_4$.

**Solution 4.6**
The simulation of the dynamic system is carried out with the initial conditions as $x_1(0) = 10\,\text{m}$, $x_2(0) = 3\,\text{m}$, $x_3(0) = 0.1\,\text{m}$, $x_4(0) = 0.5\,\text{m/s}$, $x_5(0) = 0.1\,\text{m/s}$, and $x_6(0) = 0.8\,\text{m/s}$. The simulated data are generated for 1 sec. with 0.1 sec sampling interval.

The parameter estimation was carried out using noise-free data and data with additive noise. The tuning parameters $\lambda$ and $\rho$ were kept at 0.1 and 10, respectively. Figure 4.12 shows the true and estimated system state ($x_1$ and $x_3$) for SNR = 10. Table 4.9 shows the final values and PEEN of estimated parameters for different SNR levels. The system simulation and parameter estimation are accomplished by using file ‘**parestrnn3.m**’ placed in folder ‘**Ch4RNNex6**’. Reasonably good estimation has been accomplished.

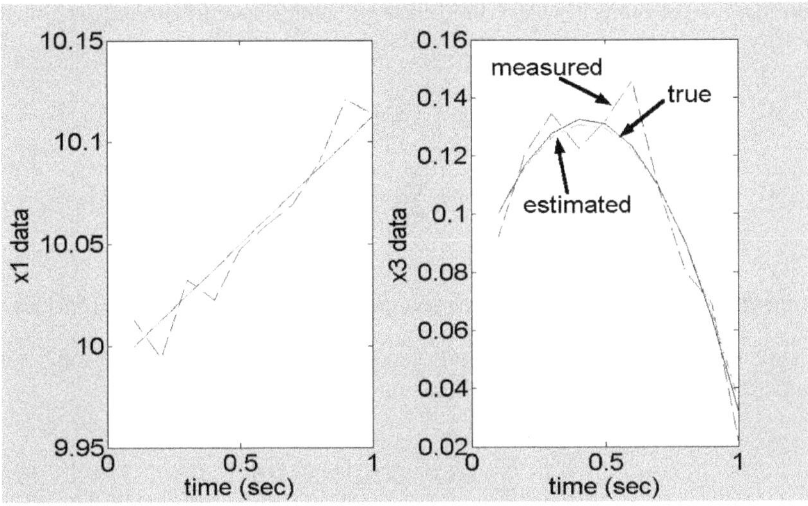


**FIGURE 4.12** True, measured, and estimated system states (Example 4.6)

**TABLE 4.9**
**Parameter estimation with RNN-S (Example 4.6)**

| Parameters | True values | Estimated values using RNN-S (HNN) method for different noise levels | | |
|---|---|---|---|---|
| | ---- | SNR = ∞ (No noise) | SNR = 10 | SNR = 2 |
| b1 | 1 | 1.0000 | 1.0000 | 1.0000 |
| b2 | 1 | 1.0003 | 1.0003 | 1.0003 |
| b3 | 1 | 1.0000 | 0.9500 | 0.7272 |
| b4 | -9.8 | -9.799 | -9.799 | -9.799 |
| PEEN | ---- | 0.003 | 0.5 | 2.74 |

### 4.3.2.3 Relationship Between Various Parameter Estimation Schemes

From section 4.3.2 we have the following important relationships

$$(a)\quad \dot{\beta}_i = -\frac{\lambda}{(f^{-1})'(\beta_i)}\,\frac{\partial E}{\partial \beta_i} \tag{4.78}$$

$$(b)\quad E(\beta) = 1/2\sum_{k=1}^{N}(\dot{x} - Ax - Bu)^T\ (\dot{x} - Ax - Bu) \tag{4.79}$$

$$(c)\quad \frac{\partial E}{\partial \beta_i} = -\left[\sum_{j=1}^{n} W_{ij}\beta_j + b_i\right] \tag{4.80}$$

From these expressions, we have the following equivalence (assuming $B$=0 in eqn. (4.79)):

$$\frac{dx}{dt} = -\frac{\partial E}{\partial \beta_i} = \left[\sum_{j=1}^{n} w_{ij}\beta_j + b_i\right] = \sum_{k=1}^{N}[\dot{x}(k) - Ax(k)]\, x^T(k)$$
$$= \sum_{k=1}^{N}\left[-\{\beta\}\, x(k)x^T(k) \ + \ \dot{x}(k)x^T(k)\right] \tag{4.81}$$

We note that for the discussion of this section only, the $x$ of $\frac{dx}{dt}$ in eqn. (4.81) is not the same as $\dot{x}, x$, etc. One can use equivalence of 1st and 5th terms, with some initial parameters $\beta(0)$, and integrate the following equation:

$$\frac{dx}{dt} = \sum_{k=1}^{N}[\dot{x}(k) - \{\beta(t)\}x(k)]x^T(k) \tag{4.82}$$

Since, the complete information required for the evaluation of the right-hand side is available for solving this equation; then compute $\beta = f(x)$, since HNN decision-making is nonlinear. Then use the new vector, $\beta$ for the next update. This procedure avoids explicit computation of weight matrix and input vector. It can be further ascertained that the role played by the sigmoid nonlinearity is somewhat similar to that played by the damping parameter in some of the gradient-based parameter estimation methods. We obtain from optimization theory, that for parameter vector, the following holds true (for non-neural based methods):

$$\frac{d\beta}{dt} = \mu(t)\sum_{k=1}^{N}[\dot{x}(k) - Ax(k)]x^T(k) \tag{4.83}$$

or equivalently

$$\beta(i+1) = \beta(i) + \mu\sum_{k=1}^{N}[\dot{x}(k) - Ax(k)]x^T(k)$$

For RNN-S (HNN), the discretization approach leads to:

$$\beta(i+1) = \beta(i) + \sum_{k=1}^{N}[\dot{x}(k) - f(x)x(k)]x^T(k); \beta = f(x) \tag{4.84}$$

Similarly, for RNN-E, the parameter estimation rule is

$$\beta(i+1) = \beta(i) + \sum_{k=1}^{N} f[\dot{x}(k) - Ax(k)]x^T(k) \tag{4.85}$$

Here $f$ could be "tanh" nonlinearity. Next, from the theory of Kalman filter, the following state estimation rule follows:

$$\hat{x}_a(k+1) = \tilde{x}_a(k+1) + K(z(k+1) - H\tilde{x}_a(k+1)) \tag{4.86}$$

Here, we presume that the state is an augmented state vector with unknown parameters $\beta$. The gradients of error w.r.t. states are implicit in the formulation of $K$. The Kalman filter is generally defined in the form of output error, which is also often known as prediction error.

From the foregoing development, the following facts emerge:

1. In Hopfield neural network (HNN), the nonlinearity directly influences the parameter vector (the state of Hopfield neural network).
2. In case of RNN-E, the nonlinearity influences the residuals directly. It can also be viewed as affecting the parameter vector indirectly.
3. In conventional parameter estimator, the factor $\mu$ affects the change in the parameter vector $\Delta\beta$, since we have $\beta(i+1) = \beta(i) + \mu\Delta\beta$.
4. The Kalman filter gain operates on the residuals and optimally helps to determine the state estimate.

We infer that nonlinearity $f, \mu$ or Kalman gain can affect the convergence of the parameter estimation algorithm. In eqn. (4.86) the inherent decision-making process is linear. Thus, the distinction lays in the way in which nonlinear/linear element affects the convergence of the algorithm, measurement errors, states, and parameters, and hence overall accuracy of the estimates. In principle, the RNN schemes developed in this Chapter can be used for parameter estimation of stable or unstable/augmented dynamical systems. The schemes are as such straightforward and require simple programming code. However, the schemes require proper use of the sigmoid nonlinearities. When formulated using the equation error, the schemes need accurate measurements of states and their derivatives. It is also possible to incorporate measurement models and formulate them in the form of output error. This will automatically extend the application of recurrent neural network-based parameter scheme to general dynamic systems.

### 4.3.4 RNN-based Estimation with Energy Terms for Speed Up

Building upon the RNN approach described in previous section, a different recurrent neural network method is presented that leverages energy function considerations to significantly accelerate the parameter estimation process. While the previous section established how Hopfield neural networks can be applied to parameter estimation using equation error formulation, this section explores a gradient-based RNN approach with adaptive learning that enhances convergence speed [5].

Consider the state space representation of a dynamical system

$$\dot{x} = Ax + Bu \tag{4.87}$$

The aim is to solve for $A$ and $B$, the system parameter matrices knowing $\dot{x}$, $x$, and $u$. The parameters $a_{ij}$ (elements of A) and $b_i$ (elements of B) to be estimated are denoted by

$$v = \left[a_{11}, a_{12}, \ldots, a_{1n}, a_{21}, \ldots, a_{2n}, \ldots, a_{n1}, a_{n2}, \ldots, a_{nn}, b_1, b_2, \ldots, b_n\right]$$

The equation error vector $e(v)$ is defined as

$$e(v) = \dot{x} - Ax - Bu$$

The cost function $E(v)$ is defined as the sum of squares over all the data points

$$E(v) = \frac{1}{2}\sum_{k=1}^{m} e^T(k)e(k)$$

$$\frac{\partial E}{\partial A} = \sum_{k=1}^{m} e(k)\left(-x(k)\right)^T$$

$$\frac{\partial E}{\partial B} = \sum_{k=1}^{m} e(k)\left(-u(k)\right)^T$$

In standard gradient-based neural networks, the learning parameter matrix $\mu$ typically remains constant, which theoretically requires infinite time for the parameter vector $v$ to converge to the true solution $v^*$. For real-time applications such as adaptive control systems or online fault detection, obtaining results within a prescribed timeframe is essential. The parameter estimation should ideally reach completion within a specified time $t_r$, such that $v(t_r) = v^*$, with the cost function $E(v)$ reaching a minimum with a specified error $\delta$ at $t = t_r$.

The key insight for achieving speed-up comes from treating the cost function $E(v)$ as an energy landscape and controlling the rate at which this energy decreases over time. By formulating an adaptive learning rate that directly relates to the energy gradient, it becomes possible to control exactly how quickly the system converges to the minimum energy state (optimal parameter values). This approach is fundamentally different from standard fixed learning rate methods, as it dynamically adjusts the step size based on the current position in the energy landscape, allowing for precise control of the convergence time. The energy-based formulation transforms what would be an asymptotic, potentially slow convergence into a process with a guaranteed finite completion time.

To achieve this accelerated convergence, the $\mu$ matrix is made adaptive during the minimization process by considering the energy landscape of the system. Let $\mu_0(t)$ denote the adaptive parameter defined as:

$$\mu_0(t) = \mu_{0max} \tag{4.88}$$

when $E(v(t)) = 0$, otherwise

$$\mu_0(t) = \frac{\mu}{tr\left\{\left(\sum_{k=1}^{m} e(k)x^T(k)\right)\left(\sum_{k=1}^{m} x(k)e^T(k)\right) + \left(\sum_{k=1}^{m} e(k)u(k)\right)\left(\sum_{k=1}^{m} u(k)e^T(k)\right)\right\}} \tag{4.89}$$

where $\mu = \dfrac{1}{\tau} > 0$ ( $\tau$ is the time constant of integration).

The time derivative of the cost function $E(v(t))$ is:

$$\frac{dE}{dt} = \sum_{s=1}^{n_p} \frac{dE}{dv_s}\frac{dv_s}{dt} = \left[\nabla E(v)\right]^T \frac{dv}{dt} \tag{4.90}$$

If $\dfrac{dv}{dt} \neq 0$, the following is obtained:

$$\frac{dE}{dt} = \mu_0(t)\sum_{s=1}^{n_p}\left(\frac{dE}{dv_s}\right)^2$$

$$= \mu_0(t)\left\{\sum_{j=1}^{n}\sum_{i=1}^{n}\left(\frac{dE}{da_{ij}}\right)^2 + \sum_{i=1}^{n}\left(\frac{dE}{db_i}\right)^2\right\}$$

$$= \mu_0(t)\left\{tr\left(\frac{\partial E}{\partial A}\right)\left(\frac{\partial E}{\partial A}\right)^T + tr\left(\frac{\partial E}{\partial B}\right)\left(\frac{\partial E}{\partial B}\right)^T\right\} \tag{4.91}$$

Expanding using the cost functions,

$$\frac{dE}{dt} = -\mu_0(t)tr\left\{\left(\sum_{k=1}^{m} e(k)x^T(k)\right)\left(\sum_{k=1}^{m} x(k)e^T(k)\right) + \left(\sum_{k=1}^{m} e(k)u(k)\right)\left(\sum_{k=1}^{m} u(k)e^T(k)\right)\right\}$$

$$= \mu < 0 \tag{4.92}$$

If $\dfrac{dv}{dt} = 0$, eqn. 34 becomes $\dfrac{dE}{dt} = 0$.

The negative value in eqn. (4.92) confirms that the energy function is always decreasing, ensuring stable convergence to the minimum [5]. The critical insight here is that by controlling parameter $\mu$, it becomes possible to precisely control the rate of energy decrease. Unlike traditional methods where convergence time depends on the initial conditions and the shape of the energy landscape, this approach allows for explicitly defining how quickly the system must reach the minimum. This is where the speed-up comes from, the energy-based approach effectively transforms an infinite-time convergence problem into a predictable finite-time solution with a guaranteed completion deadline.

The relationship in eqn. (4.92) shows that the cost function decreases linearly with time during the minimization process. This can be written in integral form as:

$$E\left(v(t)\right) = E\left(v(0)\right) - \mu t \tag{4.93}$$

For practical implementation, appropriate values for the parameters must be determined. It is assumed that the initial value $E\left(v(0)\right)$ and the final value $E(v^*)$ are known. Let us assume that at the start v(0)=0, i.e., A=0, B=0, then using the cost function equation, the value of E(v(0)) can be computed as:

$$E\left(v(0)\right) = \frac{1}{2}\sum_{k=1}^{m}\left(\dot{x}^T(k)\dot{x}(k)\right)$$

$$= \frac{1}{2}\sum_{k=1}^{m}\sum_{i=1}^{n}\left(\dot{x}_i(k)\right)^2 \tag{4.94}$$

Assuming that E(v*)=0, from eqn. (4.93):

$$tr = \frac{E\left(v(0)\right) - \delta}{\mu} \tag{4.95}$$

This equation represents the core of the energy-based speed-up mechanism. Unlike conventional algorithms where convergence time is an emergent property of the algorithm's behaviour, eqn. (4.96) allows working backward from a desired completion time to determine the necessary value of $\mu$. By choosing $\mu = \frac{E\left(v(0)\right) - \delta}{tr}$ system designers can guarantee that the parameter estimation will complete within exactly tr time units. This inversion of the traditional approach, specifying time first and deriving algorithm parameters from it, is what enables the dramatic speed-up in parameter estimation for time-critical applications.

The implementation of $\mu_0(t)$ in the adaptive block can be elegantly derived from eqns. (4.91) and (4.92):

$$\mu_0(t) = \frac{\mu}{\sum_{s=1}^{n_p} \left( \frac{\partial E}{\partial v_s} \right)^2} \tag{4.96}$$

This formulation automatically adjusts the learning rate based on the gradient magnitude, providing larger steps when gradients are small and smaller steps when gradients are large. This natural adaptation prevents overshooting near the minimum while maintaining rapid convergence in the early stages of estimation.

While this implementation offers significant convergence speed-up benefits, it requires analogue multipliers and dividers which increase hardware complexity. However, for many real-time applications where convergence time guarantees are critical, this trade-off is well justified.

The energy-based approach offers several advantages over traditional fixed-learning rate methods, including:

1. Predictable convergence time: The time required to reach the minimum can be specified in advance
2. Robustness to initialization conditions: The approach accounts for starting conditions through the energy formulation
3. Adaptive step sizing: The method naturally takes larger steps when far from thc minimum and smaller steps when close
4. Guaranteed monotonic improvement: The energy function is always decreasing, ensuring stable convergence

These characteristics make it particularly suitable for embedded systems and control applications where parameter estimation must be both fast and reliable. The energy-based normalization technique can be implemented using specialized hardware that includes multipliers and dividers, although this increases circuit complexity compared to fixed-learning rate approaches.

For practical implementation, appropriate circuit elements must be incorporated to compute the gradient norm and perform the division operation. While more resource-intensive than simpler approaches, the benefit of guaranteed convergence time justifies this additional complexity in time-critical applications such as fault detection, adaptive control, or online system identification.

### 4.3.5 RNN-based Output Error Parameter Estimation Method

The output error method represents a more comprehensive approach to parameter estimation compared to the equation error method, as it addresses situations where measurement data contain only measurement noise in the output equations without process noise in the state equations. This method is particularly suited for systems where a measurement model or output equation relates the internal states to the measured variables [5].

Consider a general dynamic system representation given by the state-space equations:

$$\dot{x} = Ax + Bu$$

$$y = Cx + Du$$

$$z = y(k) + v(k),\ k = 1,2,\ldots,m \tag{4.97}$$

Here, x is the n×1 state vector, u is the control input, y is the output vector from the model, z(k) represents the measured output at discrete time instant k, and v(k) is the measurement noise. The matrices A, B, C, and D contain the unknown parameters to be estimated. The parameter vector θ consists of all elements of these matrices.

The output error cost function is defined as:

$$E(\theta) = \frac{1}{2}\sum_{k=1}^{m}\left(z(k) - y(k)\right)^{T}\left(z(k) - y(k)\right) \tag{4.98}$$

To update the parameter estimates, the gradient descent method is employed:

$$\frac{d\theta}{dt} = -\mu(t)\nabla \mathrm{E}(\theta) \tag{4.99}$$

where $\mu(t)$ is the learning rate matrix and $\nabla \mathrm{E}(\theta)$ is the gradient of the cost function. The convergence of θ is accelerated using an adaptive learning rate matrix defined as:

$$\mu(t) = \frac{\mu_0}{tr\left\{\left(\sum_{k=1}^{m} e(k)x^{T}(k)\right)\left(\sum_{k=1}^{m} x(k)e^{T}(k)\right) + \left(\sum_{k=1}^{m} e(k)u(k)\right)\left(\sum_{k=1}^{m} u(k)e^{T}(k)\right)\right\}} \tag{4.100}$$

where e(k)=z(k)−y(k) is the output error at time k, $\mu_0$ is a positive constant, and $tr\{\cdot\}$ denotes the trace of a matrix. Using this adaptive learning rate, the parameter update equation becomes:

$$\frac{d\theta}{dt} = \mu(t)\sum_{k=1}^{m}\left(z(k) - y(k)\right)^{T}\frac{\partial y(k)}{\partial \theta} \tag{4.101}$$

The recurrent neural network architecture for the output error method processes the data iteratively to compute the model output, calculate the output error, determine the gradient of the error with respect to each parameter, and update the parameters according to the learning rule [5]. The parameters are updated after every m samples of the data using discrete integration. In this way, the well-known parameter estimation rule based on the maximum likelihood output error method can be naturally adapted to parallelization using RNN architecture. The advantage of this approach lies in its ability to handle cases where state derivatives are not directly measured or available, as the method relies only on input-output data. Furthermore, since the computational tasks can be distributed across multiple processing units in parallel computing systems, significant speed-up in obtaining parameter estimates can be achieved. This parallelization capability is particularly valuable for real-time applications such as online system identification, adaptive control, and model-based fault detection.

The output error method does not assume knowledge of the state vector or its derivatives, making it more practical for real-world applications where only input-output measurements are typically available. The method naturally handles both linear and nonlinear systems within the RNN framework. But then, the computation of sensitivity coefficients $\frac{\partial y(k)}{\partial \theta}$ is required, which for nonlinear systems can be obtained numerically using finite difference approximations when analytical derivation is complex.

### 4.3.6 RNN-based State Estimation Approach

While parameter estimation focuses on determining the unknown coefficients of a dynamic system model, state estimation addresses the problem of inferring the values of the internal states of the system from measured input and output data. Since internal states are often not directly measurable or are contaminated with noise, accurate estimation of states is essential for many applications including process control, condition monitoring, fault diagnosis, and adaptive control. RNN-based state estimation provides an alternative approach that combines the learning capability of neural networks with systematic state reconstruction from noisy measurements and input-output data [5].

The state estimation problem is formulated for discrete-time nonlinear systems described by the state-space representation:

$$x(k+1) = f\left(x(k), u(k)\right) \tag{4.102}$$

$$z(k) = h\left(x(k), u(k)\right) + v(k)$$

where x(k) is the n×1 state vector at discrete time instant k, u(k) is the control input, z(k) is the p×1 measured output, and v(k) represents the measurement noise assumed to be a zero-mean white Gaussian process. The functions f and h represent the state transition dynamics and the measurement model, respectively, which may be nonlinear and partially unknown. The objective is to obtain an estimate $\hat{x}(k)$ of the state given input and output observations.

Traditional state filtering approaches such as the extended Kalman filter assume explicit knowledge of the system model and noise statistics. But, in many practical situations such models are not available or are significantly uncertain. The RNN-based state estimation approach addresses this limitation by using neural networks to learn the system dynamics and the state update mappings directly from data, without requiring explicit knowledge of the functional forms in equation (4.102).

The RNN-based state estimator employs a two-step prediction-update structure similar to the classical Kalman filter framework. In the prediction step, the estimated state at time k is used along with the current control input to compute a predicted state at time k+1 and the corresponding predicted measurement. In the update step, the actual measured output is compared with the predicted output to generate an innovation signal, which is then used to correct the predicted state estimate and produce the final filtered state estimate. The key distinction from the Kalman filter is that the prediction and update mappings are not derived analytically but are instead learned through neural network training.

The RNN architecture for state estimation consists of multiple neural networks trained to model different functional relationships. Specifically, separate networks are developed for the state predictor $\hat{f}$, the output predictor $\hat{h}$, and the state update function. The state predictor network approximates the state transition dynamics:

$$\hat{x}(k+1 \mid k) = \hat{f}\left(x(k), u(k)\right) \tag{4.103}$$

where $\hat{x}(k+1|k)$ is the predicted state before measurement incorporation. The output predictor network approximates the measurement model:

$$\hat{z}(k) = \hat{h}\left(\hat{x}(k), u(k)\right) \tag{4.104}$$

The innovation or residual term is computed as:

$$e(k) = z(k) - \hat{z}(k) \tag{4.105}$$

The state update network then learns the nonlinear mapping from the residuals and measurement information to produce the corrected state estimate:

$$\hat{x}(k \mid k) = \hat{g}\left(e(k), z(k), \hat{x}(k \mid k-1)\right) \tag{4.106}$$

where $\hat{x}(k \mid k)$ is the learned state update function and $\hat{x}(k|k)$ is the updated state estimate after incorporating the current measurement.

The training of RNN-based state filters involves both off-line and on-line phases. During off-line training, the networks are trained using historical data where state values are either available from system simulations, models, or infrequent measurements. This initial training provides a good starting point for filter operation. Two approaches are employed in off-line training: teacher forcing, where actual measured outputs and states from the training set are used as network inputs and global feedback, where the network's own predictions are fed back as inputs to create dynamic training conditions. The global feedback approach more closely represents actual filter operation where true states are unknown.

In the on-line training phase, the filter operates in a closed-loop manner using only measured input and output data. As new measurements become available, the network weights can be updated to refine the state estimates and adapt to changes in system characteristics or operating conditions. This adaptive capability makes the approach valuable for systems experiencing parameter drift, time-varying dynamics, or conditions that were not well represented in the off-line training data. The on-line learning feature effectively decouples the final filter accuracy from the accuracy of the process model or initial training data.

A significant advantage of the RNN-based state estimation approach is its ability to handle nonlinear systems directly without requiring linearization approximations or computation of Jacobian matrices. Classical approaches like the extended Kalman filter necessitate linearization of the nonlinear functions through Taylor series expansions about operating points, which introduces approximation errors particularly when system nonlinearities are significant or when the operating point deviates substantially from the linearization point. The nonlinear activation functions in the neural networks naturally accommodate system nonlinearities without such approximations.

Furthermore, the RNN-based approach does not require explicit knowledge of noise statistics such as covariance matrices. Instead, the networks implicitly learn effective filtering characteristics from the data. This makes the approach more practical for real-world applications where noise statistics are often unknown or difficult to characterize.

The quality of state estimates depends on the richness of training data in terms of system excitation, the signal-to-noise ratio of measurements and the architecture of the neural networks. The networks must be trained on data that excites various modes of system behaviour to learn comprehensive dynamics. For complex nonlinear systems, recurrent network architectures prove more effective than feedforward networks for the predictor components, as they can better capture the temporal dependencies inherent in system dynamics. The state update network, which implements a static mapping, is typically realized as a feedforward network.

The estimated states obtained from RNN-based filters can be applied in subsequent engineering applications. In fault diagnosis, deviations of estimated state values from their nominal or expected ranges indicate potential system faults or failures. In adaptive control, the estimated states provide feedback information essential for implementing state-feedback controllers when direct state measurements are unavailable or unreliable. In sensor validation, state estimates can be compared with direct sensor measurements to detect sensor failures or bias. The integration of state estimation with parameter estimation through recurrent neural network frameworks provides a comprehensive and practical approach to system identification, monitoring, and control for nonlinear dynamic systems with unknown or uncertain models [6].

### 4.3.7 Nonlinear System Identification Using Partial RNNs

Recurrent neural networks can be broadly classified into fully recurrent neural networks and partially recurrent neural networks based on their connectivity patterns and feedback mechanisms. While fully recurrent networks have feedback connections among all neurons allowing information to flow in multiple directions throughout the network, partially recurrent neural networks employ a more restricted feedback structure with specific feedback connections combined with feedforward connections. This selective recurrence provides sufficient memory capability for dynamic system modelling while maintaining computational tractability and ease of training.

Partially recurrent neural networks have emerged as effective architectures for nonlinear dynamic system identification due to their ability to capture temporal dependencies in input-output relationships without the computational complexity associated with fully recurrent structures. The two most prominent architectures in this category are the Elman network and the Jordan network, both introduced in the late 1980s to address the limitations of purely feedforward networks in handling sequential and time-series data.

The Elman network, introduced by Jeffrey Elman in 1990, consists of an input layer, a hidden layer, an output layer, and an additional context layer. The distinguishing feature of the Elman network is that the context layer receives feedback connections from the hidden layer. At each time step, the hidden layer activations are copied to the context units with a fixed weight, typically unity. These context units then serve as additional inputs to the hidden layer at the next time step, providing the network with memory of its previous internal states. The network dynamics can be described by the following equations:

$$h(k) = f_h\left(W_{xh}x(k) + W_{ch}c(k) + b_h\right)$$

$$y(k) = f_y\left(W_{hy}h(k) + b_y\right) \tag{4.107}$$

$$c(k+1) = h(k)$$

where h(k) is the hidden layer activation vector at time k, x(k) is the input vector, y(k) is the output vector, c(k) is the context layer activation vector, $W_{xh}, W_{ch}$, and $W_{hy}$ are weight matrices connecting input-to-hidden, context-to-hidden, and hidden-to-output layers, respectively, $b_h$ and $b_y$ are bias vectors, and $f_h$ and $f_y$ are activation functions for the hidden and output layers. The context layer essentially stores the previous hidden layer state and feeds it back, allowing the network to maintain short-term memory of past activations.

The Jordan network, proposed by Michael Jordan in 1986, employs a similar architecture but with a crucial difference in the feedback mechanism. Instead of feeding back the hidden layer activations, the Jordan network feeds back the output layer activations to the context layer. Additionally, the context units in a Jordan network often include self-recurrent connections, providing a form of temporal filtering of the feedback signals. The Jordan network equations are:

$$h(k) = f_h\left(W_{xh}x(k) + W_{sh}s(k) + b_h\right)$$

$$y(k) = f_y\left(W_{hy}h(k) + b_y\right) \tag{4.108}$$

$$s(k+1) = \sigma_s\left(W_{ss}s(k) + W_{ys}y(k) + b_s\right)$$

where s(k) represents the state layer (context layer) activations, $W_{sh}, W_{ss}$ and ys are weight matrices for state-to-hidden, state-to-state, and output-to-state connections, respectively, $b_s$ is the state layer bias vector, and $\sigma_s$ is the activation function for the state layer. The self-recurrent connections characterized by $W_{ss}$ allow the context units to maintain a running weighted average of past outputs rather than simply copying them.

For nonlinear system identification, these partially recurrent networks are configured to learn the mapping from system inputs and past outputs to current outputs. Consider a nonlinear dynamic system that can be represented in discrete-time as:

$$y(k) = g\left(y(k-1), y(k-2), \ldots, y\left(k-n_y\right), u(k), u(k-1), \ldots, u\left(k-n_u\right)\right) \tag{4.109}$$

where g is an unknown nonlinear function, u(k) represents the system input, y(k) represents the system output, and $n_y$, $n_u$ specify the orders of output and input dependencies. The partially recurrent neural network is trained to approximate this function using measured input-output data.

The training of partially recurrent networks for system identification can be performed using two different configurations: the series-parallel model and the parallel model. In the series-parallel configuration, the actual measured outputs of the system y(k−1),y(k−2),... are fed back into the network inputs during training. This configuration is also known as teacher forcing and results in a feedforward structure during training, allowing the use of standard backpropagation algorithms. In the parallel configuration, the network's own predicted outputs $\hat{y}(k-1), \hat{y}(k-2), \ldots$ are fed back as inputs. This configuration more closely represents the actual operational mode where true outputs are not available, but training becomes more challenging due to error accumulation through the recurrent connections.

Modified versions of the basic Elman and Jordan networks have been developed to enhance their performance for system identification tasks. These modifications typically involve changes to the feedback connection patterns, addition of direct connections between layers, or incorporation of multiple hidden layers with different recurrent structures. The choice between Elman and Jordan architectures depends on the specific application and system characteristics. Research has indicated that Elman networks generally perform better for tasks requiring memory of internal states, while Jordan networks are more closely aligned with the structure of dynamical systems where outputs are the primary feedback signals.

The training of partially recurrent networks employs gradient-based learning algorithms. For the series-parallel configuration, standard backpropagation can be applied since the network behaves as a feedforward structure during training. For the parallel configuration, backpropagation through time or real-time recurrent learning algorithms must be used to properly account for the temporal dependencies introduced by the recurrent connections. These algorithms compute the gradients by unfolding the network through time and accumulating error contributions from multiple time steps.

One practical advantage of partially recurrent networks compared to fully recurrent networks is the reduced number of weights to be trained. The restricted feedback connectivity means fewer parameters to optimize, leading to faster training and reduced risk of overfitting. Additionally, partially recurrent networks are less prone to the vanishing and exploding gradient problems that plague fully recurrent networks during training, although these issues are not entirely eliminated.

For effective system identification using partially recurrent networks, several design considerations must be addressed: the number of hidden units must be selected to provide sufficient representational capacity without overfitting; the orders $n_y$ and $n_u$ in equation (4.106) must be chosen appropriately, either based on prior knowledge of the system or through systematic search procedures; the training data must be sufficiently rich to excite all relevant system dynamics, typically requiring persistent excitation conditions; and validation procedures using independent test data sets are essential to assess the generalization capability of the identified model.

Partially recurrent neural networks have been successfully applied to identification of various nonlinear dynamic systems including mechanical systems, chemical processes, power systems, and aerospace vehicles. Their ability to learn complex nonlinear dynamics from input-output data without requiring explicit mathematical models makes them particularly valuable for systems where first-principles modelling is difficult or infeasible. The learned models can subsequently be employed in model-based control strategies, fault detection schemes and system simulation studies [7,8].

### 4.3.8 Parameter Estimation of an Aero-elastic Aircraft

The application of parameter estimation methods to extract stability and control derivatives from flight data has been highly successful for rigid aircraft operating in linear flight regimes. However, modern high-performance aircraft with significant structural flexibility present new challenges for aeroelastic modelling and parameter identification. The fundamental difficulty lies in the fact that a complete aeroelastic aircraft model contains an order of magnitude more parameters than a rigid aircraft model, making conventional parameter estimation approaches computationally intensive and often yielding non-unique solutions. Neural network-based approaches offer an alternative methodology that circumvents many of these difficulties by learning the mapping between measured motion variables and aerodynamic coefficients directly from data without requiring explicit coupling equations or initial parameter guesses.

The longitudinal equations of motion for a flexible aircraft can be approximated for the short period mode as:

$$\dot{\alpha}-q=\frac{\rho u S}{2m}\left[C_{L_\alpha}\alpha+C_{L_q}\frac{qc}{2u}+C_{L_\delta}\delta+\sum_{i=1}^{n}\left(C_{L_{\eta i}}\eta_i+C_{L_{\dot{\eta} i}}\frac{\dot{\eta}_i c}{2u}\right)\right]$$

$$q=\frac{\rho u^2 Sc}{2I_y}\left[C_{m_\alpha}\alpha+C_{m_q}\frac{qc}{2u}+C_{m_\delta}\delta+\sum_{i=1}^{n}\left(C_{m_{\eta i}}\eta_i+C_{m_{\dot{\eta} i}}\frac{\dot{\eta}_i c}{2u}\right)\right] \tag{4.110}$$

where $\alpha$ is the angle of attack, q is the pitch rate, δ represents control surface deflection, $\eta_i$ and $\dot{\eta}_i$ are the generalized elastic deflections and their time derivatives, $C_{L_\alpha}$, $C_{m_\alpha}$ etc., are stability and control derivatives, ρ is air density, u is total velocity, S is wing area, c is wing chord, m is aircraft mass, and $I_y$ is the moment of inertia about the pitch axis. The summation accounts for contributions from multiple structural modes.

The generalized coordinates representing elastic deformations satisfy:

$$\ddot{\eta}_i+2\zeta_i\omega_i\dot{\eta}_i+\omega_i^2\eta_i=\frac{\rho u^2 Sc}{2M_i}\left[C_\alpha^{\eta_i}\alpha+C_q^{\eta_i}\frac{qc}{2u}+C_\delta^{\eta_i}\delta+\sum_{j=1}^{n}\left(C_{\eta_j}^{\eta_i}\eta_j+C_{\dot{\eta}_j}^{\eta_i}\frac{\dot{\eta}_j c}{2u}\right)\right] \tag{4.111}$$

where $\omega_i$ is the in-vacuo natural frequency of mode i, $\zeta_i$ is the modal damping ratio, $M_i$ is the modal generalized mass, and $C_\alpha^{\eta_i}$, $C_{\eta_j}^{\eta_i}$ etc., represent generalized force derivatives accounting for aerodynamic and elastic coupling. These coupled equations reveal the complexity of aeroelastic aircraft dynamics where rigid body and elastic degrees of freedom interact through aerodynamic forces [9].

Conventional parameter estimation methods such as maximum likelihood require solution of these coupled differential equations to generate model responses that are compared with measured responses in forming the cost function. This necessitates accurate initial guesses for all parameters and involves substantial computational burden. Furthermore, measurements of elastic deflections and their derivatives are typically not available from flight tests, requiring either installation of specialized instrumentation or estimation of these quantities through filtering techniques, both of which introduce additional uncertainties.

Neural network-based parameter estimation offers an alternative that avoids these complications. Feedforward neural networks trained to map motion and control variables to aerodynamic force and moment coefficients provide a black box model that implicitly captures aeroelastic effects without requiring explicit representation of elastic modes. Two methods have been developed to extract stability and control derivatives from such trained neural models: the Delta method and the Zero method.

The Delta method exploits the fundamental definition of a stability or control derivative as the rate of change of an aerodynamic coefficient with respect to a motion or control variable. After training a neural network to predict, for example, the pitching moment coefficient $C_m$ from inputs α, q, and δ, the derivative $C_{m_\alpha}$ is estimated by perturbing only the angle of attack by a small amount while holding other inputs constant. Specifically, one of the network inputs is given a small delta perturbation at each time point while all other network inputs are held at their original values. Let the predicted $C_m$ for $\alpha + \Delta\alpha$ and $\alpha - \Delta\alpha$ be denoted $C_m^+$ and $C_m^-$ respectively. The stability derivative $C_{m_\alpha}$ is then obtained as the difference between these two predictions divided by twice the perturbation magnitude. This approximation is computed at each data point in the time history, yielding a time series of derivative estimates that are then averaged to produce the final parameter estimate. The method requires no knowledge of the underlying equations of motion or coupling mechanisms.

The Zero method takes a different approach based on isolating the effect of individual variables. A stability or control derivative is viewed as follows: if one could obtain an aerodynamic coefficient as a function of only one of the motion or control variables at a time while the rest of the variables are set at zero values, then the ratio of the predicted aerodynamic coefficient to the non-zero motion or control variable will yield the corresponding stability or control derivative. For example, to estimate $C_{m_\alpha}$, the pitch rate and control deflection are set to zero while angle of attack retains its original measured value. The network prediction of $C_m$ under these conditions divided by α yields the derivative $C_{m_\alpha}$. Similarly, setting α and δ to zero while retaining q allows estimation of $C_{m_q}$. This computation is performed for each time point and the results averaged. Both methods extract parameter estimates from the trained neural model without requiring solution of differential equations or specification of initial parameter values.

For aeroelastic aircraft, the neural network aerodynamic model can be configured in two ways. In the first configuration, only rigid body motion variables and control inputs serve as network inputs. The network learns to map these quantities directly to total aerodynamic coefficients that inherently include all aeroelastic effects. Parameters extracted from such a model represent equivalent derivatives wherein contributions from all elastic modes are absorbed into effective rigid body parameters. These equivalent derivatives provide a reduced order representation suitable for control system design and simulation when detailed elastic mode information is not required.

In the second configuration, selected elastic mode deflections are included as additional network inputs along with rigid body variables. This allows extraction of not only equivalent rigid body derivatives but also generalized force derivatives associated with specific elastic modes. For instance, including the first mode deflection $\eta_1$ as an input enables estimation of derivatives $C_{L_{\eta 1}}$ and $C_{m_{\eta 1}}$ that quantify the aerodynamic forces generated by that particular mode shape. The remaining unmodelled modes still contribute to equivalent parameters for the rigid body degrees of freedom.

The concept of equivalent derivatives merits further explanation. For a flexible aircraft, substituting the steady-state elastic deflections (obtained by setting time derivatives to zero in eqn. 4.111) into eqn. (4.110) yields expressions where elastic effects are absorbed into modified stability derivatives. The resulting equivalent parameters depend on which and how many modes are retained in this algebraic manipulation. Analytical studies show that for aircraft with moderate flexibility, inclusion of only the first mode often suffices to capture the dominant aeroelastic effects in equivalent parameters, while highly flexible configurations may require additional modes for accurate representation.

An important practical advantage of the neural network approach is that it requires only measured quantities typically available from flight tests. For the longitudinal short period dynamics, the required measurements are angle of attack, pitch rate, control surface deflection, normal acceleration, and pitch acceleration. The aerodynamic coefficients are computed from these measurements using $C_L = \frac{2ma_z}{\rho u^2 S}$ and $C_m = \frac{I_y \dot{q}}{\rho u^2 Sc}$ where $a_z$ is normal acceleration. Neither elastic deflections nor their derivatives need to be measured. The neural network implicitly learns the relationship between rigid body motion and the total aerodynamic forces including elastic contributions.

Training the neural network follows standard supervised learning procedures. Input-output data pairs consisting of motion variables and computed aerodynamic coefficients are presented repeatedly while network weights are adjusted to minimize prediction error. The training continues until the mean square error between network predictions and target values falls below a specified threshold or a maximum number of iterations is reached. Network architecture parameters including the number of hidden layers, neurons per layer, learning rate and momentum factor are selected through experimentation to achieve satisfactory prediction accuracy with good generalization properties.

Once trained, the neural model provides several benefits. First, it enables parameter estimation through the Delta and Zero methods without the computational burden of solving coupled differential equations. Second, the parameters are estimated separately for each force or moment coefficient rather than simultaneously for all parameters, reducing the dimensionality of the estimation problem. Third, no initial parameter guesses are required, eliminating a potential source of bias in iterative estimation schemes. Fourth, the trained model itself can serve as an onboard aerodynamic model providing real-time predictions of forces and moments for adaptive control or health monitoring applications.

The neural network methodology for aeroelastic parameter estimation represents a pragmatic approach that balances modelling fidelity with practical constraints of available measurements and computational resources. The neural network methodology for aeroelastic parameter estimation balances modelling accuracy with practical constraints of available measurements and computational resources. By learning complex nonlinear mappings from readily measurable quantities to aerodynamic coefficients, neural networks provide a pathway to parameter estimation for flexible aircraft that would be difficult or impractical using conventional approaches requiring detailed structural models and extensive instrumentation [9].

### 4.3.9 Identification of Underwater Vehicle Models

The identification of dynamic models for complex physical systems such as underwater vehicles presents unique challenges due to the highly nonlinear and coupled nature of hydrodynamic forces and the difficulty in obtaining accurate analytical or numerical representations of certain parameters. While some model parameters can be determined through theoretical calculations or numerical methods such as computational fluid dynamics, others remain uncertain due to their dependence on operational conditions including velocity regimes, proximity to boundaries, and environmental disturbances. Neural networks provide a practical approach to identification by learning parameter relationships directly from measured data without requiring detailed knowledge of the underlying physics, while simultaneously offering the capability for online adaptation as operating conditions change.

The mathematical model of an underwater vehicle in six degrees of freedom can be expressed in vector notation as:

$$M\dot{v}+C(v)v+D(v)v+g(\eta)=\tau \tag{4.112}$$

with the kinematic equation relating body-fixed velocities to earth-fixed positions:

$$\dot{\eta}=J(\eta)v \tag{4.113}$$

where $\eta$ represents the position and orientation vector of the vehicle in the earth-fixed reference frame, $v$ is the linear and angular velocity vector in the body-fixed frame, $\dot{v}$ is the acceleration vector, M is the inertia matrix including added mass effects, $C(v)$ is the matrix of Coriolis and centripetal terms, $D(v)$ is the damping matrix, $g(\eta)$ is the vector of restoring forces and moments due to gravity and buoyancy, $\tau$ is the vector of control forces and moments, and $J(\eta)$ is the transformation matrix converting body-fixed velocities to earth-fixed velocities.

The inertia matrix consists of two components accounting for rigid body mass and added mass:

$$M = M_{RB} + M_A \tag{4.114}$$

where $M_{RB}$ represents the rigid body inertial properties including mass and moments of inertia and $M_A$ accounts for the added mass effects arising from the acceleration of the surrounding fluid. Similarly, the Coriolis matrix combines rigid body and added mass contributions:

$$C(v) = C_{RB}(v) + C_A(v) \tag{4.115}$$

The damping matrix represents various hydrodynamic resistance phenomena and can be decomposed as:

$$D(v) = D_P + D_S(v) + D_W + D_M(v) \tag{4.116}$$

where $D_P$ represents potential damping relevant in the wave zone, $D_S(v)$ accounts for linear and quadratic skin friction, $D_W$ represents wave drift damping, and $D_M(v)$ captures damping due to vortex shedding. These damping components are particularly difficult to model accurately through first principles, making them prime candidates for data-driven identification approaches.

A common simplification approximates the damping through a diagonal matrix containing linear and quadratic terms:

$$\begin{aligned} D(v) = &-diag\{X_u, Y_v, Z_w, K_p, M_q, N_r\} \\ &- diag\{X_{u|u}|u|, Y_{v|v}|v|, Z_{w|w}|w|, K_{p|p}|p|, M_{q|q}|q|, N_{r|r}|r|\} \end{aligned} \tag{4.117}$$

where the first term contains linear damping coefficients and the second term contains quadratic damping coefficients for each degree of freedom. But, this simplified representation neglects coupling between degrees of freedom and higher-order nonlinear effects that may be significant during complex manoeuvres involving simultaneous motion in multiple directions.

Traditional approaches to neural network-based system identification often employ networks in parallel with known model components, treating the neural network as a black box that compensates for all unmodelled dynamics. While this approach requires minimal prior knowledge, it results in the neural network learning a complex overall mapping that does not exploit available structural information about the system. An alternative paradigm uses neural networks selectively to model only those specific parameters or subsystems that are most uncertain, while retaining analytical or empirical models for well-understood components.

Consider the dynamic equation (4.112) rewritten to isolate the acceleration:

$$\dot{v} = M^{-1}\left[\tau - + C(v)v + D(v)v + g(\eta)\right] \tag{4.118}$$

This can be expressed in state-space form where the state vector is v and the system inputs are the control forces τ and positions η. If partial knowledge exists such that the matrices can be decomposed into known and unknown parts as $M^{-1} = M_M^{-1} + \hat{M}^{-1}$, $C(v) = C_M(v) + \hat{C}(v), D(v) = D_M(v) + \hat{D}(v)$ and $g(\eta) = g_M(\eta) + \hat{g}(\eta)$, where subscript M denotes the known model components and the hat denotes unknown components to be learned, then:

$$\begin{aligned}\dot{v} &= M_M^{-1}\left[\tau - + C_M(v)v + D_M(v)v + g_M(\eta)\right] \\ &\quad + M_M^{-1}\left[-\hat{C}(v)v + \hat{D}(v)v + \hat{g}(\eta)\right] + M^{-1}\{\cdot\}\end{aligned} \tag{4.119}$$

Rather than using a single neural network to approximate the entire right-hand side of eqn. (4.119), separate neural networks can be employed to model each uncertain component. This modular approach significantly simplifies the learning task and allows incorporation of known structural properties such as matrix symmetries, velocity dependencies and known coupling patterns.

In practical applications, parameters such as rigid body mass, moments of inertia, centre of gravity, and buoyancy can often be calculated with sufficient accuracy from vehicle geometry and mass distribution. The added mass matrix can be estimated through numerical methods or measured through specialized tests. The Coriolis matrix, being a function of the known mass matrices and measured velocities, can likewise be computed. The damping matrix however remains difficult to characterize accurately without extensive experimental testing. Therefore, a common strategy assumes all components except damping are known and applies neural networks specifically to identify the damping characteristics.

For a process plant model representing the true system dynamics and a control plant model used for prediction and control, the respective dynamic equations become:

$$\dot{v}_{ppm} = M^{-1}\left[\tau - C\left(v_{ppm}\right)v_{ppm} - D\left(v_{ppm}\right)v_{ppm} - g\left(\eta_{ppm}\right)\right] \tag{4.120}$$

$$\dot{v}_{cpm} = M^{-1}\left[\tau - C\left(v_{cpm}\right)v_{cpm} - g\left(\eta_{cpm}\right)\right] \tag{4.121}$$

where the damping term has been omitted from the control plant model initially. During operation, one-step-ahead predictions are computed using the control plant model. Assuming both models have identical states at a given instant, $v_{cpm} = v$, the difference between the accelerations reveals the effect of the damping:

$$D(v)v = M\left[\dot{v}_{cpm} - \dot{v}_{ppm}\right] \tag{4.119}$$

This relationship provides training targets for neural networks learning to approximate the damping. The product $D(v)v$ rather than the matrix $D(v)$ itself is learned, since this product represents the force contribution directly.

To avoid issues with differing output magnitudes across the six degrees of freedom, separate neural networks are typically trained for each degree of freedom rather than using a single multi-output network. Each network receives as inputs the velocity components in all six degrees of freedom plus potentially additional features such as velocity magnitudes. The output is the corresponding component of $D(v)v$. This architecture ensures balanced learning across all outputs and facilitates both offline and online training.

Offline training employs collected data from vehicle operations under various conditions. The Levenberg-Marquardt algorithm or other efficient second-order methods can be used during this phase since computational constraints are less restrictive. Online training uses gradient descent with momentum to adapt the network weights as new data become available during operation. The error signal for online training is derived from the difference between predicted and actual measurements, backpropagated through the network to adjust weights in the direction that reduces prediction error.

The trained neural networks modelling $\hat{D}(v)v$ are incorporated into the control plant model to produce:

$$\dot{v}_{cpm} = M^{-1}\left[\tau - C(v)v - g(\eta) - \hat{D}(v)v\right] \tag{4.122}$$

This augmented model can be used for feedforward control by inverting the dynamics to compute required control forces for desired accelerations:

$$\tau = m\dot{v}_{cpm} + C(v)v - g(\eta) - \hat{D}(v)v \tag{4.123}$$

The neural network approach offers several advantages for underwater vehicle identification. First, it handles the nonlinear and coupled nature of hydrodynamic damping without requiring explicit functional forms. Second, it accommodates velocity-dependent and regime-dependent behaviour that simple parametric models cannot capture. Third, online adaptation allows the model to adjust when vehicle characteristics change due to payload modifications, fouling, or damage. Fourth, neural networks inherently filter measurement noise through their generalization capability, producing smoother parameter estimates than direct least squares methods applied to noisy data.

The modular identification strategy where neural networks model specific uncertain parameters rather than the entire system dynamics exploits available knowledge effectively. Known structural properties such as the symmetry of mass matrices, coordinate frame transformations, and physically motivated coupling patterns guide the architecture design. Unknown or poorly modelled phenomena are isolated to specific neural network modules that learn from data. This combination of analytical modelling and data-driven learning achieves superior performance compared to purely analytical or purely empirical approaches [10].

### 4.3.10 Air-fuel Ratio Estimation

With increasingly stringent emissions standards, the in-cylinder air-fuel ratio must be controlled within strict limits for optimum catalyst efficiency. It is necessary to know the air-fuel ratio in order to control its value. But, there are certain problems with obtaining a direct measurement of this value. The best method of measuring air-fuel ratio is by use of an Exhaust Gas Oxygen (EGO) sensor, which measures the amount of oxygen in the gases exiting the exhaust port of the engine. From this, the air-fuel ratio of the cylinder gases before combustion can be calculated. However, since at stoichiometry or rich burning conditions there will be little or no unreacted oxygen in the exhaust gases, the EGO sensor can only accurately give lean air-fuel ratio values and this is only over a narrow band. This extremely non-linear property of such sensors, in addition to the cost per unit, would mitigate against their use in production engines.

Air-fuel ratio correlates closely to at least two other engine measurements, namely torque output and peak cylinder pressure. Thus, a good estimate of air-fuel ratio may be made from measurements of these and other variables. But then, the cost of sensors would be prohibitive. The preference would be to use only existing sensors such as throttle angle and engine speed. This can be done by considering the behaviour of the intake manifold.

The intake manifold can be represented as a closed volume with a throttle at its entrance and the engine at its exit. Air flows into the manifold through the throttle body and through it into the engine. Into this air stream is injected an amount of liquid fuel, determined by the fuel command and the injector dynamics. A certain amount of this fuel is entrained directly into the air, while the rest impinges on the manifold wall. This forms a film, from which some more fuel evaporates into the air. Thus, fuel may enter the cylinder by two routes, directly from injector into the air, or via the fuel film on the manifold wall. This complex behaviour is the cause of air-fuel ratio excursions through engine transients. For example, if the throttle steps open suddenly, the air mass flow rate increases similarly. However, the velocity of the air decreases, causing the evaporation rate of the fuel from the manifold wall to decrease. The end result is a sudden lean excursion of air-fuel ratio under proportional control.

The engine draws air from the inlet manifold into the cylinder as a result of its pumping action. The pressure in the manifold is reduced to below atmospheric, causing air to flow through the throttle. The relationship between the fuel injector command and the mass flow rate of fuel into the manifold is approximated by a first order lag:

$$\tau_f \frac{d\dot{M}_{fi}}{dt} + \dot{M}_{fi} = U_f \tag{4.124}$$

The air mass flow into the manifold at any instant is $\dot{M}_{at}$. The expression for the air mass flow rate at the engine inlet port is:

$$\dot{M}_{ap} = C_1 P_m N + C_2 N + C_3 N^2 \tag{4.125}$$

where $\dot{M}_{ap}$ is the air mass flow rate entering the engine, $P_m$ is the manifold pressure, N is the engine speed, and $C_1$, $C_2, C_3$ are constants. The mass flow of air into and out of the manifold determines the change in manifold pressure and is represented:

$$\frac{dP_m}{dt} = \frac{RT_m}{V_m}\left(\dot{M}_{at} - \dot{M}_{ap}\right) \tag{4.126}$$

where R is the gas constant, $T_m$ is the manifold temperature, and $V_m$ is the manifold volume.

The fuel is injected directly into the manifold. An amount of the fuel impinges directly onto the manifold wall, forming a film. The remaining fuel is entrained in the air, along with that which evaporates from the reservoir of film back into the air stream. The fuel entering the cylinder can be expressed as:

$$M_{fc} = (1 - X)M_{fi} + M_e \tag{4.127}$$

where X is the fraction of fuel impinging on the manifold wall, $M_e$ is the mass flow rate of fuel entering the air from the wall, and $M_{fc}$ is the mass flow rate of fuel entering the cylinder.

The fraction X is dependent on the throttle angle as follows:

$$X = 0.3 + 0.7\frac{\theta}{90^\circ} \tag{4.128}$$

where θ is the throttle angle.

The rate of evaporation of fuel from the manifold wall is represented as a first order lag:

$$\tau_e \frac{dM_e}{dt} + M_e = XM_{fi} \tag{4.129}$$

where τ $\tau_e$ is the time constant for fuel evaporation from the wall film.

Although using these equations an estimator could be developed, this would be a cumbersome method of estimating air-fuel ratio. The parameter estimation is difficult and time consuming and differences in engines would mean that this step would be necessary and separate for every engine.

An alternative approach is to treat the manifold as a black-box, in which the measured variables determine air-fuel ratio in some unknown way. The use of a neural network to identify this non-linear relationship should give a good model of the manifold behaviour, which could then be utilized as an air-fuel ratio estimator. For a multi-layer feedforward network with a set of inputs $x_i$ i, the given output y is:

$$y = F\left(\sum_i \omega_i x_i - \rho\right) \tag{4.130}$$

where $\omega_i$ represents the connection weights, ρ is a threshold value, and F(·) is the activation function such as sigmoid or hyperbolic tangent. For uniform approximation of any continuous function, only a single hidden layer is required.

The use of Autoregressive Moving Average with Exogenous inputs for linear modelling is a technique in system identification and requires time series inputs to represent system dynamics:

$$y(t)=\sum_{i=1}^{n}a_i y(t-i)+\sum_{j=1}^{m}b_j u(t-j)+\sum_{k=1}^{p}c_k y(t-k) \tag{4.131}$$

where y, u, and e are the system output, input, and noise vectors, respectively. The non-linear NARMAX representation is:

$$y(t)=f\left[y(t-1),\ldots,y(t-n),\ u(t-1),\ldots,u(t-m),e(t-1),\ldots,e(t-p)\right] \tag{4.132}$$

where y, u, and e are the system output, input, and noise vectors. The values of n, m, and p must be carefully chosen so that the system is parsimonious, that is, the number of variables is the minimum necessary to represent the system. A simpler representation is:

$$y(t)=f\left[y(t-1),\ldots,y(t-n),\ u(t-1),\ldots,u(t-m)\right]+e(t) \tag{4.133}$$

that is, the noise is additive. The application of neural networks to representing the function f follows naturally from the uniform approximation property of single hidden layer neural networks. In this case the network inputs are the vectors y and u. When the network is used to represent the system, the actual response will be given by:

$$y(t+1)=f\left[y(t),\ldots,y(t-n),\ u(t),\ldots,u(t-m)\right] \tag{4.134}$$

where y is the vector of past approximations. An alternative approach is to base the prediction only on the past values of inputs u. If it is accepted that the accuracy of using values of u alone will be less than that from using values of u and y as a result of the truncation of the series specified, then using the representation:

$$y(t+1)=f\left[u(t),\ldots,\ u(t-q)\right] \tag{4.135}$$

should be sufficient, as long as q is greater than n plus m. Two methods of training neural networks to represent these systems exist. One approach requires the use of dynamic backpropagation techniques, while the other only requires standard static backpropagation. The latter is the most correct method to use, but an alternative approach assumes the network is trained using actual outputs y, the estimation should approach the real value y(t). In this case, if the errors are small, the use of the vector as an input is acceptable. However, if the model is specified only in terms of the system inputs, both methods are equivalent.

Neural networks can be a useful tool in estimating the states of system. Starting from the situation of a known system, the so-called Observers Equation may be found. This is the representation used when only input values are considered. Having exact knowledge of this is not necessary, as only its form is needed in order to specify the network inputs. Correlation analysis can be of help in finding the order of the Observers Equation and therefore the optimum network. A neural network emulator can estimate unobserved states of linear and non-linear systems. The results for the linear case can be perfect with zero error and for a simple non-linear system minimal. A network may not be able to estimate perfectly the states of a complex non-linear system. Results provide theoretical support for the use of neural networks in the solution of the engine air-fuel ratio problem [11].

## 4.4 GENERAL DYNAMIC NEURAL NETWORKS AND NETWORK PRUNING

Neural networks used for parameter estimation face a practical problem in determining the appropriate number of neurons and connections. A network that is too small cannot learn the function properly, while a network that is too large learns the training data too well, including the noise and peculiarities of that specific dataset and fails when presented with new data. This section presents network pruning as a systematic solution to this problem.

When a neural network is trained on examples, the objective is for it to learn the underlying pattern rather than merely memorize the training data. For most real problems, providing examples of every possible input is impossible. The network must generalize from the training samples to the complete function. A fundamental principle for good generalization is to use the smallest network that can fit the data adequately. But then, finding the right size is not straightforward. Training many networks of different sizes is time consuming and very small networks often get stuck in poor solutions or take a long time to learn. The pruning approach starts with a large network that learns quickly and reliably, then removes the parts that are not needed. The large initial size allows fast learning with less sensitivity to starting conditions, while the reduced complexity of the trimmed system improves generalization.

The phenomenon of overtraining provides clear motivation for network pruning. During early training, the error on both the training set and a separate test set decreases together as the network learns the underlying function. However, if the training data is incomplete, it contains spurious patterns due to sampling. At some point, usually in later stages of learning, the network starts exploiting these artifacts in the training data. The test error then starts increasing even though the training error continues decreasing. This overtraining occurs when the network has more degrees of freedom than the number of training samples. The network fits not only the true function but also the noise in the training data. It gives correct outputs at training points but becomes inaccurate elsewhere.

Small networks offer several advantages beyond better generalization. They are faster to execute and require less memory. Their operation is easier to understand since there is less opportunity for the network to distribute functions over many nodes. This clarity can be important in critical applications where understanding the system's behaviour is essential. Formal learning theory relates the complexity of a learning system to the number of examples required to learn a particular function. If the number of examples is small relative to the system complexity, the generalization error is expected to be high. However, choosing an appropriate network architecture for a given set of examples remains challenging, as theoretical bounds do not directly specify how to select a suitable structure.

Pruning algorithms fall into two main categories. The first estimates the sensitivity of the error function to removal of network elements such as weights or nodes. Elements with the least effect on error can then be removed. The second adds penalty terms to the objective function that reward the network for choosing efficient solutions. A term proportional to the sum of all weight magnitudes, for example, favours solutions with small weights. Weights that are nearly zero do not influence the output much and can be eliminated. Sensitivity methods modify a trained network by computing sensitivities, removing low-sensitivity elements and possibly retraining. Penalty-term methods modify the cost function so that backpropagation drives unnecessary weights to zero during training, effectively removing them without explicit deletion.

A direct but computationally expensive pruning method sets each weight to zero individually and evaluates the change in error. If the error increases too much, the weight is restored, otherwise it is removed. On a serial computer, each forward propagation takes O(W) time, where W is the number of weights. This must be repeated for each weight and each of M training patterns, resulting in $O(MW^2)$ time for each pruning pass. A more conservative method evaluates the change in error for all weights and patterns, then deletes only the one weight with the least effect. This process repeats until the least change in error reaches some threshold and could take $O(MW^3)$ time. Due to these high computational costs, practical pruning algorithms employ various approximations and indirect methods to estimate sensitivity more efficiently [12].

One class of sensitivity methods approximates the importance of network elements through measures that can be computed during or immediately after training without explicit removal trials. The relevance of a hidden unit can be estimated by introducing a gating parameter α that modulates the unit's activity according to:

$$o_j = f\left(\sum_i \omega_{ji}\alpha_i o_i\right) \tag{4.136}$$

where $o_j$ is the activity of unit j, $\omega_{ji}$ is the weight from unit i to unit j, and f is the sigmoid function. When $\alpha = 0$, the unit has no influence on the network; when $\alpha = 1$, the unit operates normally. The importance of the unit can then be approximated by the derivative of the error with respect to α, evaluated at $\alpha = 1$:

$$\hat{\rho}_j = \left.\frac{\partial E^l}{\partial \alpha_i}\right|_{\alpha_i = 1} \tag{4.137}$$

This derivative can be computed efficiently using backpropagation. Units with small values of $\hat{\rho}_j$ contribute little to the network output and can be candidates for removal.

More sophisticated sensitivity calculations estimate the second derivative of the error with respect to each weight. When the weight vector W is perturbed, the change in the error is approximately:

$$\delta E \approx \sum_i s_i \delta w_i + \frac{1}{2}\sum_{i,j} h_{ij} \delta w_i \delta w_j + O\left(\|\delta W\|\right) \tag{4.138}$$

where the $\delta w_i$'s are the components of δW, $s_i = \frac{\partial E}{\partial w_i}$ are the components of the gradient of E with respect to W, and $h_{ij} = \frac{\partial^2 E}{\partial w_i \partial w_j}$ are elements of the Hessian matrix H. Since pruning is done on a well-trained network, the first term will be zero because E is at a minimum. When the perturbations are small, the last term will be negligible. Since H is a very large matrix, the simplifying assumption is made that the off-diagonal terms are zero. This leaves:

$$\delta E \approx \frac{1}{2}\sum_i h_{ii} \delta w_i^2 \tag{4.139}$$

The second derivatives $h_{kk}$ can be calculated by a modified backpropagation rule. The saliency of weight $w_k$ is then:

$$s_k = \frac{1}{2} h_{kk} w_k^2 \tag{4.140}$$

Pruning is done iteratively, i.e., train to a reasonable error level, compute saliencies, delete low saliency weights, and resume training.

Penalty-term methods modify the error function so that normal backpropagation effectively prunes the network by driving weights to zero during training. The weights may be removed when they decrease below a certain threshold; even if they are not, the network still acts somewhat like a smaller system. One approach uses the cost function:

$$C = \mu_{er} \sum_j^P \sum_i^O \left(d_{ij} - o_{ip}\right)^2 + \mu_{en} \sum_j^P \sum_i^H e\left(o_{ij}\right)^2 \tag{4.141}$$

where e is a positive monotonic function. The sums are over the set of output units O, the set of hidden units H, and the set of patterns P. The first term is the normal backpropagation error term, the second term measures the average energy expended by the hidden units. The parameters $\mu_{er}$ and $\mu_{en}$ balance the two terms. The energy expended by a unit, how much its activity varies over the training patterns, is an indication of its importance. If the unit changes a lot, it probably encodes significant information; if it does not change much, it probably does not carry much information. Qualitatively different behaviours are seen depending on the form of e. Various functions are examined which have the derivative:

$$e' = \frac{\partial e\left(o^2\right)}{\partial o_i^2} = \frac{1}{\left(1+o^2\right)^n} \tag{4.142}$$

where n is an integer. For n=0, e is linear and high and low energy units receive equal penalties. For n=1, e is logarithmic and low energy units are penalized more than high energy units. For n=2, the penalty approaches an asymptote as the energy increases so high energy units are not penalized much more than medium energy units.

A magnitude-of-weights term may also be added to the cost function, giving:

$$C = \mu_{er} \sum_j^P \sum_i^O \left(d_{ij} - o_{ip}\right)^2 + \mu_{en} \sum_j^P \sum_i^H e\left(o_{ij}^2\right) + \mu_w \sum_{ij}^W w_{ij}^2 \tag{4.143}$$

Since the derivative of the third term with respect to $w_{ij}$ is $2\mu_w w_{ij}$, this effectively introduces a weight-decay term into the backpropagation equations. Weights which are not essential to the solution decay to zero and can be removed.

Another approach minimizes the following cost function:

$$C = \sum_{k \in T} \left(t_k - o_k\right)^2 + \lambda \sum_{i \in C} \frac{w_i^2 / w_o^2}{1 + w_i^2 / w_o^2} \tag{4.144}$$

where T is the set of training patterns and C is the set of all connections. The second term represents the complexity of the network as a function of the weight magnitudes relative to the constant $w_o$. For $|w| >> w_o$, the cost of a weight approaches λ. For $|w| << w_o$, the cost is nearly zero. When λ is large, this is similar to weight-decay methods. The value of λ requires some tuning and depends on the problem. If it is too small, it will not have any significant effect, if it is too large, all the weights will be driven to zero.

A simple cost function using absolute value penalty is:

$$C = \sum_{k \in T} \left(t_k - o_k\right)^2 + \lambda \sum_{ij} \left|w_{ij}\right| \tag{4.145}$$

The second term adds $-\lambda \operatorname{sgn}(w_{ij})$ to the weight update rule. If $w_{ij} > 0$, the weight is decremented by λ; if $w_{ij} < 0$, the weight is incremented by λ.

The choice of when to stop the pruning process presents a challenge analogous to deciding when to stop training. If separate validation data are available, pruning can continue until validation error begins to increase. In sensitivity-based methods, the sensitivity values themselves may exhibit a natural threshold where they increase sharply, suggesting that further pruning would significantly degrade performance. For penalty-based methods, the balance between the error and complexity terms, controlled by the scaling parameters, determines the final network size. Some methods adapt these parameters dynamically during training to achieve desired levels of pruning without manual tuning.

One limitation of many sensitivity-based methods is their inability to detect correlated or redundant elements. Sensitivities are estimated under the assumption that $w_{ij}$ is the only weight to be deleted (or node i is the only node to be deleted). After the first element is removed, the remaining sensitivities are not necessarily valid for the smaller network. An extreme example is two nodes which cancel each other out at the output. As a pair, they have no effect on the output, but individually each has a large effect so neither will be removed. Partially correlated nodes are a less extreme, but more common example [12].

The practical application of pruning algorithms requires consideration of several factors. The initial network size should be sufficiently large to ensure that the learning problem is tractable and that the network is not trapped in poor local minima due to inadequate capacity. The pruning threshold or the penalty strength parameter must be chosen appropriately, often requiring some experimentation or validation-based selection. Training and pruning can be performed iteratively: train the network, prune low-importance elements, retrain to allow the network to adapt to the reduced architecture, and repeat until a stopping criterion is met. This iterative approach helps the network compensate for any performance degradation introduced by removing elements. Beyond improving generalization, pruned networks offer additional advantages. They typically execute faster during operation due to fewer computations, require less memory for storage, and may be easier to interpret since the learned function is distributed over fewer parameters. In applications where understanding the network's decision process is important, smaller networks can facilitate analysis and explanation of the system's behaviour.

## 4.5 STATE SPACE NEURAL NETWORKS

The identification of nonlinear dynamic systems using neural networks can follow different architectural approaches. Feedforward neural networks provide nonlinear static mappings between inputs and outputs, requiring temporal data windows containing past inputs and outputs to capture dynamic behaviour. Recurrent neural networks represent dynamics through feedback connections and signal delays, avoiding the need for explicit temporal windows. While both approaches have been successfully applied to system identification, their inherent nonlinearity and lack of correspondence with classical system descriptions make analytical tractability and stability analysis difficult. State space neural networks address these limitations by employing a recurrent architecture that directly mirrors the mathematical structure of nonlinear state space representations, enabling straightforward application of classical stability analysis tools.

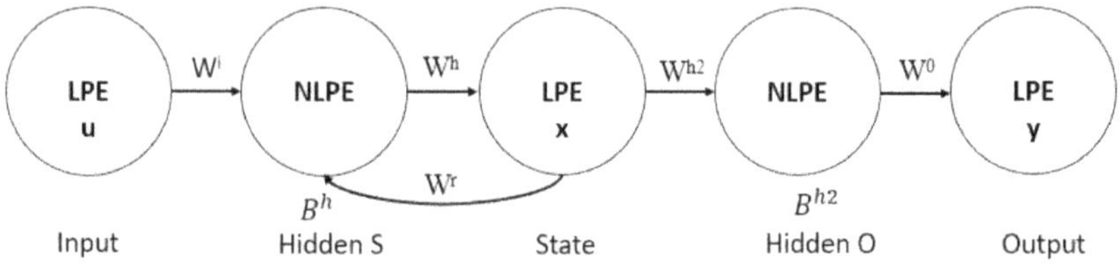


**FIGURE 4.13** Diagrammatic representation of the state space neural network architecture (adapted/modified from Zamarreno and Vega, 1998)

A deterministic nonlinear system in discrete-time state space form is represented mathematically as:

$$x(k+1) = F(x(k), u(k))$$

$$y(k) = G(x(k)) \tag{4.146}$$

where x is the state vector, u is the input vector, and y is the output vector. Both F and G are nonlinear mappings in general.

The state space neural network architecture replicates this mathematical structure through a five-layer recurrent network configuration, as illustrated in Figure 4.13. The input layer receives external input signals and distributes them to the subsequent layer. The first hidden layer, denoted $S^1$, represents the nonlinear function governing state evolution. The state layer contains neurons whose outputs directly represent the state variables, with each neuron corresponding to one state component. The second hidden layer, denoted O, represents the nonlinear function relating states to outputs. Finally, the output layer aggregates signals from the preceding hidden layer to produce the network outputs.

Figure 4.13 shows five interconnected layers: Input Layer (taking input signals), Hidden Layer (implementing nonlinear state transition function $f_1$), State Layer (containing state neurons with outputs and feedback connections with delay elements), Hidden Layer (implementing nonlinear output function $f_2$), and Output Layer (producing network outputs). LPE corresponds to Linear Processing Element and NLPE for Non-linear Processing Element (Sigmoid).

This architectural decomposition into state transition and output mappings provides two significant advantages. First, as a neural model it retains the flexibility to represent arbitrary nonlinear functions through appropriate weight configurations and activation functions. Second, as a state space model it maintains minimal external connections, establishing a natural parallelism between network inputs and outputs and the physical causes and effects in the system being modelled. The states evolve nonlinearly as functions of previous states and external inputs, while outputs are computed as functions of the current states.

The mathematical form of the state space neural network is:

$$\hat{x}(t+1) = W^h \cdot f_1\left(W^r \cdot \hat{x}(t) + W^i \cdot u(t) + B^h\right)$$

$$\hat{y}(t) = W^0 \cdot f_2\left(W^{h2} \cdot \hat{x}(t) + B^{h2}\right) \tag{4.147}$$

where $W^i, W^h, W^r, W^{h2}$, and $W^0$ are weight matrices with dimensions h×n, s×h, h×s, h2×s, and m×h2, respectively, $B^h$ and $B^{h2}$ are bias vectors with h and h2 elements, and f1 and f2 are activation functions applied elementwise. The activation functions can be linear or sigmoidal depending on the nonlinearity requirements. For single-input single-output systems, n and m equal unity, while the state dimension s and hidden layer sizes h and h2 must be determined based on system complexity.

An important operational characteristic distinguishes this architecture from other recurrent networks. The first output $\hat{y}(0)$ depends on the initial state condition $\hat{x}(0)$. Subsequently, the state $\hat{x}(1)$ is computed as a function of $\hat{x}(0)$ and $\hat{u}(0)$, then $\hat{y}(1)$ is determined from $\hat{x}(1)$ and this process repeats iteratively. Thus, state signals propagate first to produce outputs, after which states are updated based on previous states and current inputs.

The state space neural network encompasses linear systems as a special case. When the activation functions are linear and biases are zero, the functions act as identities and equation (4.139) reduces to:

$$\hat{x}(t+1) = W^h \cdot W^r \cdot \hat{x}(t+1) + W^h \cdot W^i \cdot u(t) = \Phi \cdot \hat{x}(t) + \Gamma \cdot u(t)$$

$$\hat{y}(t) = W^0 \cdot W^{h2} \cdot \hat{x}(t) = C \cdot \hat{x}(t) \tag{4.148}$$

where $F = W^h \cdot W^r$, $G = W^h \cdot W^i$, and $C = W^0 \cdot W^{h2}$ are the standard state space matrices. This connection to linear state space theory enables application of classical stability results. The stability of the linear system $\tilde{x}(t+1) = \Phi\tilde{x}(t)$ is determined by whether all eigenvalues of $\Phi$ have modulus less than unity.

For nonlinear state space neural networks, asymptotic stability of equilibrium points can be analyzed through linearization. An equilibrium point $\tilde{x}_0$ of the autonomous system satisfies:

$$\tilde{x}_0 = W^h \cdot \tanh\left(W^h \cdot \tilde{x}_0 + B^h\right) \tag{4.149}$$

when hyperbolic tangent activation functions are employed. The stability of this equilibrium is characterized by the following result. The equilibrium state $\tilde{x}_0$ is asymptotically stable if all eigenvalues of the Jacobian matrix $W^h \cdot D \cdot W^r$ lie inside the unit circle, where D is a diagonal matrix whose elements are:

$$d_{jj} = 1 - tanh^2\left(\sum_{i=1}^{s} W_r^{ji} \cdot x_{0,i} + B_h^j\right) \tag{4.150}$$

This result follows from Taylor series linearization of the nonlinear state transition around the equilibrium point. The diagonal matrix D captures the derivatives of the activation functions evaluated at the equilibrium, with elements belonging to the interval (0,1) for hyperbolic tangent functions. The product $W^h \cdot D \cdot W^r$ constitutes the Jacobian of the system at the equilibrium point and classical discrete-time stability theory applies directly.

An important observation concerns how external inputs affect stability. For non-autonomous systems with input $\tilde{u}$, the term $W^h \cdot \tilde{u}$ acts as an additional contribution to the bias, effectively creating a total bias $B_h^{new} = B^h + W^h \cdot \tilde{u}$. This modifies the diagonal elements of D and consequently changes the eigenvalues of the Jacobian. Thus the system can exhibit different stability characteristics depending on the magnitude and direction of the input, reflecting the behaviour-dependent dynamics typical of nonlinear systems.

The state space neural network formulation provides several analytical advantages beyond stability analysis. The explicit state representation allows investigation of equilibrium points and their multiplicity for given parameter values. Unlike black box recurrent networks where internal dynamics are obscured, the state variables have clear interpretations as evolving system quantities. The separation of state transition and output mappings permits independent analysis and modification of these components. Furthermore, linearization around operating points produces local linear models suitable for control system design using established linear techniques.

Training the state space neural network to identify an unknown system from input-output data requires algorithms capable of handling recurrent connections and delays. Standard backpropagation cannot be applied directly since feedback paths create temporal dependencies. Backpropagation through time unfolds the network across time steps and computes gradients by propagating errors backward through this unfolded structure. Alternative approaches include real-time recurrent learning which maintains gradient information online without unfolding. The choice of training algorithm involves trade offs between computational efficiency, convergence speed, and susceptibility to local minima. Regardless of the specific training method, the goal is to adjust the weight matrices $W^i, W^h, W^r, W^{h2}, W^0$ and bias vectors $B^h, B^{h2}$ to minimize prediction error between network outputs and target system outputs over a training dataset.

The state space neural network representation establishes a natural connection between neural network modelling and classical control theory. By preserving the mathematical structure of state space descriptions while incorporating the approximation capabilities of neural networks, this architecture enables both accurate nonlinear system identification and meaningful stability analysis. The ability to extract interpretable state variables, compute equilibrium points, and apply linearization theory makes state space neural networks particularly suitable for applications where understanding system dynamics and guaranteeing stable behaviour are essential requirements alongside accurate prediction performance [13].

## 4.6 ADAPTIVE FUNCTION LINK NETWORK IN AIRCRAFT COEFFICIENT ESTIMATION

One of the important branches of modelling and simulation is system identification and parameter estimation applied to complex dynamic systems, such as aircraft. System identification and parameter estimation are used to characterize the response of dynamic systems, either considering an input/output black box approach or assuming the equations of motion are known and estimating the unknown parameters. Certification, control law design, and flight simulation are essential tasks in the aeronautical industries that require good mathematical modelling of aircraft dynamics. More realistic aircraft models, described by aerodynamic and control coefficients, can be determined during flight test programs.

Artificial Neural Networks (ANN) are universal function modellers, able to approximate an unknown model to any desired accuracy. This characteristic has drawn the attention of scientists and engineers for the application of such tools to the problem of system identification and parameter estimation. However, the parameters (weights) of an ANN generally, have no physical meaning. Also, the learning rules used to find such parameters are nonlinear, making them costly from the computational point of view and difficult to apply in near real-time parameter estimation problems.

A network architecture called Adaptive Functional Link Network (AFLN) addresses these limitations when applied to aircraft parameter estimation. When compared with classical networks such as Multi-Layer Perceptron, there are main benefits offered by the AFLN. The nonlinear transformation of the network input is done through different types of nonlinear functions, such as polynomial functions, exponentials, and radial basis functions. This characteristic gives the AFLN enormous flexibility and makes the determination of the base functions an important phase during the network design. The network first processes the nonlinear transformation and then combines the result in a linear manner, keeping it linear with respect to the parameters. This characteristic allows the use of linear training rules, such as the Kalman filter. To improve the problem of least squares type estimators that tend to diverge when many data are processed, an adaptive procedure based on a criterion of statistical consistency is used to balance a priori information priority with that of new learning information.

The Functional Link Network (FLN) presents similar benefits, but it is not a robust estimator. If the initial parameter vector or its initial estimation error covariance matrix is far from the real parameter values, the FLN does not perform as well as an AFLN. The AFLN network was proposed to circumvent this weakness, applying a procedure to estimate the noise level adaptively in order to give the algorithm good numerical behaviour and the ability to distribute the extraction of learning information to all training data, improving the convergence and the capacity of learning by a substantial amount.

The AFLN architecture is arranged in three layers: an input layer, one hidden layer, and the output layer. The hidden layer is responsible for the mapping of the input space $\mathbb{R}^n$ to another of dimension p, that is:

$$x \in \mathbb{R}^n \rightarrow \left[h_1(x)\dots h_p(x)\right]^T \in \mathbb{R}^p \tag{4.151}$$

where $h_i(x)$, i = 1,2,..., and p is the base function set that depends exclusively on x. The output layer has m nodes, each one a linear combination leading to the following input/output mapping of the network, with $1 \leq j \leq m$:

$$\hat{f}_j(x,w) = \sum_{k=1}^{p} w_{jk} h_k(x) \tag{4.152}$$

The AFLN can approximate any continuous function with desired precision if the set of base functions h(x) selected is representative or composed by a sufficient number of high order terms. One point that should be emphasized is that the network is linear in the parameters (weights), which means that they can be determined using linear training rules [14].

An AFLN should achieve a continuous nonlinear mapping and it is assumed that the network has the following mapping and definition:

$$f : x \in C \subset \mathbb{R}^n \rightarrow y \in D \subset \mathbb{R}^m \tag{4.153}$$

$$\hat{y}(t) = \hat{f}\left(x(t), w\right) \tag{4.154}$$

where w is the weight vector (network free parameters) to be estimated from an input/output training set. During the training process, the minimization of the following functional is solved for each pair of input/output set j:

$$\begin{aligned} J_j(w_j) = {} & \frac{1}{2}\left[\left(w_j - \bar{w}_j\right)^T \bar{P}_j^{-1}\left(w_j - \bar{w}_j\right)\right] \\ & + \frac{1}{2}\sum_{t=1}^{T}\left[\left(y_j(t) - H(x(t))w_j\right)^T R_j^{-1}(t)\left(y_j(t) - H(x(t))w_j\right)\right] \end{aligned} \tag{4.155}$$

where $\bar{w}_j$ is the initial estimate vector of $w_j$, $\bar{P}_j$ is the statistic representing the quality of the initial estimation of the parameters, and $R_j$ is the statistic representing the quality of the output pattern of the training set. This is equivalent to the following stochastic linear parameter estimation problem, considering t = 1,2,...,T and j = 1, 2,...,m:

$$\bar{w}_j = w_j + \bar{e}_j$$

$$y_j(t) = H(x(t))w_j + v_j(t) \tag{4.156}$$

with the statistical properties:

$$E\left[e_j\right]=0 \quad E\left[e_j e_j^T\right]=P_j$$

$$E\left[v_j(t)\right]=0 \quad E\left[v_j(t)v_j(t)^T\right]=R_j \tag{4.157}$$

This system can be solved by a linear estimation algorithm, such as the Kalman filter algorithm, which provides the recursive solution for updating the weight estimates and their covariance matrices as new training data becomes available.

One problem of the optimal linear estimation, in particular the Kalman filter, is that it has bad numerical behaviour and observation model errors, so divergence usually occurs as many data sets are processed. These problems cause an excessive and unrealistic decrease of the estimated dispersions of the covariance in the calculated estimates; that is, a matrix of estimated covariance of the errors, in the estimates, with very small eigenvalues. In order to avoid this ill behaviour and try to keep a distributed and uniform capacity of learning, an adaptive procedure based on a criterion of statistical consistency is applied to balance a priori information priority with that of new learning information.

Considering that in typical j-th iteration for t = 1,2,…,T – 1:

$$w_j = w_{j-1} + \eta(t)$$

$$E\left[\eta(t)\right]=0 \quad E\left[\eta(t)\eta(t)^T\right]=Q(t)\delta_{tr}$$

$$Q(t)=diag\left[q_j(t): j=1,2,\ \ldots,\ n_w\right] \tag{4.158}$$

where $\delta_{tr}$ is the Kronecker delta, and $n_w$ is the number of weight parameters, the approximation of the neural weight learning using the j-th input/output training set is transformed into the following estimation problem:

$$z_j(t)=H_j\left(x(t)\right)w_j+v_j(t)$$

$$w_j(t)=w_j(t-1)+e_j(t) \tag{4.159}$$

considering $e_j(1) = e_j$ and t = 1,2,…,T.

To propagate the estimation from t to t+1:

$$w_j = w_j$$

$$P_j(t+1) = P_j(t) + Q_w(t) \tag{4.160}$$

where $P_j(t+1) = E\left[e_j(t+1)e_j(t+1)^T\right]$ and $P_j(t)$ is given by the filter algorithm. $P_j(t)$ starts with $P_j(1) = E\left[e_j e_j^T\right] = P$. The adaptation is done by adjusting the noise η(t) dispersion such as to keep the consistency to achieve distributed learning:

$$\beta E\left[v_j^2(t+1)\right] = H_j(t+1)\left[P_j(t) + Q(t)\right]H_j^T(t+1) \tag{4.161}$$

where β is to be adjusted closer to one for distributed learning. In order to use the optimal linear estimation procedure, the following associated estimation problem can be established by considering the equation above as a pseudo observation, and imposing an a priori information such that q(t):

$$q(t) = 0 + e_q$$

$$z_j^q(t+1) = q(t+1) + H_j(t+1)\beta(t+1) + v_j^q(t+1) \tag{4.162}$$

with the statistical properties:

$$E\left[e_q\right] = 0 \quad E\begin{bmatrix} e_q \\ e_q^T \end{bmatrix} = I_{n_w}$$

$$E\left[v_j^q(t+1)\right] = 0 \quad E\left[v_j^q(t+1)v_j^{qT}(t+1)\right] = R_j^q(t+1) = 0 \tag{4.163}$$

The problem above is, in fact, one of exact observation and is possible to be solved by a Kalman filter algorithm. It should be noticed that the solution gives a $\hat{q}(t)$ close to zero in magnitude. Whenever a $q_k(t)$ component is less than zero it is disregarded and taken to be zero, since a positive condition must be observed.

The neurons of the input layer of the network are formed by airspeed V, pitch velocity q, angle of attack α, elevator deflection $\delta_e$, and sideslip angle β. Each of the network neurons of the output layer, $\hat{f}_j(x,w)$, j = 1, 2, …,m is formed by ordinary polynomials:

$$CA_j = \sum_{k=1}^{p} w_{jk}h_k(x), \qquad j = 1,2,\ldots, m \tag{4.164}$$

For example, the lift force coefficient $C_L$ may be represented as a function of angle of attack, airspeed, pitch velocity, elevator deflection:

$$C_L = C_L\left(\alpha, q, V, \delta_e, \ldots\right) = w_{10} + w_{12}\Delta\alpha + w_{13}\Delta\delta_e + w_{14} q \Delta \frac{c_w}{2V} + \ldots + O_s \quad (4.165)$$

where the network parameters are represented by the aerodynamic coefficients, giving physical meaning to the network parameters (weights). This same model is used for the other coefficients such as drag coefficient $C_D$ and pitching moment coefficient $C_m$.

An important aspect of this estimation technique is the modelling of the noises inherent to the processes. For the parameters to be estimated, the initial covariance can be set as $P_{jk} \approx \left(\frac{w_{jk}}{4}\right)^2$. These values may be subject to some adjustment after the first estimates, and as guidelines the residuals are used. For parameters with initial values equal to or close to zero, the values are established based on engineering judgment. For the observation noise, the criterion adopted is the use of an expected standard deviation of the estimation [14].

The AFLN algorithm can be applied to identification and control problems. It provides parameter estimation in near real time, even with initial guesses far from reality. The algorithm is simple enough to be implemented as a function and included in the flight control computer, so the estimation process can be conducted in near real time during flight tests for certification, sensor/actuator calibration, and stability and control coefficient calculation. The adaptive procedure ensures good numerical behaviour and distributed learning capacity, making it superior to the standard FLN approach, particularly when initial parameter estimates are poor or when processing large datasets.

## 4.7 ANALYTICAL CONVERGENCE RESULTS BASED ON LYAPUNOV ENERGY FUNCTIONAL

Convergence analysis provides crucial insights into the stability and reliability of parameter estimation algorithms. A stable algorithm that converges consistently across various datasets and initial conditions is essential for obtaining robust parameter estimates, particularly in the presence of noise or uncertainties in measured data. The Lyapunov stability approach provides rigorous assurance for system convergence under any initial state condition. The attainment of Lyapunov stability depends upon the existence of a strictly positive definite Lyapunov energy function and a strictly negative definite derivative of this function. The Lyapunov-Krasovskii approach is employed to establish conditions for algorithm convergence in both feedforward and recurrent neural network-based parameter estimation methods [15].

In general, consider the error state described as:

$$e(\kappa+1) = Ge(\kappa) \quad (4.166)$$

where $e(\kappa)$ represents the estimation error at discrete time instant $\kappa$ and G is a matrix characterizing the error dynamics. A viable normalized Lyapunov energy function is defined as:

$$V(e(k)) = \frac{1}{2} e^T(k) e(k) \tag{4.167}$$

This expression represents a positive definite functional. The discrete-time equivalent of the derivative of the Lyapunov energy function is:

$$\nabla\{V(e(k))\} = V(e(k+1)) - V(e(k)) \tag{4.168}$$

Expanding this expression:

$$\nabla\{V(e(k))\} = \frac{1}{2} e^T(k+1) e(k+1) - \frac{1}{2} e^T(k) e(k)$$

$$= \frac{1}{2} e^T(k) G^T G e(k) - \frac{1}{2} e^T(k) e(k)$$

$$= \frac{1}{2} e^T(k) [G^T G - \mathrm{I}] e(k) \tag{4.169}$$

This can be rewritten as:

$$\nabla\{V(e(k))\} = -\frac{1}{2} e^T(k) [\mathrm{I} - G^T G] e(k) \tag{4.170}$$

For convergence of the error dynamics, the following condition must be satisfied:

$$|1 - G^T G| > 0 \tag{4.171}$$

which implies:

$$\|G^2\| < 1 \tag{4.172}$$

This fundamental result establishes that the error dynamics converge asymptotically to zero if the squared norm of the error propagation matrix G is bounded below unity. The specific form of G depends on the neural network architecture and training algorithm employed.

### 4.7.1 Convergence Analysis for Parameter Estimation Using FFNN

For feedforward neural network-based parameter estimation with backpropagation, let e(κ) represent the network error, defined as the difference between the desired output d(κ) and the actual network output y(κ). The network output at time κ+ 1 depends on the updated weights. The weights connecting the hidden layer to the output layer are updated according to:

$$w_{2,\kappa+1} = w_{2,\kappa} + \mu e_{2,\kappa} v_{1,\kappa}^T + \Omega\left(\Delta w_{2,\kappa}\right) \tag{4.173}$$

where μ is the learning rate, Ω is the momentum constant, $e_{2,\kappa}$ is the backpropagated error at the output layer, and $v_{1,\kappa}$ is the hidden layer output. Similarly, the weights connecting the input layer to the hidden layer are updated as:

$$w_{1,\kappa+1} = w_{1,\kappa} + \mu e_{1,\kappa} v_{0,\kappa}^T + \Omega\left(\Delta w_{1,\kappa}\right) \tag{4.174}$$

where $e_{1b,\kappa}$ is the backpropagated error at the hidden layer and $v_{0,\kappa}$ is the input vector. The backpropagated errors are computed as:

$$e_{2,\kappa} = f_1'\left(y_{2,\kappa}\right) e_\kappa$$

$$e_{1,\kappa} = f_2'\left(y_{1,\kappa}\right) w_{2,\kappa+1}^T e_{2,\kappa} \tag{4.175}$$

where $f_1'$ and $f_2'$ denote the derivatives of the activation functions at the hidden and output layers, respectively.

The error dynamics can be analyzed by examining how the error propagates from one time step to the next. Substituting the weight update equations and simplifying, the error evolution can be expressed in terms of the network parameters and activations. To make the analysis tractable, bounds on the norms of relevant quantities are introduced:

$$\|e\|_\kappa \le \varepsilon;\ \|f'(y)\| \le h;\ \|w\| \le \omega;\ \|v\| \le \upsilon \tag{4.176}$$

Through algebraic manipulation and applying these bounds with the momentum parameter set to Ω = 1, the error propagation matrix G can be characterized. After simplification, it can be shown that G satisfies:

$$G = \left|\rho\varepsilon - \frac{f_{max}}{\varepsilon}\right| \tag{4.177}$$

where $f_{max}$ represents the maximum value of terms involving the weight updates, activation function derivatives, and layer outputs, specifically:

$$f_{max} = \max\left\|\left(\omega k + \mu h \upsilon \varepsilon \times f\left(\omega \upsilon k + \mu h^2 \upsilon^2 \varepsilon\left[\omega k + \mu h \upsilon \varepsilon\right]\right)\right)\right\| \tag{4.178}$$

and ρε represents the product of bounding parameters. Applying the convergence condition from equation (4.172), we require $\|G\|^2 < 1$, which after taking the positive square root yields:

$$\rho\varepsilon < 1 + \frac{f_{max}}{\varepsilon} \tag{4.179}$$

This condition ensures asymptotic stability of the feedforward network training algorithm with backpropagation. The condition indicates that the scaled error magnitude must be bounded by the baseline error plus contributions from the network dynamics.

For the feedforward network trained with the recursive least squares (RLS) algorithm, the weight updates incorporate Kalman filter equations for improved convergence. The covariance matrix P evolves according to:

$$P(k+1) = \frac{P(k) - K(k+1)v_1(k)P(k)}{f_1} \tag{4.180}$$

where $f_1$ is a forgetting factor and the Kalman gain is computed as:

$$K(k+1) = P(k)v_1(k)\left(f_1 + v_1^T(k)P(k)v_1(k)\right)^{-1} \tag{4.181}$$

The covariance matrix is bounded by:

$$p_l I \le P(.) \le p_u I \tag{4.182}$$

Using these bounds along with $\|G^T G\| \le g^2; \|e_k\|^2 \le \varepsilon^2$, and $\|v_1\| = \|f(y)\| = y_m$ where $y_m$ is the maximum output of the activation function, the change in the Lyapunov function can be bounded as:

$$\nabla\{V(e(\kappa))\} = \varepsilon^2\left[g^2\frac{f_1 + y_m^2 p_l}{p_l} - \frac{1}{p_l}\right] \tag{4.183}$$

For convergence, we require $\nabla\{V(e(\kappa))\} < 0$, which leads to:

$$g^2\left(f_1 + y_m^2 p_l\right) < 1 \tag{4.184}$$

This condition ensures convergence of the recursive least squares-based feedforward network algorithm. All coefficients in these stability conditions are well-defined positive bounding constants that can be determined from the network architecture and training parameters [15].

### 4.7.2 Convergence Analysis for Parameter Estimation Using RNNs

For recurrent neural network-based parameter estimation, two distinct approaches are analyzed: the weight-bias method based on Hopfield neural networks and the gradient-based method. In the weight-bias method, the network dynamics are governed by:

$$\dot{x}_i(t) = \sum_{j=1}^{n} w_{ij}\beta_j(t) + b_i \tag{4.185}$$

where $\beta_j = f(x_j)$ represents the parameters to be estimated with nonlinear activation applied to the network states, $w_{ij}$ are the connection weights, and $b_i$ are the bias terms. The cost function to be minimized is:

$$E(\beta) = \frac{1}{2}\sum_{k=1}^{N} e^T(\kappa)e(\kappa) \tag{4.186}$$

where $e(\kappa) = \dot{x} - Ax - Bu$ is the equation error. From optimization theory, the gradient of the cost function with respect to the parameters satisfies:

$$\frac{\partial E}{\partial \beta_i} = -\dot{x}_i \tag{4.187}$$

The parameter update dynamics follow from the relationship $\beta_j = f(x_j)$ and the inverse function theorem, yielding:

$$\dot{\beta}_i = \frac{\lambda(\rho^2 - \beta_i^2)}{2\rho}\sum_{j=1}^{n} w_{ij}\beta_j + b_i \tag{4.188}$$

where λ is the sigmoid magnitude parameter and ρ is the sigmoid slope parameter. The error dynamics can be expressed in discrete time as:

$$e(\kappa+1) = e(\kappa) - \frac{\lambda(\rho^2 - \beta_i^2)}{2\rho e(\kappa)}(W\beta + b)^2 \tag{4.189}$$

When the following bounds are applied $\|e_\kappa\| \le \varepsilon;\ \|\beta\| \le l;\ \|W\| \le w$, the error propagation matrix can be identified as:

$$G = 1 - \frac{\lambda(\rho^2 - l^2)}{2\rho\varepsilon^2}(wl + b)^2 \tag{4.190}$$

Applying the convergence condition $\|G\|^2 < 1$ and taking the positive square root leads to the requirement:

$$\frac{\lambda(\rho^2 - l^2)}{2\rho\varepsilon^2}(wl+b)^2 > 0 \tag{4.191}$$

Since all terms are positive by definition, the final condition for stability reduces to:

$$\rho^2 > l^2 \tag{4.192}$$

This condition states that the magnitude of the sigmoid activation function must exceed the magnitude of the parameters. This is intuitively reasonable as the nonlinear activation acts as a band-pass filter for the parameters while providing saturation for large parameter variations and noise.

For the RNN gradient-based method, the nonlinearity is applied directly to the equation error rather than to the network states. The parameter update follows a gradient descent rule:

$$\beta_{\kappa+1} = \beta_{\kappa} - \mu\frac{\partial V(e(\kappa))}{\partial\beta} \tag{4.193}$$

where μ is the step size or learning rate. The gradient is computed as:

$$\frac{\partial E}{\partial\beta} = f(e(k))\frac{\partial f(e(k))}{\partial\beta} \tag{4.194}$$

where $f$ is the sigmoid activation function applied to the error. The derivative of the error with respect to the parameters can be expressed as a combination of weights, inputs, and biases. The parameter update becomes:

$$\Delta\beta = -\frac{2\mu\gamma\lambda\rho e^{-\lambda e(k)}}{\left(1+e^{-\lambda e(k)}\right)^2}(Wv+b) \tag{4.195}$$

where $f(e(k)) = \gamma$ and $v$ represents the input vector. The error dynamics evolve as:

$$e(k+1) = e(k) - \frac{2\mu\gamma\lambda\rho e^{-\lambda e(k)}}{\left(1+e^{-\lambda e(k)}\right)^2}(Wv+b)^2 \tag{4.196}$$

To establish convergence conditions, the following bounds are required:

$$\|e(k)\| \le \varepsilon,\ \|W\| \le w,\ \|\gamma\| \le \rho, \left\| \frac{e^{-\lambda e_i}}{\left(1+e^{-\lambda e_i}\right)^2} \right\| \le \frac{1}{4},\ \|b\| \le \delta \tag{4.197}$$

The last bound follows from the maximum value of the sigmoid derivative. Applying these bounds, the error propagation matrix simplifies to:

$$\mathrm{G} = 1 - \frac{\mu\lambda\rho^2}{2\varepsilon}\left(wv+\delta\right)^2 \tag{4.198}$$

Applying the convergence condition $G^2 < 1$ and simplifying yields the stability requirement:

$$\frac{\mu\lambda\rho^2}{2\varepsilon}\left(wv+\delta\right)^2 > 0 \tag{4.199}$$

Since all terms in this expression are positive, the convergence of the RNN gradient-based parameter estimation method is assured provided the learning rate μ is chosen appropriately. This condition is less restrictive than that of the weight-bias method, as it does not impose a relationship between the activation magnitude and parameter bounds [15].

The convergence conditions derived for both FFNN and RNN algorithms provide theoretical foundations for their stability and reliability in parameter estimation applications. All coefficients appearing in these conditions are well-defined positive bounding constants that can be determined from the network architecture, activation functions, and training parameters. If these conditions are satisfied, the time derivative of the Lyapunov energy function will be negative definite, ensuring that the parameter estimation algorithms converge asymptotically to the true parameter values.

## EXERCISES

1. Explain the fundamental structure common to most recursive filtering algorithms and discuss why this structure eliminates the need to store past measurements.
2. In the context of Kalman filtering, describe the roles of the process noise covariance (Q) and measurement noise covariance (R), and explain how they influence the behaviour of the filter.
3. Describe the fundamental trade-off between network size and generalization capability, and explain how pruning methods that start with large networks differ from approaches that attempt to select optimal size initially.
4. Why is the Delta method suitable for extracting system parameters using a trained feedforward neural network, and how does it relate to the fundamental definition of a derivative?

5. Explain the structural advantages that partially recurrent networks (Elman and Jordan) offer over fully recurrent networks for dynamic system identification, and describe how the choice between feeding back hidden states versus outputs affects the learning process.

## REFERENCES


1. Raol, J.R., Girija, G., and Singh, J., *Modelling and Parameter Estimation of Dynamic Systems. IEE/IET Control Series* Vol. 65, The Institution of Electrical Engineers (IEE), London, UK, Aug. 2004.
2. Astrom, K.J., "Maximum Likelihood and Prediction Error Methods", *Automatica*, Vol. 16, pp. 551–574, 1980.
3. Maine R.E., and Iliff, K.W., *Application of Parameter Estimation to Aircraft Stability and Control-the Output Error Approach, NASA RP-1168,* USA, 1986.
4. Gelb, A. (Ed.), *Applied Optimal Estimation,* MIT Press, Massachussetts, USA, 1974.
5. Raol, J.R., and Madhuranath, H., "Neural Network Architectures for Parameter Estimation of Dynamical Systems", *IEE Proceedings-Control Theory and Applications*, Vol. 143, No. 4, pp. 387–394, 1996.
6. Parlos, A.G., Menon, S.K., and Atiya, A.F., "An Algorithmic Approach to Adaptive State Filtering Using Recurrent Neural Networks", *IEEE Transactions on Neural Networks*, Vol. 12, No. 6, pp. 1411–1432, 2001.
7. Pham, D.T., and Liu, X., "Dynamic System Modelling Using Partially Recurrent Neural Networks", *Journal of Systems Engineering*, Vol. 2, No. 2, pp. 90–97, 1992.
8. Narendra, K.S., and Parthasarathy, K., "Identification and Control of Dynamic Systems Using Neural Networks", *IEEE Transactions on Neural Networks*, Vol. 1, No. 1, pp. 4–27, 1990.
9. Raisinghani, S.C., and Ghosh, A.K., "Parameter Estimation of an Aeroelastic Aircraft Using Neural Networks", *Sadhana*, Vol. 25, Part 2, pp. 181–191, April 2000.
10. Van de Ven, P.W.J., Johansen, T.A., Sørensen, A.J., Flanagan, C., and Toal, D., "Neural Network Augmented Identification of Underwater Vehicle Models", *Control Engineering Practice,* Vol. 15, pp. 715–725, 2007.
11. O'Reilly, P., and Thompson, S., "A Neural Network Air-Fuel Ratio Estimator", In *CONTROL'94* (pp. 166–172). IEE Conference Publication No. 389, 1994.
12. Reed, R., "Pruning Algorithms – A Survey", *IEEE Transactions on Neural Networks*, Vol. 4, No. 5, pp. 740–747, Sept. 1993.
13. Zamarreno, J.M., and Vega, P., "State Space Neural Network", *Properties and Application, Neural Networks*, Vol. 11, pp. 1099–1112, 1998.
14. Rios Neto, W., Curvo, M., and Góes, L.C.S., *"Aircraft Parameter Estimation Using Adaptive Functional Link Network"*, Proceedings of COBEM 2005 18th International Congress of Mechanical Engineering, ABCM, Ouro Preto, MG, Brazil, Nov. 6-11, Paper No. COBEM2005-2047, 2005.
15. Raol, J.R., Parimala, P., Reshma V., and George, S.M. *Advances in State and Parameter Estimation: Theory and Practice* (1st ed.), CRC Press, Florida, USA, 2025.

# 5 Fuzzy Logic-based Approaches

## 5.1 INTRODUCTION

Fuzzy logic (FL)-based methods have been widely applied in industrial control systems, home appliances, robotics, and aerospace engineering. A conventional control system can be regarded as a minimally intelligent system if it employs fuzzy logic or some other logic, and/or incorporates a learning mechanism. A conditional statement such as "if y exceeds a certain value, use one controller, else use another" is itself a form of logic. An AI-based system, however, must possess strong logic capability and very good learning and adaptive mechanisms.

An FL-based controller is well suited to maintain output variables and control actuation within prescribed limits, making FL one form of logic that can serve as a component in intelligent control systems. FL deals with vagueness rather than uncertainty. Since FL-based systems rely on a specified rule base, human intervention is sometimes required to tune adjustable parameters and avoid contradictions in the rules. FL-based control is most appropriate when the control system dynamics are slow and/or nonlinear, when mathematical models of the plant are unavailable, and when competent human operators are available to provide expert operational rules [1], [2].

## 5.2 CLASSICAL SET/CRISP LOGIC

Classically crisp logic (in decision and related analysis) is used. The theory of probability is based on this logic, via classical set theory. In crisp logic there are only two discrete states: Yes or No, 0 or 1, -1 or +1, and 'off' or 'on'; these are called binary or bipolar states, Figure 5.1.

A crisp logic/membership function is defined as:

$$\mu_A(u) = \begin{cases} 1 & if\ u \in A \\ 0 & if\ u \notin A \end{cases} \tag{5.1}$$

Here, $\mu_A(u)$ is a membership function that characterizes the elements $u$ of set $A$. There is no third possibility here: either a person is in the room or not, an event

DOI: 10.1201/9781003616580-5

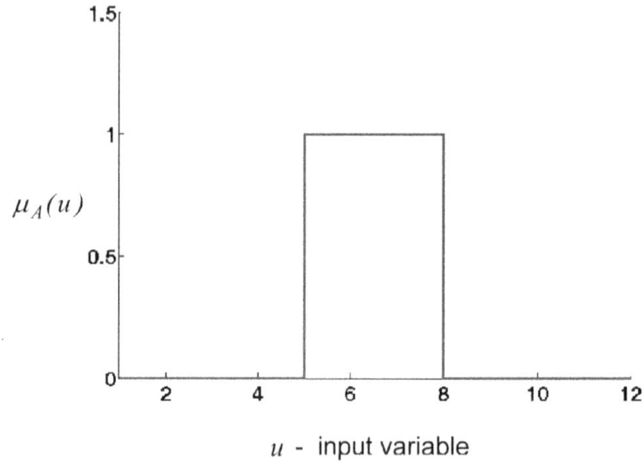


**FIGURE 5.1** Crisp set logic-Membership function

would occur or not, a light bulb is on or off, etc. This set/logic has a limitation that the information is quantized; it does not tell the state of the light if its intensity is either too high or too low, intermediate states are not defined, and not given any membership values. Despite this limitation, engineers and scientists have used the probability theory and modelled several real-life dynamic systems, and used these mathematical models for analysis, forecasting/predictions, and design and development of control systems. For the sake of uniformity this crisp logic is often called fuzzy logic type '0', and the usual fuzzy logic is called FL Type 1, because there is FL Type 2 also.

The most elementary and well known crisp set operations/operators like the intersection, the union, and the complement are represented by AND, OR, and NOT, respectively (in computer logic/Boolean logic/algebra). Let $A$ and $B$ be two subsets of $U$. The intersection of these subsets denoted by $A \cap B$ (this itself is a new resultant set) contains all the elements that are common in/between $A$ and $B$ i.e. $\mu_{A \cap B}(u) = 1$ if $u \in A$ AND $u \in B$.

The union of $A$ and $B$, denoted $A \cup B$, contains all elements in either in $A$ or $B$; i.e., $\mu_{A \cup B}(u) = 1$ if $u \in A$ **or** $u \in B$. The complement of $A$ denoted by $\bar{A}$ contains all the elements which are not in $A$, i.e. $\mu_{\bar{A}}(u) = 1$ if $u \notin A$ and $\mu_{\bar{A}}(u) = 0$ if $u \in A$.

## 5.3 MULTIVARIATE LOGIC

The real-life experiences tell us that many more than these two conditions of crisp logic are possible: the light could be dim(mer), day could be brighter with a certain degree of brightness, day could be cloudy with a certain degree of darkness, and weather could be warm, hot, more hot, or cold, too cold, etc. This calls for variation in the degree of uncertainty (called vagueness) and hence the truth and falsity (1 or 0 of the crisp logic) are the extremes of a continuous spectrum of this vagueness. This leads to a multi-valued logic and hence to FL – a theory of sets if the characteristics function is generalized to take an infinite number of values between 0 and 1 (e.g., a triangular form). In fact, the theory of possibility is based on FL (via fuzzy sets and its mathematics), in a similar way as the theory of probability is based on the crisp logic (via classical set theory).

The development of a FL system requires: i) selecting fuzzy sets and their membership functions for fuzzification process; ii) creating a rule base, with the help of domain expert, for inputs-output mapping; iii) selecting fuzzy operators in fuzzy implication and aggregation process; iv) selecting a fuzzy implication method and aggregation method; and v) selecting a defuzzification method.

### 5.3.1 Fuzzy Logic Type 1

In principle one can say that a fuzzy set is an internal extension/expansion of a crisp set, within the limits of the crisp set, hence the FL is richer than a crisp set/logic. A crisp set allows only full membership (1, an event occurred) or no membership at all (zero, event did not occur), whereas a fuzzy set allows a partial membership, of varying degree, of a member of a set (belonging to the set).

A fuzzy set A on a universe of discourse $U$ with elements $u$ is expressed as

$$A = \int \{\mu_A(u)/u\} \quad \forall\, u \in U \tag{5.2}$$

$$\text{or } A = \sum \{\mu_A(u)/u\} \quad \forall\, u \in U \tag{5.3}$$

Here, $\mu_A(u)$ is a membership function (fuzzy membership function/FMF) of $u$ on the set $A$ and provides a mapping of universe of discourse $U$ on the closed interval $[0,1]$. The $\mu_A(u)$, see Figure 5.2, is simply a measure of the degree to which or by which $u$ belongs to set $A$, i.e. $\mu_A(u): U \rightarrow [0,1]$. It should be noted here that the notation $\int$ and $\sum$ represent just a fuzzy set and do not have anything to do with usual

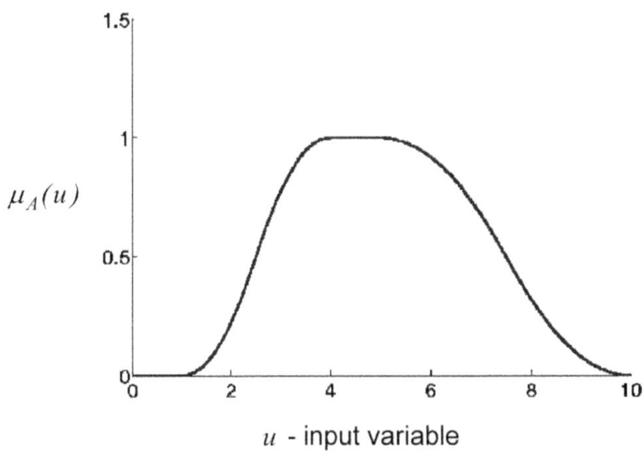


**FIGURE 5.2** A typical membership function of a fuzzy set

integration and summation interpretations. Fuzzy variable is a name of the element $u$ whose values can be considered labels of fuzzy sets, for example, TEMPERATURE as a fuzzy variable (or linguistic variable) with a numeric value ($u$) can be in the range of $U = [0,100]$. The fuzzy variable can take on different labels defined by linguistic values, such as LOW, MEDIUM, NORMAL, and HIGH, with each represented by different membership functions.

There are several types of fuzzy membership functions: Sigmoid, Gaussian (of normal probability density function-shape), Triangular, Trapezoidal, S-shaped, Π (Pi)-shaped, and $Z$-shaped.

#### 5.3.1.1 Fuzzy Logic Operators

As in the conventional crisp logic, the FL operations are also defined: the corresponding (to AND, OR, NOT), and the operators *min*, *max*, and *complement* are defined as:

$$\mu_{A\cap B}(u) = \min\left[\mu_A(u), \mu_B(u)\right] \quad \text{(intersection)} \tag{5.4}$$

$$\mu_{A\cup B}(u) = \max\left[\mu_A(u), \mu_B(u)\right] \quad \text{(union)} \tag{5.5}$$

$$\mu_{\bar{A}}(u) = 1 - \mu_A(u) \quad \text{(complement)} \tag{5.6}$$

Another way to define *AND* and *OR* operators in FL is:

$$\mu_{A\cap B}(u) = \mu_A(u)\mu_B(u) \tag{5.7}$$

$$\mu_{A\cup B}(u) = \mu_A(u) + \mu_B(u) - \mu_A(u)\mu_B(u) \tag{5.8}$$

Any consistent definition is possible in defining basic FL operations.

### 5.3.2 Fuzzy Inference Engine/system (FIE/S)

Fuzzy logic-based systems are rule-based expert systems, and are considered as one of the subcomponents of AI.

Define a fuzzy rule as 'IF $u$ is $A$, THEN $v$ is $B$'. The IF part of the rule 'IF $u$ is $A$' is called the *antecedent* or *premise*. THEN part of the rule, 'THEN $v$ is $B$' is called the *consequent* or conclusion part. The core is FIE/S that, via fuzzy implication operation, defines mapping from input fuzzy sets (variables) into output fuzzy sets. It determines the degree to which the *antecedent* is satisfied for each rule. If the *antecedent* of a given rule has more than one clause (e.g., 'IF $u_1$ is $A_1$ AND $u_2$ is $A_2$, THEN $v$ is $B$'), fuzzy operators (T-norm/S-norm) are applied to obtain one number that represents the result of the *antecedent* for that rule. FIS can take different forms depending on the manner in which inference-rule is defined. It is also possible that one or more rules may fire at the same time; in such a case, outputs of all rules are then *aggregated*, i.e.,

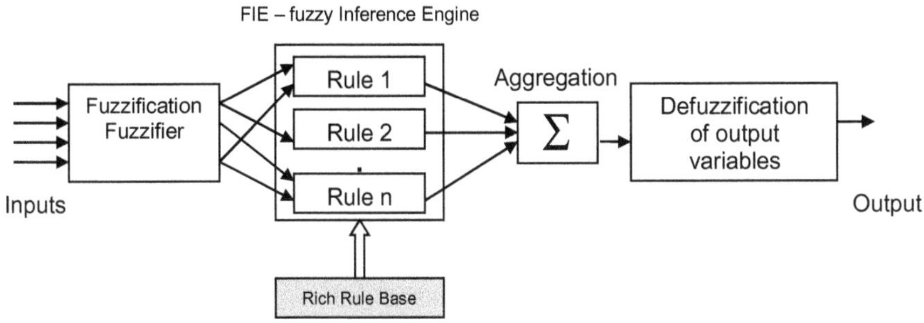


**FIGURE 5.3** FIS process

fuzzy sets that represent the output of each rule are combined into single fuzzy set. An important aspect of FIS is that fuzzy rules are fired in parallel/concurrently and the order in which firing occurs does not affect the output. Figure 5.3 shows a schematic of a FIS for multi-input/single-output system.

The fuzzifier/fuzzification maps input values into corresponding memberships which is very much essential to activate rules that are in terms of linguistic variables, i.e., the fuzzifier takes input values and determines the degree to which these numbers belong to each of fuzzy sets via membership functions.

The rule base contains linguistic rules that are provided by domain/human experts, and these rules are based on the past experience from real-life operational dynamic systems.

### 5.3.3 Fuzzy Implication Functions/methods

Due to approximate reasoning capability, FL can become an ideal tool to develop various applications that would require a logical reasoning to model imprecisely defined events. The core of FL-based system is FIS in which the fuzzy 'If…Then' rules are processed using fuzzy implication methods (FIM) to finally obtain output as fuzzy sets. A FIM plays a crucial role in the successful design of the FIS/FLS. It becomes necessary to select an appropriate FIM from the existing methods. If there is any new implication method, then it should satisfy some of the intuitive criteria of GMP/GMT so that this FIM can be fitted into the logical development of any system using FL.

In any FIS, fuzzy implication provides mapping between input and output fuzzy sets so that fuzzified inputs can be mapped to desirable output fuzzy sets. Basically, a fuzzy IF-THEN rule, provided by a domain expert, is interpreted as a fuzzy implication. Consider a simple rule such as:

$$\text{IF } u \text{ is } A, \text{ THEN } v \text{ is } B \tag{5.9}$$

Here, 'IF $u$ is $A$' is known as antecedent or premise and 'THEN $v$ is $B$' is consequent part of the Fuzzy rule. Crisp variable $u$, fuzzified by set $A$ in universe of discourse $U$, is an input to an inference engine whereas crisp variable $v$, represented by set $B$ in universe of discourse $V$, is an output from an inference engine. The fuzzified output of an inference engine, computed using sup-star composition is given by:

$$B = RoA \tag{5.10}$$

Here, '$o$' is a compositional operator and '$R$' is a Fuzzy relation in the product space $UxV$. Eqn. (5.10) in the form of the membership function is given by

$$\mu_B(v) = \mu_R(u,v) o \mu_A(u) \tag{5.11}$$

A Fuzzy implication, denoted by $\mu_{A\to B}(u,v)$, is also a kind of relation that provides mapping between input and output; hence, eqn. (5.11) can be re-written as

$$\mu_B(v) = \mu_{A\to B}(u,v) o \mu_A(u) \tag{5.12}$$

There are seven standard ways/interpretation of the Fuzzy IF-THEN rule to define the fuzzy implication process [1]:

i) Fuzzy conjunction (FC)

$$\mu_{A\to B}(u,v) = \mu_A(u) * \mu_B(v) \tag{5.13}$$

Here, '*' represents T-norm operator. This operator is a general symbol of fuzzy 'AND' function or operation.

ii) Fuzzy disjunction (FD)

$$\mu_{A\to B}(u,v) = \mu_A(u) \dot{+} \mu_B(v) \tag{5.14}$$

Here, '$\dot{+}$' represents S-norm operator. This operator is a general symbol of fuzzy 'OR' function or operation.

iii) Material implication (MI)

$$\mu_{A\to B}(u,v) = \mu_{\bar{A}}(u) \dot{+} \mu_B(v) \tag{5.15}$$

Here, $\mu_{\bar{A}}(u)$ is a *Fuzzy complement* of $\mu_A(u)$. It can be seen that the S-norm operator symbol is used here.

iv) Propositional calculus (PC)

$$\mu_{A\to B}(u,v) = \mu_{\bar{A}}(u) \dot{+} \mu_A(u) * \mu_B(v) \tag{5.16}$$

It is noted here that PC utilizes both the operators 'AND' and 'OR', i.e. it uses T-norm and S-norm.

v) Extended propositional calculus (EPC)

$$\mu_{A\to B}(u,v)=\mu_{\bar{A}}(u)x\mu_{\bar{B}}(v)\dot{+}\mu_B(v) \tag{5.17}$$

The EPC utilizes complement and S-norm.

vi) Generalization of modus ponens (GMP)

$$\mu_{A\to B}(u,v)=\sup\left\{c\in[0,1],\mu_A(u)*c\le\mu_B(v)\right\} \tag{5.18}$$

vii) Generalization of modus tollens (GMT)

$$\mu_{A\to B}(u,v)=\inf\left\{c\in[0,1],\mu_B(v)\dot{+}c\le\mu_A(u)\right\} \tag{5.19}$$

By employing different combinations of T-norms and S-norms [1], one can get a variety of ways to interpret the fuzzy IF-THEN rules, i.e., a number of different fuzzy implications can be derived. However, not all fuzzy implications or interpretations might satisfy completely the intuitive criteria of GMP/GMT. The most commonly used specific fuzzy implications which satisfy one or more of these intuitive criteria are:

i) Mini-operation rule of fuzzy implication (MORFI–Mamdani) is derived by applying standard intersection (SI) operator of T-norms:

$$R_{MORFI}=\mu_{A\to B}(u,v)=\min\left(\mu_A(u),\mu_B(v)\right) \tag{5.20}$$

ii) Product-operation rule of Fuzzy implication (PORFI–Larsen) is derived by applying algebraic product (AP) operator of T-norms:

$$R_{PORFI}=\mu_{A\to B}(u,v)=\mu_A(u)\mu_B(v)) \tag{5.21}$$

iii) Arithmetic rule of Fuzzy implication (ARFI–Zadeh/Lukasiewicz) is derived by using bounded sum (BS) operator of S-norms and the complement operator:

$$\begin{aligned}R_{ARFI}=\mu_{A\to B}(u,v)&=\mu_{\bar{A}}(u)\dot{+}\mu_B(v)\\&=\min\left(1,\mu_{\bar{A}}(u)+\mu_B(v)\right)\\&=\min\left(1,1-\mu_A(u)+\mu_B(v)\right)\end{aligned} \tag{5.22}$$

iv) Max-min rule of Fuzzy implication (MRFI–Zadeh) is derived by using standard intersection (SI) operator of T-norms, standard union (SU) operator of S-norms, and fuzzy complement operator:

$$R_{MRFI} = \mu_{A\to B}(u,v) = \mu_{\bar{A}}(u) \dot{+} \mu_A(u) * \mu_B(v)$$
$$= \max\left(\mu_{\bar{A}}(u), \mu_A(u) * \mu_B(v)\right)$$
$$= \max\left(1 - \mu_A(u), \mu_A(u) * \mu_B(v)\right)$$
$$= \max\left(1 - \mu_A(u), \min\left(\mu_A(u), \mu_B(v)\right)\right) \quad (5.23)$$

v) Standard sequence of Fuzzy implication (SSFI) is derived by using bounded difference or product (BD) operator of T-norms:

$$R_{SSFI} = \mu_{A\to B}(u,v) = \sup\left\{c \in [0,1], \mu_A(u) * c \le \mu_B(v)\right\}$$
$$= \sup\left\{c \in [0,1], \max\left(0, \mu_A(u) + c - 1\right) \le \mu_B(v)\right\}$$

$$= \begin{cases} 1 \ \textit{if} \ \mu_A(u) \le \mu_B(v) \\ 0 \ \textit{if} \ \mu_A(u) > \mu_B(v) \end{cases} \quad (5.24)$$

vi) Boolean rule of Fuzzy implication (BRFI) is derived by using standard union (SU) operator of S-norms and fuzzy complement operator:

$$R_{BRFI} = \mu_{A\to B}(u,v) = \mu_{\bar{A}}(u) \dot{+} \mu_B(v)$$
$$= \max\left(\mu_{\bar{A}}(u), \mu_B(v)\right)$$

$$= \max\left(1 - \mu_A(u), \mu_B(v)\right) \quad (5.25)$$

vii) Goguen's rule of Fuzzy implication is derived by using 'algebraic product' operator of T-norms:

$$R_{GRFI} = \mu_{A\to B}(u,v) = \sup\left\{c \in [0,1], \mu_A(u) * c \le \mu_B(v)\right\}$$
$$= \sup\left\{c \in [0,1], \mu_A(u) c \le \mu_B(v)\right\}$$

$$= \begin{cases} 1 \ \textit{if} \ \mu_A(u) \le \mu_B(v) \\ \dfrac{\mu_B(v)}{\mu_A(u)} \ \textit{if} \ \mu_A(u) > \mu_B(v) \end{cases} \quad (5.26)$$

Apart from these seven methods there are a few more ways to perform fuzzy implication operations that require a combination of both the T-norms and S-norms.

### 5.3.4 Fuzzy Composition Algorithms

Fuzzy operators T-norm and S-norm can also be used in rule optimization for combining two rules into one [1]. This operation is known as 'Sup (max) – Star (T-norm) composition'. Consider fuzzy sets $A$, $B$, and $C$ with elements $u$, $v$, and $w$ in the UOD $U$, $V$, and $W$ respectively. Consider the following sequential fuzzy conditional rules:

$$R = \text{IF } A \text{ THEN } B \text{ (rule 1) ; } S = \text{IF } B \text{ THEN } C \text{ (rule 2)}$$

Here, $R$ and $S$ are the fuzzy relation in $UXV$ and $VXW$ respectively. It is possible to combine two rules into a single rule by absorbing the intermediate result $B$ and finding the relationship between antecedent and the ultimate consequent directly, i.e.,

$$RoS = \text{IF } A \text{ THEN } C \text{ (new rule)}$$

The composition of $R$ and $S$ is a fuzzy relation denoted by $RoS$ (where $o$ implies *rule composition*), i.e., a fuzzy relation defined on $UXW$. In general a fuzzy composition is defined as

$$RoS = \left\{\left[(u,w), \sup_v (\mu_R(u,v) * \mu_S(v,w))\right], u \in U, v \in V, w \in W\right\} \quad (5.27)$$

Here, sup means supremum and is denoted as $\cup$, i.e., union or max (standard union), and * represents triangular norms by either of i) standard intersection, ii) algebraic product, iii) bounded difference, or iv) drastic intersection. There are different methods to compute composition of fuzzy relations.

#### 5.3.4.1 Method 1: Max-min Composition (Mamdani)

This is defined as:

$$\begin{aligned} \mu_{RoS}(u,w) &= \max_v \min(\mu_R(u,v), \mu_S(v,w)) \\ &= \cup_v (\mu_R(u,v) \cap \mu_S(v,w)) \end{aligned} \quad (5.28)$$

If fuzzy relations $R$ and $S$ are given by relational matrices:

| | | | | | |
|---|---|---|---|---|---|
| | 0.1 | 0.2 | 0.0 | 1.0 | (R :1st row all columns) |
| | 0.9 | 0.2 | 0.8 | 0.4 | (S: 1st column all rows) |
| min | 0.1 | 0.2 | 0.0 | 0.4 | $(\mu_R(u,v) \cap \mu_S(v,w))$ |
| max | 0.4 | | | | $(\cup_v)$ |

After completing the operations for all rows of $R$ with all columns of $S$, $RoS$ can be represented; it should be noted that in eqn. (5.28) (for *min*) the basic definition of standard intersection is used,

#### 5.3.4.2 Method 2: Max-product Composition (Larsen)

This is defined as:

$$\mu_{RoS}(u,w) = \max_v \underbrace{(\mu_R(u,v) \bullet \mu_S(v,w))}_{\text{algebric product}} \tag{5.29}$$

Consider the same example as mentioned in method 1. The Max-Product composition operation for 1st row (all columns) of fuzzy relation $R$ with 1st column (all rows) of fuzzy relation $S$ can be given as:

| | | | | | |
|---|---|---|---|---|---|
| | 0.1 | 0.2 | 0.0 | 1.0 | (R :1st row all columns) |
| | 0.9 | 0.2 | 0.8 | 0.4 | (S: 1st column all rows) |
| • | 0.09 | 0.04 | 0.0 | 0.4 | $(\mu_R(u,v) \bullet \mu_S(v,w))$ |
| max | 0.4 | | | | $(\cup_v)$ |

After completing the operations for all rows of $R$ with all columns of $S$, $RoS$ can be represented; it should be noted that in eqn. (5.29) (for '•'), the basic definition of algebraic product is used.

### 5.3.5 Defuzzification Methods

The defuzzification process is used to obtain a crisp value from/using the fuzzy outputs from an inference engine. Consider the discrete aggregated fuzzy output set $B$:

$$v = [0\ 1\ 2\ 3\ 4\ 5\ 6\ 7\ 8\ 9\ 10\ 11\ 12\ 12.6\ 13.6\ 14.6\ 15.6]$$

$$\mu_B(v) = [0\ 0.2\ 0.4\ 0.4\ 0.4\ 0.6\ 0.8\ 0.8\ 0.8\ 0.6\ 0.6\ 0.6\ 0.6\ 0.6\ 0.4\ 0.2\ 0] \tag{5.30}$$

The different ways to get defuzzified output are as follows:

i) **Centroid method or centre of gravity (CoG) or centre of area (CoA)**
The defuzzifier determines the centre of gravity (centroid) $v'$ of fuzzy set $B$ and uses that value as the output of the fuzzy logic system. For a continuous aggregated fuzzy set, the centroid is given by

$$v' = \frac{\int_S v\mu_B(v)dv}{\int_S \mu_B(v)dv}, \text{ where } S \text{ denotes the support of } \mu_B(v) \tag{5.31}$$

In discrete fuzzy sets, the centroid can be given as

$$v' = \frac{\sum_{i=1}^{n} v(i)\mu_B(v(i))}{\sum_{i=1}^{n} \mu_B(v(i))} \tag{5.32}$$

Using the numerical values from equations (5.29) and (5.30) in equation (5.32) results in:

$$\sum_{i=1}^{n} v(i)\mu_B(v(i)) = 0*0+1*0.2+2*0.4+3*0.4+4*0.4+5*0.6+6*0.8$$
$$+7*0.8+8*0.8+9*0.6+10*0.6+11*0.6+12*0.6$$
$$+12.6*0.6+13.6*0.4+14.6*0.2+15.6*0 = 64.72$$

$$\sum_{i=1}^{n} \mu_B(v(i)) = 0+0.2+0.4+0.4+0.4+0.6+0.8+0.8+0.8$$
$$+0.6+0.6+0.6+0.6+0.6+0.4+0.2+0 = 8$$

$$v' = \frac{64.72}{8} = 8.09$$

**ii) Maximum-decomposition method**

In this method, the defuzzifier examines the aggregated fuzzy set and chooses the output $v$ for which $\mu_B(v)$ is the maximum. In the example case, finding the defuzzified value for discrete output fuzzy set $B$ is not advisable due to maximum value of $\mu_B(v) = 0.8$ is at more than one point i.e. at $v = 6,7,8$.

**iii) Centre of maxima (CoM) or mean of maximum (MoM)**

In a multimode fuzzy region (more than one rule fires), the centre-of-maxima technique finds the highest plateau and then the next highest plateau. The mid-point between the centres of these plateaus is selected as the output of fuzzy logic system (FLS). In this case the defuzzified value for discrete output fuzzy set $B$ can be computed as follows: since the maximum value of $\mu_B(v) = 0.8$ is at$v = 6,7,8$, the defuzzified value computed as mean of maximum is equal to $(6 + 7 + 8)/3 = 7$.

**iv) Smallest of maximum (SoM)**

The smallest value of $v$ at which the aggregated output fuzzy set has the maximum membership grade. Since the maximum value of $\mu_B(v) = 0.8$, is at $v = 6,7,8$, the defuzzified value computed as smallest of maximum is equal to $\min(v = 6,7,8) = 6$.

**v) Largest of maximum (LoM)**

The largest value of $v$ at which the aggregated output fuzzy set has the maximum membership grade. Since the maximum value of $\mu_B(v) = 0.8$ is at $v = 6,7,8$, the defuzzified value computed as largest of maximum is equal to $\max\,(v = 6,7,8) = 8$.

**vi) Height defuzzification**

In this method, the defuzzifier first evaluates $\mu_{Bi}(v_i^{'})$ at $v_i^{'}$ and then computes the output of the FLS given by

$$v' = \frac{\sum_{i=1}^{m} v_i' \mu_{B_i}(v_i')}{\sum_{i=1}^{m} \mu_{B_i}(v_i')} \tag{5.33}$$

Here, m represents the number of output fuzzy sets $B_i$ obtained after fuzzy implication and $v_i^{'}$ represents the centroid of fuzzy region $i$. In the example case, the number of output fuzzy sets is one, i.e., $m = 1$. Hence, the defuzzified output with $m = 1$ yields the value 8.09.

### 5.3.6 Relationships between Fuzzy Logic Operators

An effort has been made to define some interconnectivity and relationships between T-norm and S-norm operators; the various relations between fuzzy logic operators (FLOR) [1] are given as follows:

FLOR 1:

$$T_{BD}(x, y) = \max(0, x + y - 1)$$

$$T_{BD}(x, y) = \max(0, S_{AS}(x, y) + x \bullet y - 1), \text{since } S_{AS}(x, y) = x + y - xy$$

$$T_{BD}(x, y) = \max(0, S_{AS}(x, y) + T_{AP}(x, y) - 1), \text{since } T_{AP}(x, y) = xy \tag{5.34}$$

FLOR 2:

$$S_{BS}(x, y) = \min(1, x + y)$$

$$S_{BS}(x, y) = \min(1, S_{AS}(x, y) + xy)$$

$$S_{BS}(x, y) = \min(1, S_{AS}(x, y) + T_{AP}(x, y)) \tag{5.35}$$

FLOR 3:

$$T_{AP}(x, y) = x + y - S_{AS}(x, y), \text{since } S_{AS}(x, y) = x + y - xy \tag{5.36}$$

FLOR 4:

$$T_{SI}(x,y) = x + y - S_{SU}(x,y) \tag{5.37}$$

$$Assume,\ x = 0.2\ \&\ y = 0.8$$

$$S_{SU}(x,y) = \max(x,y) = \max(0.2, 0.8) = 0.8$$

$$RHS = 0.2 + 0.8 - 0.8 = 0.2$$

$$LHS = T_{SI}(x,y) = \min(x,y) = \min(0.2, 0.8) = 0.2$$

$$LHS = RHS$$

$T_{SI}(x,y) = T_{AP}(x,y) + S_{AS}(x,y) - S_{SU}(x,y)$ (by putting FLOR 3 in FLOR 4)

FLOR 5:

$$T_{BD}(x,y) = x + y - S_{BS}(x,y) \tag{5.38}$$

### 5.3.7 Steps of Fuzzy Inference Process

Consider the i-th fuzzy rule (with more than one part in antecedent) for a MISO (Multi-Inputs Single-Output) system defined as

$$R^i : \text{IF } u \text{ is } T_u^i \text{ AND } v \quad T_v^i \text{ THEN } w \text{ is } T_w^i \tag{5.39}$$

Here, $u$, $v$, and $w$ are the fuzzy or linguistic variables whereas $T_u$, $T_v$, and $T_w$ are their linguistic values (LOW, HIGH, LARGE, etc.). In order to get the crisp output using a Fuzzy Inference System (FIS), the following steps are needed.

Step 1: fuzzify the inputs $u$, and $v$ using membership functions ($\mu^i(u)$, and $\mu^i(v)$) for i-th rule. Basically it means define appropriate FMFs.

Step 2: Since antecedent part of every rule has more than one clause, fuzzy logic operator is used to resolve the antecedent to a single number between 0 and 1 that gives degree of support (or firing strength) for i-th rule. The firing strength can be expressed by

$$\alpha^i = \mu^i(u) * \mu^i(v) \tag{5.40}$$

Here, * represents triangular norm. The most popular T-norms used are standard intersection and algebraic product:

$$\alpha^i = \min(\mu^i(u), \mu^i(v)) \text{ or } \alpha^i = \mu^i(u) \bullet \mu^i(v) \tag{5.41}$$

Step 3: Apply implication method to shape the consequent part (the output fuzzy set) based on the antecedent. The input to the implication process is a single number ($\alpha$) given by the antecedent, and the output is a fuzzy set. Commonly used methods are: i) min-operation rule of fuzzy implication (Mamdani), and ii) product operation rule of fuzzy implication:

$$\mu^i(w)' = \min(\alpha^i, \mu^i(w)) \tag{5.42}$$

$$\mu^i(w)' = \alpha^i \bullet \mu^i(w) \tag{5.43}$$

Step 4: Since more than one rule (i.e., more than one output fuzzy set) can be fired at a time, it is essential to combine the corresponding output fuzzy sets into a single composite fuzzy set; this process of combining is known as *aggregation*. The inputs to the aggregation process are outputs of the implication process and the output of the aggregation process is a single fuzzy set that represents the output variable. The order in which rules are fired does not affect the aggregation process. The most commonly used aggregation method is the *max* (standard union) method. Suppose rule 3 and rule 4 are fired then the composite output fuzzy set could be expressed as

$$\mu(w) = \max(\mu^3(w)', \mu^4(w)') \tag{5.44}$$

Step 5: In order to get crisp value of output variable $w$, a defuzzification process is used. The input to this process is the output from the aggregation process, and the output is a single crisp number.

## 5.4 PROPERTIES OF INTERPRETATIONS OF FUZZY IF-THEN RULES

The GMP and GMT are two ideal inference rules in our day-to-day reasoning or thought process. To compute the consequences (so that they can be compared with the consequences of GMP) when the fuzzy implication methods are applied in the fuzzy inference process, the following formula is used:

$$B' = RoA' \tag{5.45}$$

Here, '$o$' is known as the compositional operator, represented using sup-star with 'sup' as supremum and 'star' as T-norm operator. In the present case standard union or 'max' operator for 'sup' and standard intersection or 'min' operator for 'star' have been used. In terms of membership function eqn. (5.45) is given by

$$\mu_{B'}(v) = \sup_{u \in U} \left\{ \mu_{A \to B}(u,v) * \mu_{A'}(u) \right\} \text{ or}$$

$$y = \mu_{B'}(v) = \sup_{u \in U} \left\{ \min\left[ \mu_{A \to B}(u,v), \mu_{A'}(u) \right] \right\} \tag{5.46}$$

Here, $\mu_{A \to B}(u,v)$ is a fuzzy implication method and $x = \mu_{A'}(u)$ is the premise 1 of GMP, Table 5.1 containing any one of the following:

$$\mu_{A'}(u) = \mu_A(u), \mu_{A'}(u) = \mu^2{}_A(u), \mu_{A'}(u) = \sqrt{\mu_A(u)}, \mu_{A'}(u) = 1 - \mu_A(u) \quad (5.47)$$

Similarly in GMT to compute the consequences (so that they can be compared with consequences of GMT) when the fuzzy implication methods are applied in the fuzzy inference process, the following formula is used;

$$A' = RoB' \quad (5.48)$$

In terms of membership function:

$$\mu_{A'}(u) = \sup_{v \in V} \left\{ \min \left[ \mu_{A \to B}(u,v), \mu_{B'}(v) \right] \right\} \quad (5.49)$$

Here, $\mu_{B'}(v)$ is the premise 1 of GMT, Table 5.2 containing any one of the following:

$$\mu_{B'}(v) = 1 - \mu_B(v), \mu_{B'}(v) = 1 - \mu^2{}_B(v), \mu_{B'}(v) = 1 - \sqrt{\mu_B(v)}, \mu_{B'}(v) = \mu_B(v)$$
$$(5.50)$$

### 5.4.1 Forward- and Backward-chain Logic Criteria

The intuitive criteria of Generalized Modus Ponens (GMP) and Generalized Modus Tollens (GMT) of forward and reverse logic are described in this section. These are two important rules that can be also used in FL for approximate reasoning or inference [1].

#### 5.4.1.1 Generalized Modus Ponens

This is a forward-driven inference rule defined by following the modus operandi [8]:

Premise 1: $u$ is $A'$

Premise 2: IF $u$ is $A$ THEN $v$ is $B$ (5.51)

Consequence: $v$ is $B'$

Here, $A'$ and $A$ are input fuzzy sets. $B'$ and $B$ are output fuzzy sets, and $u$ and $v$ are the variables corresponding to input and output fuzzy sets, respectively. The fuzzy set $A'$ of premise 1 can have values: $A$, *very* $A$, *more or less* $A$, and *not* $A$. The linguistic values 'very' and 'more or less' are hedges and are defined in terms of their membership grade as $\mu(.)^2$ and $\mu(.)^{1/2}$, respectively. Here, (.) denotes fuzzy sets $A$ or $B$.

**TABLE 5.1**
**GMP Intuitive criteria – a forward chain-inference rule**

| GMP Criteria | u is A′ (premise 1) | v is B′ (consequence) |
|---|---|---|
| C1 | u is A | v is B |
| C2-1 | u is very A | v is very B |
| C2-2 | u is very A | v is B |
| C3-1 | u is more or less A | v is more or less B |
| C3-2 | u is more or less A | v is B |
| C4-1 | u is not A | v is unknown |
| C4-2 | u is not A | v is not B |

The criteria of GMP, relating premise 1 and the consequence for any given premise 2, are given in Table 5.1. There are seven criteria under GMP and each can be related to our everyday reasoning. If a fundamental relation between '*u* is *A*' and '*v* is *B*' is not strong in premise 2 then the satisfaction of criterion C2-2 and C3-2 is allowed.

#### 5.4.1.2 Generalized Modus Tollens

This is a backward goal-driven inference rule defined by following procedure [8]:

$$\text{Premise 1}: v \text{ is } B'$$

$$\text{Premise 2: IF } u \text{ is } A \text{ THEN } v \text{ is } B \tag{5.52}$$

$$\text{Consequence: } u \text{ is } A'$$

The values that fuzzy set $B'$ of premise 1 could have are: *not B*, not *very B*, not *more or less B*, and *B*. The linguistic values such as 'not very' and 'not more or less' are known as hedges and are defined in terms of their membership grade as $1-\mu(.)^2$ and $1-\mu(.)^{1/2}$, respectively. The criteria of GMT, relating premise 1 and consequence for any given premise 2, are given in Table 5.2.

## 5.5 FUZZY LOGIC ALGORITHMS AND FINAL COMPOSITION OPERATIONS

A fuzzy logic-based system can model any continuous function or system. The quality of approximation depends on the quality of rules that are formed by an expert. Fuzzy engineering is a function approximation with fuzzy systems, and it rests on the mathematics of function approximation (FA) and statistical learning theory (SLT). The basic unit of the fuzzy algorithm is an 'If then' rule: 'IF the water in the washing machine is dirty, THEN add more detergent powder'. Thus, the fuzzy system is a set of 'IF THEN' rules that maps input sets like 'dirty water' to output sets like 'more

**TABLE 5.2**
**GMT Intuitive criteria – a backward chain inference rule**

| GMT Criteria | v is B′ (premise 1) | u is A′ (consequence) |
|---|---|---|
| C5 | v is not B | u is not A |
| C6 | v is not very B | u is not very A |
| C7 | v is not more or less B | u is not more or less A |
| C8-1 | v is B | u is unknown |
| C8-2 | v is B | u is A |

detergent powder'. Overlapping rules are used to define polynomials- and richer-functions. A set of possible rules would be:

Rule 1: IF the air is cold THEN set the air conditioning motor speed to stop
Rule 2: IF the air is cool THEN set the motor speed to slow
Rule 3: IF the air is just right THEN set the motor speed to medium
Rule 4: IF the air is warm THEN set the motor speed to fast
Rule 5: IF the air is hot THEN set the motor speed to blast

This set gives the first-cut fuzzy system and more rules can be guessed and added by experts or by learning new rules adaptively from training data sets obtained from the system. ANNs can be used to learn the rules from the data. The fuzzy engineering mainly deals with (firstly defining these rules, then) tuning such rules and adding new rules or pruning old rules. In an additive fuzzy system each input partially fires all rules in parallel, and the system acts an associative processor as it computes the output F(x). The FL system then combines the partially fired 'then' part fuzzy sets in a sum and converts this sum to a scalar or vector output. Thus, a match-and-sum fuzzy approximation can be viewed as a generalized AI (artificial intelligent) expert system or as a (neural-like) fuzzy associative memory. The additive fuzzy systems belong to the proven universal approximators for rules that use fuzzy sets of any shape and are computationally simple.

Fuzzy variable's values can be considered labels of fuzzy sets: temperature → fuzzy variable → linguistic values such as low, medium, normal, high, very high, etc. This leads to membership values on the universe of discourse. The dependence of a linguistic variable on another variable is described by means of a fuzzy conditional statement: R: If S1 (is true), then S2 (is true) Or S1 → S2, more specifically 1) IF the load is small THEN torque is very high, and 2) IF the error is negative-large, THEN output is –ve large. A composite conditional statement in a FA would be like R1: IF S1 THEN (IF S2 THEN S3) is equivalent to R1: IF S1 THEN R2 AND R2: IF S2 THEN S3. FA is formed by combining 2 or 3 fuzzy conditional statements: IF speed error is negative large THEN (IF change in speed error is NOT (negative large Or negative medium) THEN change in fuel is positive large),…, Or,..., Or,…, etc.

The composition operation can be realized using various combinations of T-norm and Supremum operators. Composition operation using MAX-SI or MAX-AP combination is easy to implement as compared to the other possible combinations. The accuracy of the final result would mostly depend on composition operation/fuzzy implication functions. Hence, there could be possibility of having better compositional output by logically selecting the different fuzzy implication functions and combining them with appropriate T-norm and supremum operators, requiring further study.

## 5.6 FUZZY LOGIC TYPE 2

We can regard the T1FL (Type I FL) as analogous to a probabilistic system only through the first moment, the mean. Then the so-called T2FL Type II FL is analogous to a probabilistic system via first and second moments [2]. Thus, the T2FL provides an additional design degree of freedom (DOF) in Mamdani and TSK fuzzy logic systems. The provision of this extra DOF is useful when such FL-based systems are used in cases when lots of uncertainties (and that too of different types) are present. In that case the resulting T2FLS can potentially provide better performance than a T1FL. This additional design DOF would be very useful for modelling sensors/communications channels that would be affected by vagueness and impreciseness, and then provide efficient and realistic DF and decision (DeF) fusion systems.

The T2FL can handle uncertainties by modelling them and minimizing their effects on the control system's performance. When these uncertainties disappear T2FL reduces to T1FL. The T2FL consists of the fuzzifier (with T2FL MFs), the inference mechanism/engine (FIS, If...Then...rules, and FIF for operations on fuzzy sets that are characterized by MFs), the Type Reducer (TR), and finally the defuzzification operation (DFO). The O/P (output processor) of a T2FLS has two components: i) T2FSs are transformed into T1FSs by means of TR, and then ii) the T1FS is transformed into a crisp number by means of DFO, i.e., into T0FS.

The T1FL does not directly handle rule uncertainties for it uses T1FSs (that are certain to some extent). T2FL is very useful where it is difficult to determine an exact MF for a FS. Thus, the T2FL can be used to handle the uncertainties in the rules themselves (if so desired) and even measurement uncertainties. As such a T2FL provides more DOF for design than a T1FL, and this is because its T2FSs are described by more parameters than are T1FS (in fact the MFs for the T2FS have one extra DOF, i.e., a third dimension than the T1FL MFs). This aspect itself opens up to new possibilities of representing and modelling more or several types of uncertainties in measurements as well as in the fuzzy rule base, thereby enlarging the scope of the T2FL in applications to MSDF including situational awareness (SA) and wireless sensor networks (WSN).

### 5.6.1 Type 2 and Interval Type 2 Fuzzy Sets

As we have seen that in T1FLS the MFG for each element is a crisp value in the range [0, 1], whereas the T2FL is characterized by a 3D MF and a footprint of uncertainty (FOU) (Figure 5.4), the MFG for each element of this set is a FS in [0, 1] itself. This T2FS is bounded from below by the lower MF (LMF), and from above by the upper

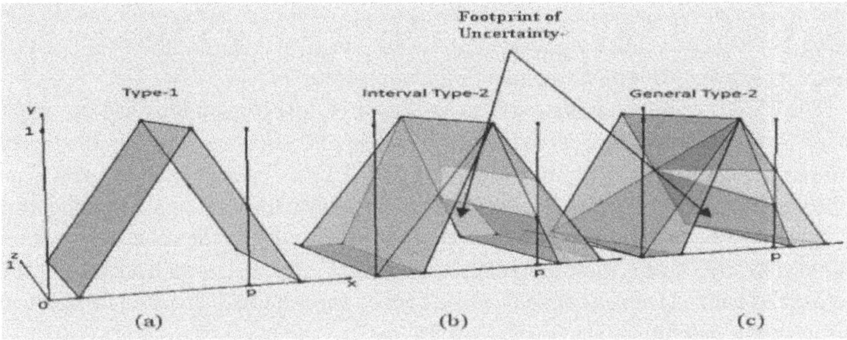


**FIGURE 5.4** Types of FL membership functions: (a) Type-1 FL/FS (T1FL); (b) interval type-2 FL/FS (IT2FL) and FOU; (c) general type-2 FL/FS (T2FL) and FOU (adapted/modified from Raol et al., 2023)

MF (UMF). The area between the LMF and UMF is the FOU. Thus, we see that the new and third dimension of T2FS and the FOU provide additional degrees of freedom that make it possible to directly model and treat the uncertainties. This is the IT2FL that uses the T2FS in either their inputs or outputs and thus provides the potential to have a suitable framework to handle the uncertainties in real-world problems. The T2FSs encompass a large number of T1FSs. We can quickly realize that in T2FLS the DOM (degree of membership) is in itself fuzzy (it has a range of values rather than one value), and is represented by a secondary membership function (SMF). If the SMF is at its maximum of *1* at every point, we call it an interval type 2 set (IT2FS, most people use this IT2FL and hence we continue to denote it as IT2FLS). The SMF in T1FS has only one value in its domain corresponding to the primary membership function/value (PMF/V) at which the SMF grade equals *1*. Thus, in T1FS for each value of $x$, there is no uncertainty associated with the PMV.

In IT2FLSs (still we may use the notation T2FS) as the primary membership takes values within the interval $[a, b]$, where each point in this interval is having an associated SMF of *1*. Hence, there is maximum uncertainty associated with the SMF. However, in the general type 2 fuzzy sets (T2FS per se are not discussed further in the present book), the uncertainty represented by SMF can be modelled with any degree between T1FLSs and IT2FSs, for example by the triangular SMF. Since, the IT2FS-MFs are fuzzy and have a FOU, they can model and handle the linguistic and numerical uncertainties associated with the I/O of the FL control/systems (FLS). The merit is that the use of IT2FS will result in the reduction of the rule base when compared to using T1FS. This is because the uncertainty represented in the FOU in IT2FL allows us to cover the same range as T1FS with a smaller number of labels. The additional DOFs given by the FOU enables IT2FS to produce outputs that cannot be achieved by T1FSs (with the same number of MFs). In fact, a IT2FS gives rise to an equivalent T1 MFG that is negative or larger than unity. Each I/O will be represented by a large number of T1FSs that are embedded in the T2FSs, and hence the use of a large number of T1FSs to describe the I/O variables permits a detailed description of the analytical control. This addition of the extra levels of classification made possible by using T2FS gives a much smoother control response. In addition, the IT2FL can be thought of as a collection of many different embedded T1FSs.

The T2FL is computationally intensive, however, this computation can be simplified a lot by using IT2FL that enables us to design a FLS in a lesser time and with less effort. Hence, the majority of FLSs are based on IT2FL. Traditionally the T1FL is not visualized as a 3D set. However, its conception and visualization in 3D is straightforward when we consider each of its points to be associated with the value *1* in the third dimension. This means that a complete confidence is attached to each achieved MFG associated with a given crisp input. The extent of uncertainty/distribution linked with the degree is a (whitish) patch in Figure 5.4.

### 5.6.2 Interval Type 2 Fuzzy Logic Mathematics

If we blur the T1FS MF, for example, to say a triangular MF, then by shifting the points on the triangle either to the left or to the right, we can obtain the T2MF as seen in Figure 5.4. This blurring need not necessarily be of the same amounts, or it could be regular, irregular, or uniform, the latter being preferred. Then, at a specific value of $x$, say $x^j$, there is no single value for the MF; instead it takes on values wherever the vertical line intersects the blur. All these values need not be weighted the same, so we assign an amplitude distribution to all of these points. As a result of this process we obtain 3-D MF-a T2MF, this characterizes a T2FS. We have the following definition of the T2FS [3]:

$$\tilde{A} = ((x,u), \mu_{\tilde{A}}(x,u)); \;\; \forall x \in X, \forall u \in J_x \subseteq [0,1] \tag{5.53}$$

Here, $\tilde{A}$ is the T2FS, and it is characterized by a T2MF (Type 2 membership function), $\mu_{\tilde{A}}(x,u))$ *where* $x \in X$, *and* $u \in J_x \subseteq [0,1]\}$. Also, we have $0 \le \mu_{\tilde{A}}(x,u) \le 1$. T2FS can also be expressed as:

$$\tilde{A} = \int_{x\in X} \int_{u\in J_x} \mu_{\tilde{A}}(x,u)/(x,u) \;\; J_x \subseteq [0,1] \tag{5.54}$$

Here, the integral signs denote the union-operator over all the permissible $x$ and $u$, and for the discrete UOD (DUOD) the integral sign is replaced by the summation (sigma) sign. We see two conditions: i) $\forall u \in J_x \subseteq [0,1]$ is in line with the T1FL constraint, i.e., $0 \le \mu_{\tilde{A}}(x,u) \le 1$. This is because when the uncertainty disappears a T2FL should degenerate to T1FS, i.e., T2MF should reduce to T1MF; and ii) the condition $0 \le \mu_{\tilde{A}}(x,u) \le 1$ is also in line with the fact that the amplitudes of MF should be within 0 and 1, both inclusive. We have now the following definition for the IT2FS [3]:

$$\tilde{A} = \int_{x\in X} \int_{u\in J_x} 1/(x,u), \quad J_x \subseteq [0,1]\} \tag{5.55}$$

We emphasize here, that IT2FS/IT2FL are more prevalent and usable in FLS/FL control and are easier to handle than T2FL. Next, for each value of $x$, say $x'$, the 2D plane, whose axes are u and the $\mu_{\tilde{A}}(x',u)$, is called a vertical slice of the $\mu_{\tilde{A}}(x,u)$. Then, a secondary MF (SMF) is considered as a vertical slice of $\mu_{\tilde{A}}(x,u)$, it is $\mu_{\tilde{A}}(x=x',u)$ for $for\, x' \in X, \forall u \in J_{x'} \subseteq [0,1]$ and is given as follows:

$$\mu_{\tilde{A}}(x=x',u) \equiv \mu_{\tilde{A}}(x') = \int_{u \in J_{x'}} 1/u, \quad J_{x'} \subseteq [0,1] \tag{5.56}$$

Invoking the concept of secondary sets, one can interpret IT2FS as the union of all the secondary sets. Thus, one can express $\tilde{A}$ in a vertical-slice manner as follows:

$$\tilde{A} = (x, \mu_{\tilde{A}}(x)); \forall\, x \in X \tag{5.57}$$

Or,

$$\tilde{A} = \int_{x \in X} \mu_{\tilde{A}}(x)/x = \int_{x \in X} [\int_{u \in J_x} 1/u]/x; \quad J_x \subseteq [0,1] \tag{5.58}$$

Next, we define the domain of a SMF. It is called the primary membership (PMF) of $x$. Thus, $J_x$ is the PM of $x$, and $J_x \subseteq [0,1]; \forall x \in X$. The amplitude of a SMF is called the SMFG (-grade), and these for an IT2FS are all equal to 1. In case $X$ and $J_x$ are discrete then we can express the right most part of equation (5.58) as [6]:

$$\tilde{A} = \sum_{x \in X}\left[\sum_{u \in J_x} 1/u\right]/x, \quad = \sum_{i=1}^{N}\left[\sum_{u \in J_{x_i}} 1/u\right]/x_i,$$

$$= \left[\sum_{k=1}^{M_1} 1/u_{1k}\right]/x_1 + \ldots + \left[\sum_{k=1}^{M_N} 1/u_{Nk}\right]/x_N \tag{5.59}$$

In equation (5.59) the '+' sign denotes union operation, and we see that $x$ is discretized into $N$ values, and at each of $x$ value $u$ is also discretized into $M_i$ values. The discretization along each $u_{ik}$ is not necessarily the same; if it is the same then $M_1=M_2=\ldots=M_N=M$. The uncertainty in the PMF of an IT2FS, $\bar{A}$, has the bounded region and is called the FOU (foot-print of uncertainty), and this FOU is the union of all the PMFs expressed as follows:

$$FOU(\tilde{A}) = \bigcup_{x \in X} J_x \tag{5.60}$$

This FOU is a vertical-slice representation of the FOU. This is so because each of the PMs is a vertical slice. The FOU is a complete description of an IT2FS, since the

SMFGs of an IT2FS do not convey any new information. The upper and lower MFs (UMF. LMF) are next defined as:

$$\begin{aligned}\bar{\mu}_{\tilde{A}}(x) &\equiv \overline{FOU(\tilde{A})}; \quad \forall x \in X \\ \underline{\mu}_{\tilde{A}}(x) &\equiv \underline{FOU(\tilde{A})}; \quad \forall x \in X\end{aligned} \tag{5.61}$$

Thus, we have $J_x = [\underline{\mu}_{\tilde{A}}(x), \bar{\mu}_{\tilde{A}}(x)]$; $\forall x \in X$ for IT2FS. For DUODs $X$, and $U$, an embedded IT2FS has $N$ elements. The set of embedded such functions/FSs is given as:

$$\tilde{A}_e = \sum_{i=1}^{N} [1/u_i]/x_i, \quad u_i \in J_{x_i} \subseteq U = [0,1] \tag{5.62}$$

The set is given as:

$$A_e = \sum_{i=1}^{N} u_i/x_i, \quad u_i \in J_{x_i} \subseteq U = [0,1] \tag{5.63}$$

In equation (5.63), we have $\prod_{i=1}^{N} M_i A_e$ embedded functions. We have now the representation theorem [3]: for an IT2FS with discrete $X$, and $U$, we have the following expressions:

$$\tilde{A} = \sum_{j=1}^{n_A} \tilde{A}_e^j \tag{5.64}$$

With $(j=1,2,...,n_A)$ and

$$\tilde{A}_e^j = \sum_{i=1}^{N} [1/u_i^j]/x_i, \quad u_i^j \in J_{x_i} \subseteq U = [0,1] \tag{5.65}$$

$$\text{with } n_A = \prod_{i=1}^{N} M_i \tag{5.66}$$

In equation (5.66), $M_i$ denotes the discretization levels of SMF variables. Equation (5.64) gives a wavy slice of the set $\tilde{A}$. We make the following observations: i) the MF of a IT2FS is 3D wavy; ii) one can take all the possible wavy slices and take their union to reconstruct the original 3D MF; iii) then, the same points that occur in different wavy slices, only occur once in the set-theoretic union; iv) then it means collecting all of the embedded IT2FSs into a bunch of such T2FS; and v) equivalently one can collect all of the embedded T1FS into a bunch of such T1FSs. We have also the following relationship from equation (5.64):

$$\tilde{A} = 1/FOU(\tilde{A}) \tag{5.67}$$

With the following expressions for the set:

$$FOU(\tilde{A}) = \sum_{j=1}^{n_A} A_e^j = \begin{cases} \{\underline{\mu}_{\tilde{A}}(x),...,\overline{\mu}_{\tilde{A}}(x)\}; & \forall x \in X_d \\ [\underline{\mu}_{\tilde{A}}(x),\overline{\mu}_{\tilde{A}}(x)]; & \forall x \in X \end{cases} \tag{5.68}$$

$$\text{With } A_e^j = \sum_i^N u_i^j / x_i, \quad u_i^j \in J_{x_i} \subseteq U = [0,1] \tag{5.69}$$

In equation (5.68) $X_d$ is the DUOD, and $X$ is the continuous UOD, and the equation gives an interval set of functions implying that it contains an uncountable number of functions that would completely fill the space between $\overline{\mu}_{\tilde{A}}(x) - \underline{\mu}_{\tilde{A}}(x)$; $\forall x \in X$.

## EXERCISES

1. Consider a crisp set A defined on universe of discourse U={0,1,2,3,4,5} such that

$$\mu_A(u) = \begin{Bmatrix} 1 & u \geq 3 \\ 0 & u < 3 \end{Bmatrix}$$

   Construct a fuzzy set B on the same universe to represent the linguistic variable 'Medium'. Use a triangular membership function with peak at $u = 3$ and base spanning from $u = 1$ to $u = 5$. Plot and compare $\mu_A(u)$ and $\mu_B(u)$.
2. A fuzzy rule is defined as:

   *IF Temperature is HIGH AND Humidity is LOW THEN Fan Speed is FAST.*

   Given input values: Temperature = 32°C and Humidity = 40%, with membership grades

$$\mu_{HIGH}(32)=0.7,\ \mu_{LOW}(40)=0.5.$$

   (a) Compute the rule of firing strength using (i) min operator and (ii) algebraic product.
   (b) Comment on the difference between the two results.
3. For a discrete aggregated fuzzy output set B given by:

$$v = [0,1,2,3,4,5],\ \mu_B(v) = [0,0.2,0.5,0.7,0.9,0.4]$$

Compute the defuzzified output using:
(i) Centroid (Centre of Gravity) method,
(ii) Mean of Maximum (MoM) method,
(iii) Smallest of Maximum (SoM).

4. Two fuzzy rules are given:
   - R: IF Load is SMALL THEN Torque is HIGH
   - S: IF Torque is HIGH THEN Speed is FAST
   - Using **max-min composition**, derive the composite rule R∘S. Illustrate with assumed membership values for SMALL Load, HIGH Torque, and FAST Speed.
5. Explain how an Interval Type-2 Fuzzy Set (IT2FS) differs from a Type-1 fuzzy set. Construct an example of a triangular membership function for 'Medium Temperature' with uncertainty in the base endpoints ([18,22]to [28,32]). Show how the footprint of uncertainty (FOU) is formed, and discuss its advantage in handling measurement uncertainty.

## REFERENCES

1. Raol, J.R., *Multi-Sensor Data Fusion with MATLAB*, CRC Press, Taylor & Francis Group, Florida, USA, 2010.
2. Raol, J.R., *Data Fusion Mathematics-Theory and Practice,* CRC Press, Taylor & Francis Group, Florida, USA, 2016.
3. Raol, J.R., Lakshmi, S., and Kashyap, S.K., *Situation Assessment in Aviation-Bayesian Networks and Fuzzy Logic Based Approaches,* CRC Press, Taylor & Francis Group, Florida, USA, 2023.

# 6 System Identification and Estimation with Fuzzy Logic/system

## 6.1 INTRODUCTION

Fuzzy logic (FL) has found increasing application in decision fusion, in addition to its use in mathematical modelling. Decision fusion, also referred to as symbol level fusion, operates at a higher level than both kinematic level and image level fusion processes, where a symbol represents a decision [1]. Such symbolic input in the form of a decision requires the fusion process to describe both logical and statistical inference. Since the fusion of symbolic information necessitates reasoning and inference under uncertainty, computational mechanisms drawn from artificial intelligence such as FL are well suited for handling symbolic data and information.

## 6.2 FUZZY LOGIC-BASED KALMAN FILTER

The Kalman filter is one of the most promising algorithms for recursive estimation of states of linear and non-linear systems. The accuracy of the filter is based on: i) the accuracy of the mathematical model of the actual dynamic system and measurement device (because of the random uncertainties), and ii) its tuning parameters Q (process noise covariance matrix) and R (measurement noise covariance matrix).

The (time varying, especially in the case of the EKF) gain of a KF decides how much weighting should be given to the present measurement (in fact to the residuals): i) if the measurement data are highly contaminated with noise then automatically less weight is assigned to that data and the filter depends on the model of the target (i.e., state propagation equation), and ii) if measurement data are less noisy then more weight is given to that data and estimated state is a combination of state predicted (through target model) and observation data.

Thus, the KF has an inherent decision-making capability that helps in soft-switching between the process model and measurement model (via appropriate weight assignments via Kalman gain and P, Q, R). This decision of soft-switching is based on Kalman gain which depends on the relative value of tuning parameters/matrices Q and R. Measurement noise variance reflects the noise level in measurement data, higher R (in terms of the norm of the matrix) means data are very noisy and vice versa. In the case of very high R, the filter takes a decision to have much less faith in measurement data by assigning a low weight, through Kalman gain equation,

DOI: 10.1201/9781003616580-6

to the correction part of the state update. This can also be interpreted by saying that the filter relies highly on the process model. Similarly for very low R, the filter highly relies on the measurement model. The foregoing soft decision-making ability (of the Kalman filter) has come about from the optimization procedure while deriving the KF, and it can be viewed as a part of artificial intelligence (Appendix 1C), like one of the PATs, i.e., process, action, or technique (Chapter 1).

FL assists in modelling the conditions that are inherently imprecisely defined and FL-based methods in the form of approximate reasoning provide decision support and an expert system with reasonably good reasoning capabilities – this being called a FL-type 1 system. FL can also be used for tuning Kalman filters. The algorithms can be developed by considering the combination of fuzzy logic and KF. The proper combination of FL and KF-based approaches can be used to obtain improved accuracy and performance in tracking as well as in data fusion systems. In such systems the FL can be considered as aiding soft decision-making in the filtering process because of the use of fuzzy 'IF… Then' rules in making some judgment on the use of, say residuals, in navigating the prediction/filtering in the direction of achieving accurate results in either the tracking process, feature selection/detection/matching, or multisensor data fusion: this aspect in a sense can be considered as a 'system that acts like human', one component of AI.

In this section two schemes based on i) KF, and ii) Fuzzy Kalman filter (FKF) are studied for target tracking applications and their performances evaluated. The concept of FL is extended to state-vector (level) data fusion for similar sensors. The performances of fuzzy logic-based fusion methods are compared with the conventional fusion method (also state vector fusion, SVF) to track a manoeuvring target. FL concept is combined with KF filter at a measurement update level. The equations for the FKF are the same as those of KF except for the following equation:

$$\hat{X}(k+1/k+1) = \tilde{X}(k+1/k) + KB(k+1) \tag{6.1}$$

Here, $B(k+1)$ is regarded here as an output of the fuzzy logic-based process variable (FLPV) and is, in general, a nonlinear function of the innovations 'e' of the KF. It is assumed that position (in x-y axes) measurements of the target are available. The FLPV vector consists of the modified innovation sequence for x and y axes:

$$B(k+1) = \begin{bmatrix} b_x(k+1) & b_y(k+1) \end{bmatrix} \tag{6.2}$$

To determine FLPV vector the innovation vector $e$ is first separated into its x and y components, $e_x$ and $e_y$. The target motion in each axis is assumed to be independent. The FLPV vector for the x direction can be developed and then it is generalized to include y direction. This vector consists of two inputs, $e_x$ and $\dot{e}_x$, and a single output$b_x(k+1)$, where $\dot{e}_x$ is computed by

$$\dot{e}_x = \frac{\{e_x(k+1) - e_x(k)\}}{T} \tag{6.3}$$

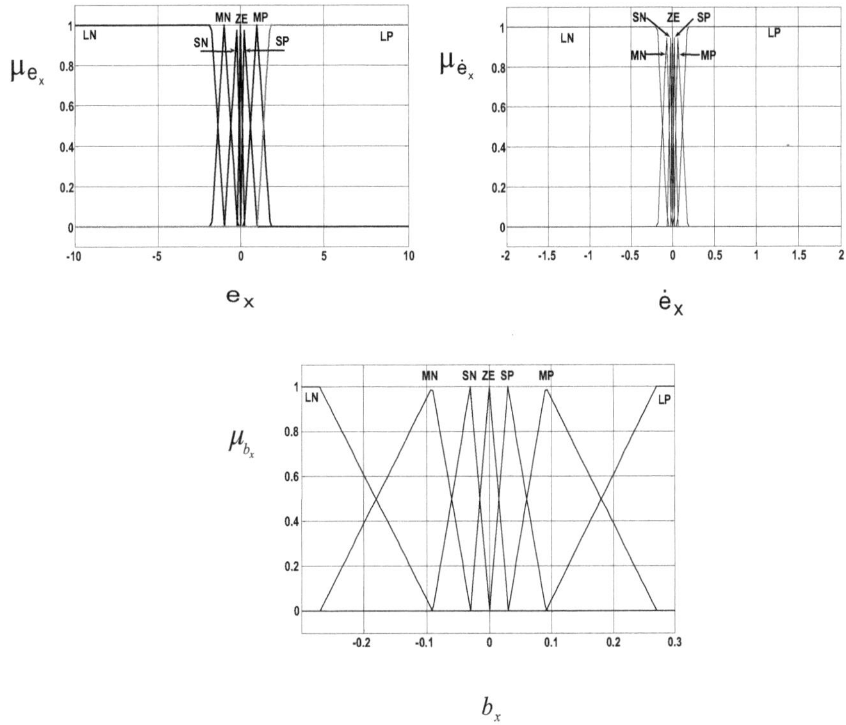


**FIGURE 6.1** FMFs-Fuzzy membership functions for error-input and its finite difference and for the output FLP variable

Here, T is the sampling interval in seconds, and the expression of equation (6.3) can be extended to y-direction, and even z-direction if required.

### 6.2.1 Fuzzy Logic-based Process

FLP is obtained via a FIS. In order to develop the FLP the antecedent membership functions that define the fuzzy values for inputs $e_x$ and $\dot{e}_x$, and the membership function for the output $b_x$ are shown in Figure 6.1. The linguistic variables/labels to define membership functions are LN – large negative, MN – medium negative, SN – small negative, ZE – zero error, SP – small positive, MP – medium positive, and LP – large positive. The rules for the inference in FIS, in general are created based on the experience and intuition of the domain expert, one such rule being:

$$\text{IF } e_x \text{ is LP AND } \dot{e}_x \text{ is LP THEN } b_x \text{ is LP} \tag{6.4}$$

This signifies the fact that having $e_x$ and $\dot{e}_x$ with large positive values indicates an increase in innovation sequence at a faster rate. Then the future value of $e_x$ (and hence $\dot{e}_x$) can be reduced by increasing the present value of $b_x(\approx Z - H\,\tilde{X})$ with a large magnitude. This will generate 49 rules to implement FLP. The output $b_x$ at any

TABLE 6.1
Features of FIS for FKF

| FIM | Mamdani |
|---|---|
| AND operator | min |
| OR operator | max |
| Fuzzy implication | min |
| Aggregation | max |
| Defuzzification | centroid |

instant of time can be computed using: i) the inputs $e_x$ and $\dot{e}_x$, ii) input membership functions, iii) 49 rules, iv) FIS (fuzzy inference system/engine), v) aggregator, and vi) defuzzification. The properties/features of the fuzzy operators/fuzzy implication methods used are given in Table 6.1.

### 6.2.2 Comparison of KF and FKF

The simulated data for x-axis position are generated using CAM (constant acceleration model, [1], [2]) with process noise increment and with $T = 0.1$ s; total number of scans being $N = 100$. The simulation uses the parameter values and related information as: i) initial states of target $(x, \dot{x}, \ddot{x})$ are (0 m, 100 m/s, 0 m/s$^2$), respectively, and ii) process noise variance $Q = 0.0001$. The CAM model is given as:

$$F = \begin{bmatrix} 1 & T & T^2/2 \\ 0 & 1 & T \\ 0 & 0 & 1 \end{bmatrix} \tag{6.5}$$

$$G = \begin{bmatrix} T^3/6 & T^2/2 & T \end{bmatrix} \tag{6.6}$$

The target state equation is given as

$$X(k+1) = FX(k) + Gw(k) \tag{6.7}$$

Here, $k$ is the scan number and $w$ is white Gaussian process noise with zero mean and covariance matrix $Q$. The measurement equation is given as

$$Z_m(k) = HX(k) + v(k) \tag{6.8}$$

$$H = \begin{bmatrix} 1 & 0 & 0 \end{bmatrix} \tag{6.9}$$

Here, $v$ is white Gaussian measurement noise with usual meaning ($R = \sigma^2$; $\sigma$ is the standard deviation of noise with a value of 10 meters). The initial conditions, $F$, $G$,

$H$, $Q$, and $R$ for both the filters, KF and FKF, are the same. The initial state vector $\hat{X}(0/0)$ is close to true initial states. The KF and FKF algorithms were coded in MATLAB. The results for both the filters are compared in terms of true and estimated states, and states errors with bounds at every scan number. It was observed that FKF performs much better compared to KF. The consistency checks on these filters were performed using the normalized cost function (CF) computed using the formula:

$$CF = \frac{1}{N}\sum_{k=1}^{N} e(k)S(k)^{-1}e(k)^{T} \tag{6.10}$$

Here, $e$ is the innovation sequence vector, and $S$ is innovation covariance matrix. The filter performance is deemed consistent if its normalized CF, eqn. (6.10) is equal to the dimension of the vector of observables. The CF for KF (=2.94) was found to be very close to the theoretical value of 3, whereas for FKF its CF value 2.85 was slightly different from the theoretical value 3. However, it is still comparable with the KF value and is not much different from the theoretical number. Hence, the FKF can be treated as an approximately consistent filter. The performance of both the filters in terms of states errors was also evaluated and the FKF showed better performance than KF. The procedure of this section validated the application of the FKF for target tracking and the FKF is either comparable to or gives better performance than the KF [1]; the details of the results are not given here for the sake of brevity, however, further applications of both the filters are evaluated and some results are given in the next section.

### 6.2.3 Comparison of KF and FKF for Manoeuvring Target-tracking

In order to use FKF for tracking a manoeuvring target, re-design of the FLPV in order to capture the various possible manoeuvre modes of the target is required. This involves [1]: i) proper selection of membership functions of I/O; ii) tuning of the membership functions; iii) selection of fuzzy operators (e.g. T-norm & S-norm [1]); and iv) selection of fuzzy implication, aggregation, and defuzzification techniques. We here use MATLAB-based functions such as 'genfis1()' to create initial FLP vector and 'anfis()'

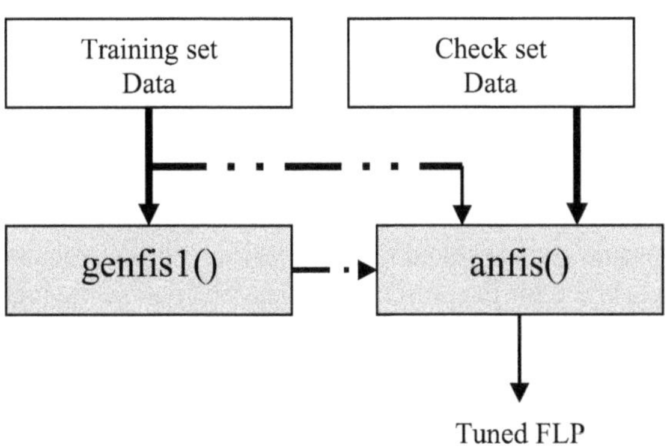


**FIGURE 6.2** FLP-Fuzzy Logic Process tuning scheme

to tune it. The required training and check(ing) data are obtained from true and measured target positions. Figure 6.2 depicts the procedure to get tuned FLP vector.

#### 6.2.3.1 Training Set and Check Set Data

The target states are simulated using a 3 DOF kinematic model with process noise acceleration increments and additional accelerations. Measurement data are obtained using with a sampling interval of 1 second, and total of 150 scans are generated. The data simulation is done with: i) initial states $(x, \dot{x}, \ddot{x}, y, \dot{y}, \ddot{y})$ of target as (100, 30, 0, 100, 20, 0); ii) process noise variance $Q = 0.1$; it is assumed that $Q_{xx} = Q_{yy} = Q$; and iii) the measurement noise variance $R = 25$. It is also assumed that $R_{xx} = R_{yy} = R$. The target has an additional acceleration of $(x_{acc}, y_{acc})$ at scans 25 and 100 and an acceleration of $(-x_{acc}, -y_{acc})$ at scans 50 and 75. Data simulation is carried out with process noise vector $w$ (a 2x1 vector) modified to include these additional accelerations at the specified scan points in order to induce a specific manoeuvre [1]:

$$\left.\begin{aligned} w(1) &= guass()*\sqrt{Q_{xx}} + x_{acc} \\ w(2) &= guass()*\sqrt{Q_{yy}} + y_{acc} \end{aligned}\right\} \tag{6.11}$$

$$\left.\begin{aligned} w(1) &= guass()*\sqrt{Q_{xx}} - x_{acc} \\ w(2) &= guass()*\sqrt{Q_{yy}} - y_{acc} \end{aligned}\right\} \tag{6.12}$$

At the other scan points the vector $w$ is simply defined without these additional accelerations terms. Accelerations $x_{acc} = -9*9.8m/s^2$ and $y_{acc} = 9*9.8m/s^2$ are used in the above equations and the function $guass()$ is used to generate Gaussian random numbers with mean 0 and variance 1. First the initial FLPV is created for the x-axis and tuned using inputs $u_x^1, u_x^2$ and output $o_x$ obtained using:

$$u_x^1(k) = z_x(k) - x(k) \tag{6.13}$$

$$u_x^2(k) = \frac{u_x^1(k) - u_x^1(k-1)}{T} \tag{6.14}$$

$$output_x(k) = m\,u_x^1(k) \tag{6.15}$$

Here, $x$ and $z_x$ are true and measured target x-position respectively. $m$ is the unknown parameter and it is $m = 2$ for the present case. The first half of the total simulated data is taken for training and the remaining half as the check-set data. The same procedure is followed to get tuned FLPV for y-axis, then the trained FLP vector is plugged in to FKF and its performance is compared with the KF for the two cases discussed next.

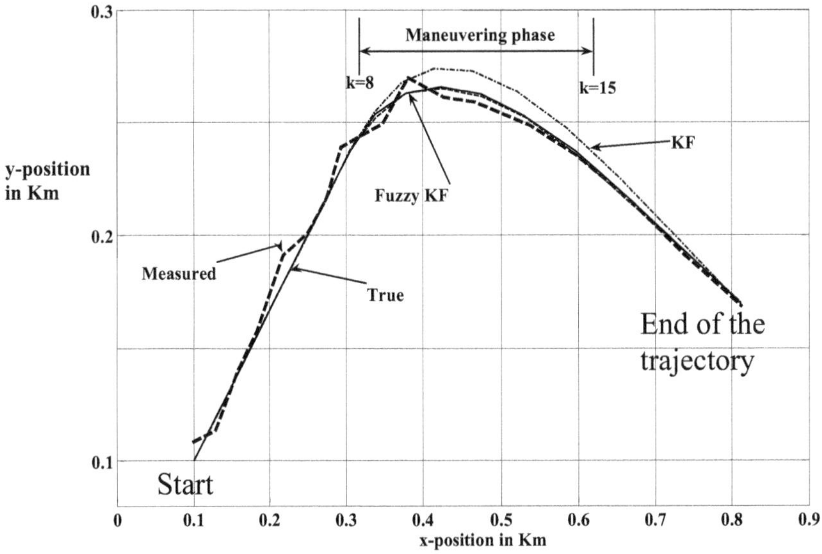


**FIGURE 6.3** True, measured and estimated x-y positions of the target for Mild manoeuvre-KF, FKF

### 6.2.3.2 Mild and Evasive Manoeuvre Data

The mild manoeuvre data (MM) are generated with minor modifications in the acceleration injection points, and a total of 17 scans are generated. Accelerations are injected at scans 8 ($x_{acc} = 6m/s^2$ and $y_{acc} = -6m/s^2$) and 15 ($x_{acc} = -6m/s^2$ and $y_{acc} = 6m/s^2$) only. The evasive manoeuvre data (EM) are generated with the same points for acceleration injection but with a manoeuvre magnitude of $40 * 9.8m/s^2$ (instead of $9 * 9.8m/s^2$). The results are obtained for 100 Monte-Carlo simulation runs with the initial state vectors of KF and FKF kept the same and close to the initial true states. Initial state error covariance matrices for both the filters are unity values. In Figure 6.3 the measured, true, and estimated x-y target positions for MM data are shown. The estimated trajectories using KF and FKF compare reasonably well with the true ones. There are some discrepancies in the manoeuvring phase of the flight where FKF exhibits better performance than KF. Similar observations are made for the case of EM data (the detailed results are not shown here). However, Figure 6.4 shows the comparison of RSSPE, RSSVE, and RSSAE for both the filters and these errors for the KF are found to be somewhat large compared to those for FKF [1]. Thus, the feasibility of the use of fuzzy logic-based KF has been established for target tracking applications. Further applications of the use of fuzzy logic-based KFs for sensor data fusion can be found in [1].

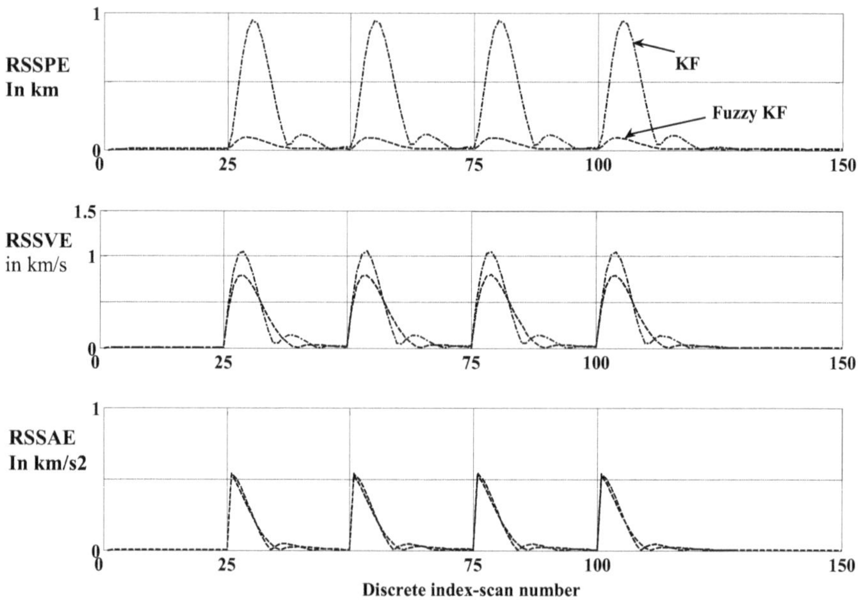


**FIGURE 6.4** RSSPE, RSSVE, and RSSAE plots for KF, FKF for EM data

## 6.3 FUZZY CLUSTERING METHOD (FCM)

FCM is also known as soft clustering or soft k-means, and is a clustering technique where each data point can belong to more than one cluster. Unlike hard clustering, where each data point is assigned to a single cluster, fuzzy clustering assigns membership grades to each data point, indicating the degree to which it belongs to each cluster.

### 6.3.1 Fuzzy C-means Clustering

The Fuzzy C-means clustering (FCMC) algorithm is most widely used. The FCMC algorithm is similar to the k-means (FKMC) algorithm, but allows data points to belong to multiple clusters with varying degrees of membership. The Algorithm Steps are: i) Initialization: Randomly assign coefficients to each data point for being in particular clusters; ii) Centroid calculation: Compute the centroid for each cluster, weighted by the degree of belonging of each data point; iii) Membership update: For each data point, update its coefficients of belonging to the clusters; and iv) Iteration: Repeat the centroid calculation and membership update steps until the convergence is reached (i.e., the change in coefficients is below a given threshold).

Illustrative example of implementing FCMC using the skfuzzy library in Python:

```
import numpy as np
import skfuzzy as fuzz
```

```
# Generate some example data
np.random.seed(0)
data = np.random.rand(100, 2)

# Define the number of clusters
n_clusters = 3

# Apply fuzzy c-means clustering
cntr, u, u0, d, jm, p, fpc = fuzz.cluster.cmeans (
data.T, n_clusters, 2, error=0.005, maxiter=1000,
init=None)

# Predict cluster membership for each data point
cluster_membership = np.argmax(u, axis=0)

# Print the cluster centers
print('Cluster Centers:', cntr)

# Print the cluster membership for each data point
print('Cluster Membership:', cluster_membership)
```

### 6.3.2 Applications

FCM has various applications: i) Image Segmentation: Grouping pixels with similar properties for better image analysis; ii) Pattern Recognition: Identifying patterns in large datasets by grouping similar data points; iii) Marketing: Segmenting customers based on preferences and purchasing behaviour; iv) Medical Diagnosis: Grouping patients with similar symptoms for better diagnosis; and v) Environmental monitoring: Identifying areas of environmental concern by grouping similar indicators.

### 6.3.3 Merits and Demerits

Merits are: i) Flexibility: Allows overlapping clusters, useful for complex data structures; ii) Robustness: More robust to outliers and noise; and iii) Interpretability: Provides a nuanced understanding of data relationships.

Demerits are: i) Complexity: Computationally more expensive than traditional clustering algorithms; and ii) Model Selection: Choosing the right number of clusters and membership functions would be challenging.

### 6.3.4 Parameter Estimation Using Fuzzy Logic

The steps of the fuzzy identification/parameter estimation process are: i) I/O data collection; ii) model structure selection, this is confined to a choice of the order of the system's dynamics, and complexity of the fuzzy model; iii) model parameter estimation; and iv) model validation

The system dynamics selected can be NARX (nonlinear autoregressive with exogenous input [3]), where the relation between past input/output data and predicted out has the form:

$$y(k+1)=F(y(k),...,y(k-n_y+1),u(k),...,u(k-n_u+1)) \tag{6.16}$$

where $n_{y,}$ and $n_u$ are related to the system's order. The complexity of the fuzzy model is mainly connected to the number of fuzzy rules that are used to define the model itself. The number of fuzzy rules is essentially related to the choice of the number of clusters that are used to partitioned I/O data. Data clustering is a technique to process and classify a set of observed data according to similarity by minimizing the following objective function:

$$J(X;U;V)=\sum_{i=1}^{c}\sum_{k=1}^{n}(\mu_{ik})^m\, d^2(v_i,x_k) \tag{6.17}$$

Here, c is the number of fuzzy clusters, and $X=[x_1,...,x_n]\in R^{pxn}$ is the matrix-containing data set composed of n observations with each consisting of p measured variables. $U=[\mu_{ik}]$ is the fuzzy partition matrix, where $\mu_{ik}\in[0,1]$ is the membership function of datum $x_k$ to cluster $i$ and $d(v_i,v_k)$ is the distance. The parameter m>1 is the fuzziness index: when m–>1 the cluster tend to be crisp, whereas m–> infinity the partition becomes maximum fuzzy and $\mu_{ik}->1/c$.

Development of rule-based fuzzy models: The development of the rule base is required for parameter estimation of the system, for example, an aircraft's lift and momentum coefficients. It can be done based on the Takagi-Sugeno (TS) models:

If u is $A_i$ then $y_i=a_i^T u+b_i$, where u is the input variable, $y_i$ is the output variable, and index i denotes i-th rule in the rule base (no. of rules is equals to no. of clusters).

$A_i$ is the antecedent fuzzy set of the i-th rule, $a_i$ is a parameter vector, and $b_i$ is a scalar offset. In general $A_i$ can be extracted by fuzzy partition matrix U by the projection method, whereas $a_i$ and $b_i$ can be estimated from the data by means of the least squares methods.

**Example 6.1:** Shot period dynamics, and lift and moment coefficients of an aircraft are expressed as

$$\begin{aligned}\dot{\alpha}-q&=-\frac{\rho VS}{2M}C_L\\ \dot{q}&=\frac{\rho V^2 S\bar{c}}{2I_y}C_m\ ......\end{aligned} \tag{6.18}$$

$$\begin{aligned}C_L&=C_{L0}+C_{L\alpha}\alpha+C_{Lq}\frac{\bar{c}}{2V}q+C_{L\delta_e}\delta_e\\ C_m&=C_{m_0}+C_{m\alpha}\alpha+C_{m_q}\frac{\bar{c}}{2V}q+C_{L\delta_e}\delta_e\ ............\end{aligned} \tag{6.19}$$

**TABLE 6.2**
**Parameter Estimation Results**

| | | Noise 1% | | Noise 5% | |
|---|---|---|---|---|---|
| **Parameters** | **True value** | **Estimated value** | **Standard deviation** | **Estimated value** | **Standard deviation** |
| $C_{L\alpha}$ | 2.92 | 2.901 | 0.007 | 2.765 | 0.031 |
| $C_{Lq}$ | 14.70 | 13.616 | 0.769 | 9.648 | 2.941 |
| $C_{L\delta_e}$ | 0.43 | 0.414 | 0.010 | 0.275 | 0.048 |
| $C_{m\alpha}$ | -1.67 | -1.669 | 0.002 | -1.601 | 0.008 |
| $C_{mq}$ | -34.74 | -34.411 | 0.206 | -31.598 | 0.756 |
| $C_{m\delta_e}$ | -2.5 | -2.507 | 0.003 | -2.439 | 0.012 |
| $C_{L0}$ | 0.05 | 0.053 | 0.003 | 0.069 | 0.019 |
| $C_{m_0}$ | 0.10 | 0.099 | 0.001 | -0.092 | 0.008 |

Estimate the aerodynamic coefficients of eqn. (6.19).

**Solution 6.1** For model identification the number of clusters is set as c = 3 (found after trial and error). The parameter estimation is done using the so-called zero method. The resulting TS model for the Lift coefficient is: If is $A_i$ and is $B_i$ and is $C_{i\,(6.20)}$

$$\text{then: } C_L = a_{i1}.q + a_{i2}.\alpha + a_{i3}.\delta_e + b_i \tag{6.20}$$

Here, i = 1,2,3 are three clusters. A similar expression for $C_m$ is valid. Table 6.2 shows estimated aerodynamic coefficients with a fuzzy identification method for various levels of noise. We see that the results are very satisfactory.

## 6.4 ANFIS – ADAPTIVE NEURO FUZZY INFERENCE SYSTEM

The Adaptive Neuro Fuzzy Inference System (ANFIS) is a hybrid soft-computing approach that combines the transparency of fuzzy inference with the adaptive learning capability of neural networks. It was introduced by Jang in 1993 [4] and has since been widely used for system identification, control, and fault detection of nonlinear systems. ANFIS represents a powerful computational intelligence approach that leverages the interpretability of fuzzy logic and the learning capabilities of neural networks to create a system that can handle uncertainty while adapting to new data.

The strength of ANFIS lies in its ability to automatically learn both the parameters of the membership functions (premise parameters) and the parameters of the rule consequents (consequent parameters), while still retaining the interpretability of fuzzy rules. Unlike pure neural networks that operate as black boxes, ANFIS maintains the transparency of fuzzy systems while incorporating the learning capabilities of neural networks. This makes ANFIS particularly attractive for applications where both accuracy and interpretability are important.

### 6.4.1 Structure of ANFIS

ANFIS is usually built upon the Takagi-Sugeno (TS) fuzzy model [5], which provides a systematic approach to generating fuzzy rules from input-output data. The TS framework allows ANFIS to represent complex nonlinear relationships through a collection of local linear models, each valid in a specific region of the input space. Consider a system with two inputs, x and u, and one output, y. A first-order TS fuzzy model with two rules can be written as:

$$\text{Rule 1: IF x is } A_1 \text{ AND u is } B_1\text{, THEN } y_1 = p_1x + q_1u + r_1$$

$$\text{Rule 2: IF x is } A_2 \text{ AND u is } B_2\text{, THEN } y_2 = p_2x + q_2u + r_2 \tag{6.21}$$

where $p_i$, $q_i$, $r_i$ are consequent parameters to be learned, and $A_i$, $B_i$ are fuzzy sets characterized by their membership functions.

The ANFIS architecture consists of five functional layers, each performing specific operations in the fuzzy inference process. The network structure is functionally equivalent to a Sugeno-type fuzzy rule base, but the parameters are determined through a learning process using input-output training data.

**Layer 1 – fuzzification layer:** This layer performs the fuzzification operation by converting crisp inputs into fuzzy sets using membership functions. Each node in this layer corresponds to a linguistic label of one of the input variables. Every node i in this layer is an adaptive node with outputs:

$$O_1^i = \mu_{\alpha i}(x) \quad \text{for i} = 1{,}2$$

$$O_1^i = \mu_{\beta i-2}(u) \quad \text{for i} = 3, 4 \tag{6.22}$$

where x and u are the inputs to the node, and $A_i$ and $B_{i-2}$ are linguistic labels associated with this node. The membership functions $\mu_{\alpha i}(x)$ and $\mu_{\beta i-2}(u)$can be any appropriate parameterized membership function, such as the generalized bell function:

$$\mu_{\alpha i}(x) = \frac{1}{1 + \left[\left(\frac{x - c_i}{a_i}\right)^2\right] b_i} \tag{6.23}$$

where {ai, bi, ci} is the parameter set that changes the shape of the membership function. The parameters in this layer are referred to as premise parameters.

**Layer 2 – product layer:** Every node in this layer is a fixed node whose output is the product of all the incoming signals. This layer implements the fuzzy AND operation, typically using the algebraic product as the T-norm operator. The firing strength of each rule is obtained from the membership functions:

$$w_1 = \mu_{\alpha 1}(x) \cdot \mu_{\beta 1}(u)$$

$$w_2 = \mu_{\alpha 2}(x) \cdot \mu_{\beta 2}(u) \tag{6.24}$$

Each node output represents the firing strength of the associated rule. So, Any T-norm operator that performs the AND operation can be used in this layer.

**Layer 3 – normalized layer:** Every node in this layer is a fixed node that calculates the ratio of the i-th rule's firing strength to the sum of all rules' firing strengths. The normalized firing strengths are:

$$\bar{w}_1 = \frac{w_1}{w_1 + w_2}$$

$$\bar{w}_2 = \frac{w_2}{w_1 + w_2} \tag{6.25}$$

This normalization ensures that the sum of all normalized firing strengths equals unity, which is a requirement for the proper functioning of the Sugeno-type inference system.

**Layer 4 – defuzzification layer:** Every node in this layer is an adaptive node with a node function that computes the contribution of each rule to the overall output. The output of each node is the product of the normalized firing strength and the individual rule output:

$$O_4^i = \bar{w}_i f_i = \bar{w}_i \left(p_i x + q_i u + r_i\right) \tag{6.26}$$

where $\{p_i, q_i, r_i\}$ are the parameter set of this node, referred to as consequent parameters. These parameters define the linear function in the consequent part of the Sugeno-type fuzzy rule.

**Layer 5 – summation layer:** The single node in this layer is a fixed node that computes the overall output as the summation of all incoming signals. The final ANFIS output is then given by:

$$y = \bar{w}_1 f_1 + \bar{w}_2 f_2 = \frac{w_1 f_1 + w_2 f_2}{w_1 + w_2} \tag{6.27}$$

This represents the final output of the ANFIS network, which is a weighted average of the individual rule outputs. Figure 6.5 shows the complete ANFIS network architecture that implements the mathematical framework described above.

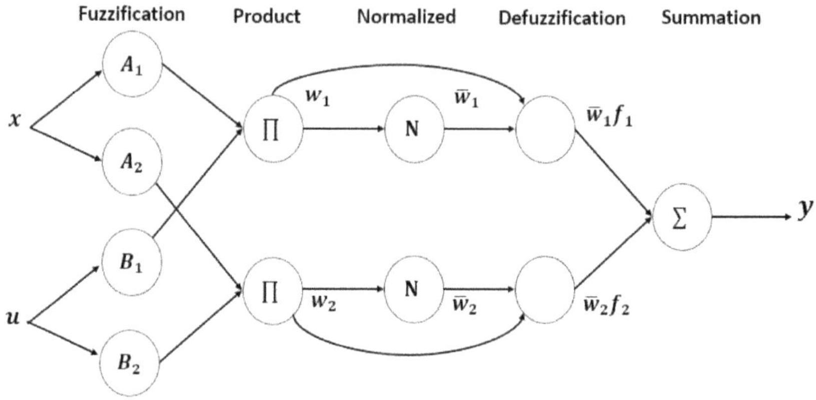


**FIGURE 6.5** ANFIS five-layer architecture for two-input Takagi-Sugeno fuzzy system

**Source: Adapted/modified from Jang, 1993**

### 6.4.2 Learning Algorithm

The learning in ANFIS is carried out using a hybrid optimization approach [4] that combines the gradient descent method and the least-squares estimate to identify the parameters of Sugeno-type fuzzy systems. The hybrid learning procedure is composed of a forward pass and a backward pass, which makes the learning process significantly more efficient than using gradient descent alone.

**Forward pass (Least Squares Method):** In the forward pass, the input patterns are propagated through the network, and the consequent parameters ($p_i$, $q_i$, $r_i$) are identified using least-squares estimation while the antecedent parameters (membership function parameters) are held fixed. The ANFIS output can be expressed as a linear combination of the consequent parameters. If we collect all the consequent parameters into a vector θ, then:

$$f = A\theta \tag{6.28}$$

where A is a matrix whose elements are functions of the input variables and the antecedent parameters, and f is the ANFIS output vector. The optimal consequent parameters can be obtained using the least-squares formula:

$$\theta^* = \left(A^T A\right)^{-1} A^T f \tag{6.29}$$

**Backward pass (Gradient Descent):** In the backward pass, the error rates propagate backward, and the premise parameters of membership functions are tuned using backpropagation while the consequent parameters remain fixed. The update formula for any antecedent parameter α is:

$$\alpha(t+1) = \alpha(t) - \eta\left(\frac{\partial E}{\partial \alpha}\right) \tag{6.30}$$

where $\eta$ is the learning rate and E is the error measure, typically the sum of squared errors between the desired and actual outputs.

This combined scheme allows ANFIS to capture nonlinear input-output mappings effectively while maintaining the interpretability of fuzzy rules. The hybrid approach ensures fast convergence compared to pure gradient descent methods because the least squares method provides the optimal solution for the linear parameters in one step.

### 6.4.3 Advantages of ANFIS

ANFIS offers several significant advantages over conventional neural networks and pure fuzzy systems. First, it provides universal approximation capability, meaning it can approximate any continuous function on a compact set to arbitrary accuracy, given a sufficient number of rules and appropriate membership functions. The Takagi-Sugeno framework enables ANFIS to represent complex nonlinear relationships through local linear models, with membership functions determining the boundaries and overlap between regions.

Second, ANFIS maintains interpretability through its fuzzy rule structure, allowing users to understand the reasoning process behind the system's decisions. The linguistic nature of the rules makes it possible to incorporate expert knowledge and to explain the system's behaviour in terms that are meaningful to domain experts.

Third, the hybrid learning algorithm ensures computational efficiency and fast convergence. The separation of linear and nonlinear parameter optimization allows each type of parameter to be updated using the most appropriate method, resulting in better convergence properties than pure gradient-based approaches.

Fourth, ANFIS can handle both numerical data and linguistic information, making it suitable for applications where expert knowledge needs to be incorporated along with numerical data. The system can start with expert-defined rules and membership functions and then refine these through learning from data.

### 6.4.4 System Identification Capability

ANFIS is particularly well-suited for system identification tasks where the goal is to build a model of an unknown system based on input-output data. The parameter estimation process in ANFIS involves determining both the structure and parameters of the fuzzy model. Structure identification includes determining the number of membership functions for each input variable, the type of membership functions, and the number of rules. Parameter identification involves estimating the parameters of the membership functions (premise parameters) and the coefficients of the linear functions in the rule consequents (consequent parameters).

The membership functions determine the boundaries and overlap between these regions, providing a smooth transition between local models. This characteristic makes ANFIS effective for modelling systems with multiple operating regimes or systems that exhibit different behaviours in different regions of the operating space. The hybrid learning algorithm efficiently handles both types of parameters, making ANFIS a practical tool for real-world system identification problems.

Applications of ANFIS in system identification include modelling nonlinear dynamics of aircraft and spacecraft, process control in chemical plants and power systems, fault detection and isolation in sensors and actuators, and multisensor data fusion for tracking problems. The versatility and effectiveness of ANFIS have made it a popular choice for a wide range of engineering applications where accurate modelling of complex nonlinear systems is required.

### 6.4.5 Fault Detection, Identification, and Accommodation for Sensors and Actuators

Fault detection and identification (FDI) plays a critical role in modern engineering systems, especially in aerospace, robotics, and process industries, where failures of sensors or actuators can compromise safety and performance. ANFIS is particularly useful for FDI because it can learn fault-free system behaviour from data and then detect deviations caused by faults, while maintaining the interpretability of the fault detection logic through fuzzy rules.

The ANFIS-based fault detection framework for sensors and actuators operates through a systematic approach that combines the learning capabilities of neural networks with the interpretability of fuzzy logic. This hybrid approach is especially advantageous in fault detection applications because it can handle the nonlinear characteristics and uncertainties inherent in real-world systems while providing meaningful insights into the nature of detected faults.

#### 6.4.5.1 System with Faults

A general nonlinear dynamic system can be described as:

$$x(k+1) = f\left(x(k), u(k)\right) + w(k) \tag{6.31}$$

$$y(k) = g\left(x(k)\right) + u(k) \tag{6.32}$$

where x(k) is the state vector, u(k) is the input vector, y(k) is the measured output vector, and w(k), v(k) are process and measurement noise, respectively, assumed to be zero-mean white Gaussian processes.

When faults occur in the system, additional terms appear in the system equations. For sensor faults, the measured output becomes:

$$y_f(k) = y(k) + f_s(k) \tag{6.33}$$

For actuator faults, the effective input to the system becomes:

$$u_f(k) = u(k) + f_a(k) \tag{6.34}$$

where $f_s(k)$ and $f_a(k)$ represent sensor and actuator faults, respectively. These fault signals can take various forms including bias faults (constant offsets), drift faults (gradual degradation), scaling faults (gain changes), and complete component failures.

#### 6.4.5.2 Residual Generation

The fundamental principle of ANFIS-based fault detection is to establish a baseline model of normal system behaviour and then monitor deviations from this baseline. An ANFIS model is trained under fault-free conditions to produce an estimated output ŷ(k) based on current and past inputs and outputs. The ANFIS model takes the form:

$$\hat{y}(k) = ANFIS\left(u(k-1),\ldots,\ u(k-n_u), y(k-1),\ldots,\ y(k-n_y)\right) \tag{6.35}$$

where $n_u$ and $n_y$ are the input and output memory orders, respectively, which determine how many past values are used for prediction.

During operation, the residual signal is defined as:

$$r(k) = y_f(k) - \hat{y}(k) \tag{6.36}$$

Under fault-free conditions, the residual should remain small and exhibit only the characteristics of modelling errors and noise. When a fault occurs, the residual will deviate significantly from its normal behaviour, providing the basis for fault detection.

#### 6.4.5.3 ANFIS-based FDI Procedure

The complete ANFIS-based fault detection, identification, and accommodation procedure consists of several interconnected stages:

**Training phase:** The first stage involves collecting comprehensive fault-free input-output data covering the entire operating envelope of the system. This data is used to train the ANFIS model to represent normal system behaviour. The training process uses the hybrid learning algorithm to optimize both the premise parameters (membership function parameters) and consequent parameters (linear function coefficients) to minimize the prediction error on the training dataset.

**Residual monitoring:** During online operation, the trained ANFIS model continuously generates predicted outputs, and residuals are computed using equation (6.36). The residual signals are then analyzed for fault detection. A simple threshold-based approach declares a fault when:

$$\|r(k)\| > \tau \tag{6.37}$$

where τ is a predetermined threshold. More sophisticated approaches use statistical methods or fuzzy logic to define adaptive thresholds that account for varying operating conditions and noise levels.

**Fault classification:** Once a fault is detected, the next step is to identify the specific type and location of the fault. This is accomplished by analyzing the patterns in the residual signals and applying fuzzy classification rules. Different fault types produce characteristic residual patterns that can be learned and recognized by the ANFIS system. For example, sensor bias faults produce step-like changes in residuals, while drift faults result in gradual trends.

**Fault accommodation:** The final stage involves implementing corrective actions to maintain system performance despite the presence of faults. The accommodation strategy depends on the type and severity of the identified fault and may include switching to redundant sensors, reconfiguring the control system, or modifying control parameters to compensate for actuator degradation.

#### 6.4.5.4 Mathematical Formulation for Sensor Fault Detection

For sensor fault detection, separate ANFIS models can be developed for each sensor or group of sensors. Consider a multi-sensor system where the i-th sensor is suspected of being faulty. An ANFIS model is trained to predict the output of the i-th sensor using the outputs of the remaining healthy sensors and the system inputs:

$$\hat{y}_i(k) = ANFIS_i\left(u(k), y_j(k), j \neq i\right) \tag{6.38}$$

The residual for the i-th sensor is then:

$$r_i(k) = y_i(k) - \hat{y}_i(k) \tag{6.39}$$

A fault in the i-th sensor is detected when the corresponding residual exceeds its threshold while residuals for other sensors remain within normal bounds.

#### 6.4.5.5 Actuator Fault Detection and Accommodation

For actuator fault detection, the approach is modified to account for the fact that actuator faults affect the system dynamics rather than the measurements directly. The ANFIS model is trained to predict the expected system response based on the applied control inputs as in eqn. (6.35).

When an actuator fault occurs, the actual system response will deviate from the predicted response because the effective input to the system differs from the commanded input. The residual analysis can then be used to identify which actuator is faulty and estimate the fault magnitude.

For accommodation of actuator faults, the control system can be reconfigured to redistribute the control effort among the remaining healthy actuators. If the fault magnitude can be estimated, the control commands can be modified to compensate for the reduced actuator effectiveness:

$$u_{compensated,i}(k) = \frac{u_i(k)}{1 - \varepsilon_i} \tag{6.40}$$

where $\varepsilon_i$ represents the estimated effectiveness loss of the i-th actuator.

#### 6.4.5.6 Performance Evaluation

The performance of ANFIS-based FDI systems is evaluated using several key metrics. Detection performance is assessed through the detection rate, which measures the percentage of actual faults that are correctly detected, and the false alarm rate, which quantifies the frequency of false fault declarations during normal operation. The detection delay, measuring the time between fault occurrence and detection, is crucial for time-critical applications.

Identification performance is evaluated through classification accuracy, which measures the percentage of detected faults that are correctly classified by type and location. Confusion matrices provide detailed insight into the system's ability to distinguish between different fault types and help identify areas where the classification performance can be improved.

#### 6.4.5.7 Advantages and Implementation Considerations

The ANFIS-based approach to fault detection offers several significant advantages. The system can handle nonlinear dynamics and complex relationships between system variables that are difficult to capture with traditional linear methods. The interpretability of fuzzy rules allows system operators to understand the reasoning behind fault detection decisions, which is crucial for building confidence in the system and facilitating maintenance actions.

The adaptive learning capability enables the system to improve its performance over time as more data becomes available and the hybrid learning algorithm ensures efficient parameter optimization. However, successful implementation requires careful consideration of several factors including the quality and completeness of training data, proper selection of ANFIS structure parameters, and appropriate threshold setting for fault detection.

In practice, ANFIS-based FDI systems have been successfully applied in aircraft flight control systems for detecting sensor bias and actuator effectiveness loss, in process plants for identifying nonlinear sensor faults such as pressure sensor drifts and in robotics for monitoring joint sensor and actuator health. These applications have demonstrated that ANFIS provides a robust, data-driven and interpretable framework for ensuring the reliability and safety of complex engineering systems.

## 6.5 A QUALITATIVE-FUZZY FRAMEWORK FOR BLACK-BOX SID-SYSTEM IDENTIFICATION

Black-box system identification represents one of the most challenging problems in control engineering, where the mathematical structure of the system is completely unknown and only input-output data are available. Traditional parametric identification methods often fail in such scenarios due to insufficient prior knowledge about system dynamics and the presence of nonlinearities that cannot be adequately captured by linear models [6]. The qualitative-fuzzy framework addresses this challenge by systematically integrating expert knowledge expressed in linguistic terms with measured data through fuzzy logic mechanisms.

The fundamental principle underlying this approach recognizes that domain experts often possess valuable qualitative understanding of system behaviour, even when precise mathematical relationships are unavailable. This qualitative knowledge typically manifests as linguistic descriptions such as 'when the input increases rapidly, the system responds with moderate delay' or 'under high-load conditions, the response becomes sluggish'. The framework formalizes such expert knowledge through fuzzy rules and combines it with numerical optimization techniques to achieve accurate system identification.

The qualitative-fuzzy identification framework operates through a multi-stage process that begins with knowledge acquisition from domain experts and culminates in a validated fuzzy model. The process involves qualitative knowledge extraction, fuzzy rule construction, parameter optimization, and model validation. The initial phase involves systematic extraction of expert knowledge about the system behaviour. This knowledge is typically expressed in the form of conditional statements that relate input conditions to expected output responses. For a dynamic system with input u(k) and output y(k), qualitative relationships are encoded through rules of the form:

$$R_i: \quad IF\ u(k)\ is\ A_i\ AND\ \Delta u(k)\ is\ B_i\ THEN\ y(k+1) is\ C_i \tag{6.41}$$

where $\Delta u(k) = u(k) - u(k-1)$ represents the input rate of change and $A_i, B_i, C_i$ are linguistic terms such as 'Small', 'Medium', 'Large', 'Increasing', 'Decreasing', etc. The linguistic terms are defined through membership functions μ(·) that capture the imprecise nature of human reasoning.

The linguistic knowledge is formalized into a structured rule base. For multi-input systems, the rule structure extends to accommodate all relevant variables. A general rule for an n-input system takes the form:

$$R_i: \quad IF\ x_1\ is\ A_1^i\ AND\ x_2\ is\ A_2^i \ldots\ldots\ AND\ x_n\ is\ A_n^i\ \ THEN\ y\ is\ B^i \tag{6.42}$$

where each $A_j^i$ represents a linguistic term for the j-th input variable in the i-th rule and $B^i$ defines the corresponding output linguistic value. The total number of rules is typically much smaller than the exhaustive combination of all linguistic terms due to the sparse nature of expert knowledge.

The inference mechanism combines individual rules to produce the overall system output using the Mamdani inference method. The aggregated output membership function is computed as:

$$\mu_c(y) = \max_i \left[ \min\left( \mu_{A_1^i}(x_1), \mu_{A_2^i}(x_2), \ldots\ldots, \mu_{A_n^i}(x_n), \mu_{B^i}(y_n) \right) \right] \tag{6.43}$$

The final crisp output is obtained through defuzzification using the centroid method:

$$y^* = \int \frac{y\mu_c(y)dy}{\mu_c(y)dy} \tag{6.44}$$

where the integrals are computed over the output universe of discourse.

The framework incorporates a learning mechanism that refines the initial expert knowledge using available input-output data. The optimization process minimizes the prediction error while preserving the linguistic interpretability of the rules. The objective function is formulated as:

$$E = \sum_{k=1}^{N} \left[ y(k) - \hat{y}(k) \right]^2 + \lambda R \tag{6.45}$$

where N is the number of data points, $\hat{y}(k)$is the model prediction, $\lambda$ is a regularization parameter, and R represents a penalty term that prevents the learning process from distorting the original qualitative knowledge structure. The regularization term typically takes the form:

$$R = \sum_{i} \left\| \theta_i - \theta_i^0 \right\|^2 \tag{6.46}$$

where $\theta_i$ represents the current parameters and $\theta_i^0$ denotes the initial expert-defined parameters. This ensures that the optimization process remains close to the original expert knowledge while improving the quantitative accuracy.

The parameter optimization can be performed using gradient-based methods or evolutionary algorithms. The gradient of the objective function with respect to the membership function parameters can be computed using the chain rule:

$$\frac{\partial E}{\partial \theta_i} = \sum_{k=1}^{N} 2\left[ y(k) - \hat{y}(k) \right] \frac{\partial \hat{y}(k)}{\partial \theta_i} + 2\lambda\left( \theta_i - \theta_i^0 \right) \tag{6.47}$$

The qualitative-fuzzy framework offers several distinct advantages over traditional identification methods. It can incorporate prior expert knowledge systematically, handle nonlinear systems effectively, provide interpretable models and work with limited data by leveraging qualitative information. The approach is particularly valuable when mathematical models are difficult to derive from first principles and when system behaviour exhibits complex nonlinear characteristics that are better understood through experiential knowledge.

## 6.6 A FUZZY IDENTIFICATION METHOD FOR NONLINEAR SYSTEMS

Nonlinear system identification presents significant challenges in control engineering due to the complex relationships between inputs and outputs that cannot be captured by traditional linear models. The fuzzy identification method provides a systematic approach to model nonlinear systems by partitioning the input space into local regions and approximating the system behaviour within each region using simple models. This approach leverages the universal approximation capability of fuzzy systems to represent complex nonlinear mappings while maintaining interpretability through linguistic rules.

The fuzzy identification method is based on the concept that any complex nonlinear system can be decomposed into a collection of simpler subsystems, each valid in a specific operating region. The boundaries between these regions are defined through fuzzy sets, allowing for smooth transitions and overlap between neighbouring regions. This decomposition is particularly effective for systems that exhibit different behaviours under different operating conditions, such as aircraft dynamics at various flight regimes or chemical processes under different temperature and pressure conditions.

The mathematical foundation of fuzzy identification rests on the Takagi-Sugeno (TS) fuzzy model structure. For a nonlinear system with inputs $x = \left[x_1, x_2, \ldots, x_n\right]^T$ and output y, the TS fuzzy model consists of rules of the form:

$$R_i: \quad IF\ x_1\ is\ A_1^i\ AND\ x_2\ is\ A_2^i \ldots\ldots\ AND\ x_n\ is\ A_n^i\ \ THEN\ y_i\ is\ f_i\left(x\right) \tag{6.48}$$

where $A_j^i$ are fuzzy sets defining the antecedent part of the i-th rule and $f_i\left(x\right)$ is a function that describes the local relationship between inputs and output in the i-th region. The function $f_i\left(x\right)$ can be linear or nonlinear, depending on the complexity of the local behaviour. For many practical applications, linear local models are sufficient:

$$f_i\left(x\right) = a_1^i x_1 + a_2^i x_2 + \ldots + a_n^i x_n + b^i \tag{6.49}$$

where $a_j^i$ and $b^i$ are the parameters of the i-th local linear model.

The overall output of the fuzzy system is computed as a weighted average of the individual rule outputs:

$$\hat{y} = \frac{\sum_{i=1}^{m} w^i\left(x\right) f_i\left(x\right)}{\sum_{i=1}^{m} w^i\left(x\right)} \tag{6.50}$$

where m is the number of rules and $w^i\left(x\right)$ is the degree of activation of the i-th rule, calculated as:

$$w^i\left(x\right) = \prod_{j=1}^{n} \mu_{A_j^i}\left(x_j\right) \tag{6.51}$$

Here, $\mu_{A_j^i}\left(x_j\right)$ represents the membership degree of $x_j$ in the fuzzy set $A_j^i$.

The identification process involves determining both the structure and parameters of the fuzzy model. Structure identification includes determining the number of rules, the type and parameters of membership functions, and the complexity of local models. Parameter identification involves estimating the parameters of both the membership functions (premise parameters) and the local models (consequent parameters).

One effective approach for structure identification is the clustering-based method, which partitions the input-output data into clusters and derives fuzzy rules from these clusters. The fuzzy C-means clustering algorithm can be extended for this purpose. Given a set of input-output data pairs $\{(x_k, y_k), k = 1,2,\ldots,N\}$, the objective function to be minimized is:

$$J = \sum_{i=1}^{c}\sum_{k=1}^{N}\left(u_{ik}\right)^{m}\left\|z_k - v_i\right\|^2 \tag{6.52}$$

where $z_k = \left[x_k^T, y_k\right]^T$ is the augmented data vector, $v_i$ is the centre of the i-th cluster, $u_{ik}$ is the membership degree of data point $z_k$ to cluster i, c is the number of clusters, and m >1 is the fuzziness parameter.

The cluster centres are updated using:

$$v_i = \frac{\sum_{k=1}^{N}\left(u_{ik}\right)^{m} z_k}{\sum_{k=1}^{N}\left(u_{ik}\right)^{m}} \tag{6.53}$$

The membership degrees are updated according to:

$$u_{ik} = \frac{1}{\sum_{j=1}^{c}\left(\frac{\left\|z_k - v_i\right\|}{\left\|z_k - v_j\right\|}\right)^{2/m-1}} \tag{6.54}$$

Once the clusters are obtained, fuzzy rules are derived by projecting the cluster centres onto the input and output spaces. The antecedent membership functions are typically Gaussian functions centred at the projected cluster centres:

$$\mu_{A_j^i}\left(x_j\right) = exp\left(-\frac{\left(x_j - c_{ij}\right)^2}{\sigma_{ij}^2}\right) \tag{6.55}$$

where $c_{ij}$ and $\sigma_{ij}$ are the centre and width of the membership function for the j-th input variable in the i-th rule.

For consequent parameter estimation, the weighted least squares method is commonly employed. For a given input $x_k$, the degree of activation of each rule is calculated and the parameters of the local models are estimated to minimize:

$$J_c = \sum_{k=1}^{N}\sum_{i=1}^{m} w_i\left(x_k\right)\left(y_k - f_i\left(x_k\right)\right)^2 \tag{6.56}$$

When local linear models are used, equation (6.56) becomes a weighted linear regression problem that can be solved analytically.

The iterative optimization process alternates between updating the premise parameters (membership function parameters) and the consequent parameters (local model parameters). The premise parameters are typically updated using gradient descent:

$$c_j^i(t+1) = c_j^i(t) - \eta_1 \frac{\partial J}{\partial c_j^i} \tag{6.57}$$

$$\sigma_j^i(t+1) = \sigma_j^i(t) - \eta_2 \frac{\partial J}{\partial \sigma_j^i} \tag{6.58}$$

where $\eta_1$ and $\eta_2$ are learning rates and $J$ is the overall identification error.

The validation of the identified fuzzy model involves testing its performance on independent data not used during the identification process. Common validation metrics include the mean squared error (MSE), normalized root mean squared error (NRMSE), and correlation coefficient between predicted and actual outputs. The model's ability to generalize to new operating conditions is crucial for its practical applicability.

Advanced variants of fuzzy identification include recursive algorithms for online identification, where the model parameters are updated continuously as new data become available. The recursive least squares (RLS) algorithm can be adapted for updating consequent parameters:

$$P(k+1) = P(k) - \frac{P(k)\phi(k)\phi(k)^T P(k)}{\lambda + \phi(k)^T P(k)\phi(k)} \tag{6.59}$$

$$\theta(k+1) = \sigma_{ij}(t) + P(k+1)\phi(k)\left(y(k+1) - \phi(k)^T \theta(k)\right) \tag{6.60}$$

where $\phi(k)$ is the regression vector, $\theta(k)$ contains the consequent parameters, $P(k)$ is the covariance matrix, and $\lambda$ is the forgetting factor.

## 6.7 FUZZY WEIGHTED INPUT ESTIMATION FOR THE INPUT FORCE ON THE PLATE STRUCTURE

Input force estimation in structural dynamics represents a challenging inverse problem where unknown forces must be determined from limited, noisy response measurements. This problem is particularly relevant for plate structures subjected to dynamic loading, where traditional inverse methods often fail due to ill-conditioning and measurement uncertainties. Liu and Chen [7] developed a fuzzy weighted estimation approach that addresses these challenges by combining multiple specialized estimators through an intelligent fusion mechanism.

The inverse force estimation problem can be formulated mathematically as recovering the force vector f(k) from the measured structural response vector z(k), where the relationship is governed by the structural transfer function matrix H(ω):

$$Z(\omega) = H(\omega)F(\omega) + V(\omega) \tag{6.61}$$

where $Z(\omega)$, $F(\omega)$and $V(\omega)$represent the Fourier transforms of the response, force, and measurement noise vectors, respectively. Direct inversion of this relationship is often impossible due to insufficient measurements, noise contamination, and the ill-conditioned nature of $H(\omega)$.

The fuzzy weighted approach operates multiple specialized estimators in parallel, each designed for different force characteristics. Let $\hat{f}_j(k)$ represent the force estimate from the j-th estimator, where j∈{1,2,…,M}. The individual estimators include:

**(i) Regularized least squares estimator** for smooth, slowly varying forces:

$$\hat{f}_1(k) = \left(H^T H + \alpha I\right)^{-1} H^T z(k) \tag{6.62}$$

where $\alpha$ is the regularization parameter and $I$ is the identity matrix.

**(ii) Kalman filter-based estimator** for stochastic force processes:

$$\hat{f}_2(k) = \hat{f}_2(k+1) + K(k)\left[z(k) - H\hat{f}_2(k+1)\right] \tag{6.63}$$

where $K(k)$ is the Kalman gain matrix.

**(iii) Wavelet-based estimator** for impulsive or transient forces:

$$\hat{f}_3(k) = W^{-1}\left(Threshold\left(Wz(k)\right)\right) \tag{6.64}$$

where $W$ represents the wavelet transform operator.

The fuzzy inference system assigns time-varying weights $w_j(k)$to each estimator based on three key indicators derived from the measured response characteristics. The **spatial** response distribution indicator $\rho_s(k)$characterizes the spatial pattern of structural responses:

$$\rho_s(k) = \frac{max\left(z(k)\right)}{mean\left(z(k)\right)} \tag{6.65}$$

High values of $\rho_s(k)$ indicate localized responses typical of impact forces, while low values suggest distributed loading.

The **frequency content indicator** $\rho_f(k)$ analyzes the spectral characteristics of the measured signals:

$$\rho_f(k) = \frac{\int_0^{\omega_c} |Z(\omega)|^2 \, d\omega}{\int_0^{\infty} |Z(\omega)|^2 \, d\omega} \tag{6.66}$$

where $\omega_c$ represents a cutoff frequency that separates high and low-frequency content.

The **estimation consistency measure** $\rho_{c,j}(k)$ evaluates how well each estimator's force prediction matches the measured response:

$$\rho_{c,j}(k) = exp\left( -\frac{\left\| z(k) - H\hat{f}_j(k) \right\|^2}{\sigma^2} \right) \tag{6.67}$$

where $\sigma^2$ represents the expected noise variance.

The fuzzy rules governing weight assignment take the linguistic form: *'IF spatial distribution is Localized AND frequency content is High AND consistency is Good THEN assign High weight to impulse estimator'*. These rules are mathematically implemented through membership functions and fuzzy inference mechanisms similar to those described in previous sections.

The final force estimate is computed as the weighted combination:

$$\hat{f}(k) = \sum_{j=1}^{M} \omega_j(k) \hat{f}_j(k) \tag{6.68}$$

subject to the normalization constraint:

$$\sum_{j=1}^{M} \omega_j(k) = 1 \tag{6.69}$$

Validation studies on plate structures subjected to various loading scenarios (impact, distributed pressure, and harmonic excitation) have demonstrated that this fuzzy weighted approach provides superior accuracy and robustness compared to single-estimator methods, particularly under challenging conditions involving measurement noise, limited sensor coverage and varying force characteristics.

## 6.8 SYSTEM IDENTIFICATION OF NONLINEAR SYSTEMS BY ADAPTIVE FUZZY TAKAGI-SUGENO MODEL

Real-world nonlinear systems often exhibit time-varying characteristics due to changing operating conditions, component aging, or environmental variations. Angelov and Filev [8] developed an adaptive fuzzy Takagi-Sugeno approach that extends the static framework with online learning mechanisms for both parameter and structure adaptation, enabling continuous model improvement as new data become available.

The adaptive T-S model maintains the basic rule structure while incorporating time-varying parameters:

$$R_i: \ IF\ x_1(k) is\ A_1^i(k) AND\ x_2(k) is\ A_2^i(k) \ldots AND\ x_n(k) is\ A_n^i(k)$$

$$THEN\ y_i(k+1) = \left(p^i(k)\right)^T x(k) + q^i(k) \tag{6.70}$$

where both the antecedent fuzzy sets $A_n^i(k)$ and the consequent parameters $p^i(k), q^i(k)$ are adapted online based on incoming data.

The adaptation mechanism operates on two levels: parameter adaptation and structural adaptation.

(i) Parameter adaptation continuously updates the consequent parameters using recursive weighted least squares (RWLS):

$$p^i(k+1) = p^i(k) + P^i(k+1)x(k)\bar{w}^i(k)e(k) \tag{6.71}$$

$$P^i(k+1) = \frac{1}{\lambda}\left[P^i(k) - \frac{P^i(k)x(k)x^T(k)P^i(k)\bar{w}^i(k)}{\lambda + x^T(k)x(k)P^i(k)\bar{w}^i(k)}\right] \tag{6.72}$$

where $e(k) = y(k) - \hat{y}(k)$ represents the prediction error, $\bar{w}^i(k)$ is the normalized firing strength of the i-th rule, $\lambda \in [0.95,1]$ is the forgetting factor that enables tracking of time-varying parameters, and $P^i(k)$ is the covariance matrix for the i-th rule.

(ii) Antecedent parameter adaptation updates membership function centres and spreads using gradient descent with momentum:

$$c_j^i(k+1) = c_j^i(k) - \eta_1 \frac{\partial E(k)}{\partial c_j^i(k)} + \mu_1\left[c_j^i(k) - c_j^i(k-1)\right] \tag{6.73}$$

$$\sigma_j^i(k+1) = \sigma_j^i(k) - \eta_2 \frac{\partial E(k)}{\partial \sigma_j^i(k)} + \mu_2\left[\sigma_j^i(k) - \sigma_j^i(k-1)\right] \tag{6.74}$$

where $\eta_1, \eta_2$ are learning rates, $\mu_1, \mu_2$ are momentum coefficients, and the partial derivatives are computed using the chain rule of differentiation.

(iii) Structural adaptation addresses the challenge of unknown optimal model complexity by dynamically adding or removing rules. The rule addition criterion is based on the potential of new data points to form distinct clusters:

$$P(k) = \frac{1}{1+\sum_{i=1}^{m} exp\left(-\alpha\left\|x(k)-c^{i2}\right\|\right)} \tag{6.75}$$

A new rule is added when:

$$P(k) > P_{\text{threshold}} \text{ AND } e^2(k) > \varepsilon_{\text{add}} \tag{6.76}$$

where $P_{threshold}$ and $\varepsilon_{add}$ are predetermined thresholds for potential and prediction error, respectively.

Rule pruning removes rules with consistently low activation levels or redundant behaviour. A rule i is considered for removal when:

$$\frac{1}{N_{\text{window}}} \sum_{j=k-N_{window}+1}^{k} \bar{w}^i(j) < w_{\text{min}} \tag{6.77}$$

where $N_{window}$ defines the evaluation window length and $w_{min}$ is the minimum acceptable average firing strength.

The complete adaptive algorithm processes each new data point through the following recursive procedure:

1. Evaluate rule activations using current membership functions.
2. Compute model output using equation (6.70) and prediction error e(k).
3. Update consequent parameters using RWLS (equations 6.71–6.72).
4. Update antecedent parameters using gradient descent (equations 6.73–6.74).
5. Assess structural adaptation needs using criteria (6.76) and (6.77).
6. Add new rules or prune existing ones if necessary.

This adaptive framework has demonstrated effectiveness in diverse applications including chemical process control with catalyst aging, robotic systems with configuration-dependent dynamics, and biomedical systems requiring patient-specific adaptation. Experimental results show that the adaptive T-S model successfully maintains accuracy while tracking system changes, providing a practical solution for nonlinear system identification in time-varying environments.

## EXERCISES

1. How does the fuzzy Kalman filter enhance the decision-making capability of the standard Kalman filter? Explain the role of the Fuzzy Logic Process Variable and why FKF performs better for manoeuvring target tracking.
2. Distinguish between fuzzy C-means clustering and traditional hard clustering methods. Why is the concept of membership degrees particularly important for system identification applications?

3. Describe the hybrid learning algorithm in ANFIS and explain how it combines the advantages of neural networks and fuzzy logic systems. What makes this approach computationally efficient for nonlinear system modelling?
4. How does ANFIS-based fault detection work for sensors and actuators? Explain the residual generation process and how different fault types can be identified through pattern analysis.
5. Adaptive fuzzy Takagi–Sugeno models extend the basic fuzzy modelling approach. What makes the adaptive version suitable for time-varying systems and how does it maintain balance between accuracy and interpretability?

## REFERENCES


1. Raol, J.R., *Multisensor Data Fusion with MATLAB,* CRC Press, Taylor & Francis Group, Florida, USA, 2010.
2. Raol, J.R., and Naidu, V.P.S., *Mathematical Modelling of Aerospace Dynamic Systems with Practical Applications*, CRC Press, Taylor & Francis Group, Florida, USA, 2025.
3. Raol, J.R., Girija, G., and Singh, J., *Modelling and Parameter Estimation of Dynamic Systems. IEE/IET Control Series* Vol. 65, The Institution of Electrical Engineers (IEE), London, United Kingdom, Aug. 2004.
4. Jang, J.S.R., "ANFIS: Adaptive-Network-Based Fuzzy Inference System", *IEEE Transactions on Systems, Man, and Cybernetics*, Vol. 23, No. 3, pp. 665–685, 1993.
5. Takagi, T., and Sugeno, M., "Fuzzy Identification of Systems and Its Applications to Modeling and Control", *IEEE Transactions on Systems, Man, and Cybernetics*, Vol. 15, No. 1, pp. 116–132, 1985.
6. Wang, L.X., *A Course in Fuzzy Systems and Control,* Prentice Hall PTR, Upper Saddle River, NJ, USA, 1997.
7. Liu, F., and Chen, G., "Fuzzy System Identification Using a Robust Learning Algorithm", *IEEE Transactions on Fuzzy Systems*, Vol. 11, No. 6, pp. 809–820, 2003.
8. Angelov, P.P., and Filev, D.P., "An Approach to Online Identification of Takagi-Sugeno Fuzzy Models", *IEEE Transactions on Systems, Man, and Cybernetics, Part B: Cybernetics*, Vol. 34, No. 1, pp. 484–498, 2004.

# 7 Concept and Theory of Genetic Algorithms (GAs)

## 7.1 EVOLUTION IN NATURE

The natural systems have evolved over millions and millions of years. They have gone through iterations over many generations and in the process have become very robust, especially to their many different environments. Due to its strong evolutionary 'experience', Nature offers good solutions whenever robustness is called for. Biological systems are generally more robust, efficient, and flexible compared to the most sophisticated artificial systems. Artificial systems have to learn from biological systems to improve their performance and carry out their daily-required functions for longer period of time and with greater efficiency.

Genetic algorithms are subcomponents of the evolutionary algorithms which in turn are part of the grand evolution in nature; and this evolution refers to the gradual change in the heritable characteristics of biological populations (all kinds of species) over successive generations; the latter is put in action by mechanisms like natural selection and genetic drift, which in turn result in the diversification/variation and adaptation of life forms. By definition evolution is the process by which populations of living organisms change over time (often huge), and these changes are passed down through inheritance to subsequent generations.

It operates on the basis of three mechanisms: i) **Natural selection:** Individuals with traits that are better suited to their environment are more likely to survive and reproduce, passing on those advantageous traits (behaviours or tendencies to acquire that behaviours/BTAB) to their offspring; ii) **Genetic drift:** Random fluctuations in the frequency of alleles (different versions of a gene) within a population occur, especially common in small populations; and iii) **Gene flow:** The movement of genes (considered as a unit of life) between populations, which can introduce new variations and alter allele frequencies.

Some glaring examples of evolved species are: a) **Giraffe's long neck:** Giraffes with slightly longer necks had an advantage in reaching food sources, and over time, this trait (BTAB) became more common in the population; b) **Antibiotic resistance in bacteria:** Bacteria can evolve resistance to antibiotic medicines through natural selection, where bacteria with mutations that make them (more) resistant survive and reproduce more easily.


DOI: 10.1201/9781003616580-7

Key Concepts in evolution are: i) **Variation:** Individuals within a population exhibit differences in their traits/BTAB; ii) **Heritability:** Traits/BTAB are passed down from parents to offspring; iii) **Adaptation:** Organisms evolve traits/BTAB that help them survive and reproduce in their environment; and iv) **Biodiversity:** Evolution is the driving force behind the incredible diversity of life on Earth.

The evolutionary biology is the study of evolution encompassing a wide range of topics, including the origin of life, the diversification of species, and the mechanisms of evolutionary change. Thus, evolution is a process that results in changes in the genetic material of a population over time; and it reflects the adaptations of organisms to their changing environments and can result in altered genes, novel traits/BTAB, and new species.

## 7.2 EVOLUTIONARY ALGORITHMS

Evolutionary algorithms (EAs) are nature-inspired optimization techniques that mimic biological evolution, using mechanisms like selection, crossover, and mutation to find optimal solutions to problems.

Some key concepts in EAs are: i) **Inspiration from nature:** Evolutionary algorithms draw inspiration from the principles of natural selection and evolution, where populations of individuals evolve over generations, with the fittest individuals surviving and reproducing; ii) **Population-based:** These algorithms work with a population of potential solutions (individuals) rather than a single solution; and iii) **Iterative process:** The algorithms iteratively refine the population through generations, using evolutionary operators to create new, and preferably robust solutions based on the existing ones.

**The Evolutionary Operators** are: i) **Selection:** Individuals are selected based on their fitness (how well they solve the problem); ii) **Crossover:** Selected individuals are combined to create new offspring, inheriting characteristics from both parents; iii) **Mutation:** Random changes (mutations) are introduced into the offspring to explore, and perhaps expand the solution space; iv) **Fitness function:** A fitness function evaluates the quality of each individual in the population, guiding the evolutionary process towards better solutions; and v) **Termination criteria:** The algorithm is stopped when a satisfactory solution is found or a predetermined number of generations have been reached.

The examples of evolutionary algorithms are: i) **Genetic Algorithms (GAs):** A well-known type of evolutionary algorithm that uses a population of encoded solutions (chromosomes) and applies genetic operators to evolve towards better/robust solutions; ii) **Genetic Programming (GP):** A variant of GA that evolves computer programs as solutions; iii) **Evolution Strategies (ES):** Another type of EA that focuses on evolving parameters of a solution, often used in optimization problems; iv) **Differential Evolution (DE):** A fast and powerful EA that is often used for engineering optimization problems; and v) **Cellular Evolutionary Algorithms (cEAs):** EAs where individuals interact with their neighbours, leading to a more localized and potentially parallel evolution.

Evolutionary algorithms are used in various fields: i) **Optimization:** Finding the best and robust solution to a problem, such as minimizing costs or maximizing profits; ii) **Machine Learning:** Training machine/computer learning models by optimizing their parameters; iii) **Robotics:** Designing robot controllers and behaviours; iv) **Drug Discovery:** Identifying potential drug candidates; and v) **Financial Modelling:** Optimizing investment strategies.

## 7.3 GENETIC ALGORITHMS

Genetic algorithms are search methods inspired by nature's evolutionary systems. They can be used to obtain global and robust solutions to many optimization problems in science, engineering, economics, psychology and biology. Genetic algorithms are based on some of the principles that govern the natural systems.

Genetic algorithms are computational optimization schemes with an approach that seems rather unconventional. The algorithms solve optimization problems imitating nature in the way it has been working for millions of years on the evolution of life forms. Inspired by the biological systems, genetic algorithms adopt the rules of natural selection and genetics to attain robustness. Acting on the premise of survival of the fittest, a population or samples of feasible solutions is combined in a manner similar to the combination of chromosomes in a natural genetic system. The fitter population members pass on their structures as genes in far greater measure than their less fit members do. As the generations evolve, the net effect is evolution of the population towards an optimum (species, solution, etc.). Genetic algorithms operate by combining the information present in different possible solutions in a way that a better solution is obtained in next/future generations. The terms used in the study of genetic algorithms are given in Table 7.1.

### 7.3.1 Operations in a Typical Genetic Algorithm

#### 7.3.1.1 Chromosomes

They represent encoding of information in a string of finite length and each chromosome consists of a string of bits (binary digit; 0 or 1). Or it could be a symbol from a set of more than two elements. Generally, for function optimization, chromosomes are constructed from binary strings as seen from the following table:

**TABLE 7.1**
**Genetic algorithm and natural genetic system**

| Natural Genetic System (NGS) | Genetic Algorithm (GA) |
|---|---|
| Chromosomes | String of numbers |
| Gene | Feature or detection |
| Allele | Feature value |
| Locus | String position |
| Genotype | Structure |
| Phenotype | Parameter set, alternative form, a decoded structure |

| Parameter value | String |
|---|---|
| 6 | 000110 |
| 34 | 100010 |

The long stretches of DNA that carry the genetic information needed to build an organism are called chromosomes. The chromosomes consist of genes. Each gene represents a unit of information and it takes different values. These values are called alleles at different locations called loci. The strings, composed of features or detectors, assume values like 0 or 1, which are located at different positions in the string. The total package or system is called the genotype or structure. The phenotype results when interaction of genotype with environment takes place.

#### 7.3.1.2 Population and Fitness

Genetic algorithms operate on population of possible solutions with chromosomes. The population members are known as individuals. Each individual is assigned a fitness-value based on the objective function, or cost function. Better individuals (solutions) have higher fitness values and weaker ones have lower fitness-values.

#### 7.3.1.3 Initialization and Reproduction

By randomly selecting information from the search space and encoding it, a population of possible initial solutions is created. Reproduction is a process in which individual strings are copied as per their fitness values. Thus, the strings with a greater fitness-value have higher probability of contributing one or more offsprings to the next generation.

#### 7.3.1.4 Crossover

In a crossover, a site is selected randomly along the length of the chromosomes, and each chromosome is split into two pieces at the crossover site. The new ones are formed by joining the top piece of one chromosome with the tailpiece of the other.

#### 7.3.1.5 Mutation

It is a small operation in which a bit in a string is changed at a random location. The main idea is to break monotony and add a bit of novelty. This operation would help gain information not available to the rest of the population. It lends diversity to the population.

#### 7.3.1.6 Generation

Each iteration in the optimization procedure is called a generation. In each generation pairs are chosen for crossover operation, the fitness is determined, and mutation is carried out during crossover operation (during or after has a subtle distinction). With these operations performed, a new population evolves which is carried forward.

#### 7.3.1.7 Survival of the Fittest

The individuals may be fitter or weaker than some other population members. Hence, the members must be ranked as per their fitness value. In each generation, the weaker members are allowed to wither out and the ones with good fitness values take part in the genetic operations. The net result is the evolution of the population towards the global optimum.

#### 7.3.1.8 Cost function, Decision Variables, and Search Space

In most practical optimization problems, the goal is to find optimal parameters to increase the production and/or to reduce the expenditure/loss. That is to get maximum profit by reorganizing the system and its parameters that affect the cost function. Since, this in effect, reflects on the cost, it is represented by the cost function. A carefully devised and convergent computational algorithm would eventually find an optimum solution to the problem. The parameters of the system that decide the cost are termed decision variables. The search space is a Euclidean space in which parameters take different values and each point in the space is a probable solution.

**Example 7.1: Simple Genetic Algorithm illustration**

A simple genetic algorithm is described, which will use binary coding technique.

Step 1: Create population of N samples from a chosen search space – denoting the decision variables
Step 2: Produce series of 0s and 1s to create chromosomes – i.e., encoding the decision variables
Step 3: Calculate the cost function values and assign fitness (values) to each member
Step 4: Sort the members accordingly to their respective fitness values
Step 5: Carry out crossover operation taking two chromosomes at a time
Step 6: Mutate the chromosomes with a given probability of mutation
Step 7: Retain the best members of the population and remove the weaker members based on their fitness values
Step 8: Replace old generation by new one and repeat steps 3 to 8.

Let us consider the problem of maximizing the function:

$$f(x) = x^2 - 64x + 100$$

Here, $x$ varies from 0 to 63. The function '$f$' has a maximum value of 100 at x=0. The decision variables are coded in strings of finite length. We can encode the variables as a binary string of length 6. We create an initial population with 4 samples by randomly selecting them from the interval 0 to 63 and encode each sample. A binary string of length 6 can represent any value from 0 to 63; $(2^6-1)$. Four encoded samples in the initial population are: 5 (000101); 60 (111100); 33 (100001); 8 (001000). These individuals are sorted according to their fitness values- they are arranged in the descending order of their fitness values. For simplicity, mutation is not used. Also, the problem that could have been solved using the conventional approach is used to illustration of GA operations for simplicity. For the present example, the fitness value is the same as the value of cost function and these individuals are sorted accordingly to their fitness values:

| No. | X | String | Fitness value |
|---|---|---|---|
| 1 | 60 | 111100 | -140 |
| 2 | 5 | 000101 | -195 |
| 3 | 8 | 001000 | -348 |
| 4 | 33 | 100001 | -923 |

Next, the crossover is randomly selected, and in the first generation, the 1st and 2nd strings are crossed over at site 3 to get two new strings:

| Crossover site | New strings | Fitness of new strings |
|---|---|---|
| 111! 100 | 111101 | -83 |
| 000! 101 | 000100 | -140 |

Similarly, the 3rd and 4th strings are crossed over at site 2, to get:

| Crossover site | New strings | Fitness of new strings |
|---|---|---|
| 00! 1000 | 000001 | 37 |
| 10! 0001 | 101000 | -860 |

Sorting these new individuals one gets:

| No. | X | String | Fitness value |
|---|---|---|---|
| 1 | 1 | 000001 | 37 |
| 2 | 61 | 111101 | -83 |
| 3 | 4 | 000100 | -140 |
| 4 | 40 | 101000 | -860 |

It is now seen that in one generation fitness is improved from -140 to 37 (f(1) > f(60)). The weakest member of the population is replaced by the fittest member of the previous population; string 101000 that has fitness -860 is replaced by string 111100, whose fitness is -140. In the 2nd generation, the 1st and 2nd strings are crossed over at site 1 to obtain the following:

| Crossover site | New strings | Fitness of new strings |
|---|---|---|
| 0! 00001 | 011101 | -915 |
| 1! 11101 | 100001 | -923 |

Similarly, the 3rd and 4th strings are crossed over at site 3 to obtain:

| Crossover site | New strings | Fitness of new strings |
|---|---|---|
| 000! 100 | 000100 | -140 |
| 111! 100 | 111100 | -140 |

We replace the weakest member by the fittest member of the previous population (string 100001 with fitness value of -923 is replaced by the string 000001 with fitness value of 37). The sorting results in:

| No. | X | String | Fitness value |
|---|---|---|---|
| 1 | 1 | 000001 | 37 |
| 2 | 4 | 000100 | -140 |
| 3 | 60 | 111100 | -140 |
| 4 | 29 | 011101 | -915 |

In the 3rd generation, the process of crossover at site 4 is carried out (not shown here). The new set of strings in the population, after replacement of the weakest by the fittest member is given as:

| No. | X | String | Fitness value |
|---|---|---|---|
| 1 | 0 | 000000 | 100 |
| 2 | 1 | 000001 | 37 |
| 3 | 61 | 111101 | -83 |
| 4 | 5 | 000101 | -195 |

We see that as the genetic algorithm progresses from one generation to the next, the improved solutions evolve. At x=0, f(x)=100, the desired result.

### 7.3.2 Stopping Strategies for Genetic Algorithms

One needs to know as to where and when to stop the genetic algorithm iterations. If the population size is fixed, then more generations might be needed for the convergence of a genetic algorithm to an optimal solution. One way is to track the fitness value for no further improvement. As the algorithmic steps progress, a situation would occur wherein we need a large number of generations to bring about a small improvement in the fitness-value. One can define predetermined number of generation/iterations to solve the problem. Also, insignificant change in the norm of estimated parameters can be tracked for a few consecutive iterations before stopping the search. It must be possible to do an effective search if one exploits some important similarities in the coding used in genetic algorithms. Another way is to evaluate the gradient of the cost function and use the conventional approaches for assessing the quality of the estimates for their convergence to true values. It is possible to use

GA with gradient-based approach for evaluating the estimation accuracy as is done for OEM (Chapter 3). Again, here as is true with all the other parameter estimation methods, the matching of time histories of the measured data and model responses is a necessary and not the sufficient condition. An increase in the number of samples would generally increase the success rate.

### 7.3.3 Genetic Algorithms without Coding of Parameters

Genetic algorithms become more complex because of coding the chromosomes, especially for more complex problems. In the problems of science and engineering, we come across real numbers. Thus, we need to use real numbers and still use genetic algorithms on these numbers for solving optimization problems. Major change is in the crossover and mutation operations. Averaging the two samples, i.e., the two sets of parameter values can perform the crossover operation. After the crossover, the best individual is mutated. In mutation, a small noise is added.

Assume that two individuals have $\beta_1$ and $\beta_2$ as numerical values of the parameters. Then after crossover, we obtain the new individual as $(\beta_1 + \beta_2)/2$. For mutation we have $\beta_3 = \beta_1 + \varepsilon * v$, where ε is a constant and $v$ is a number chosen randomly between -1 and 1. Thus, all the genetic algorithm operations can be performed by using real numbers like 4.8904, etc, without coding the samples. This feature is extremely well suited for several engineering applications: parameter estimation, control, optimization, and signal processing [1].

## 7.4 SCHEMES OF PARALLELIZATION OF GENETIC ALGORITHMS

The genetic algorithms are powerful and yet very simple strategies for optimization problems. They can be used for multi-modal, multi-dimensional and multi-objective optimization problems, not only in science and engineering, but also in business and related fields. However, despite the fact that the computations required in genetic algorithm operations are very simple, they become complex as the number of iterations grows. This will put heavy demand on the computational power. Often, the procedures can be parallelized and the power of parallel computers can be used. Since genetic algorithms can work on population samples simultaneously, their natural parallelism can be exploited to implement them on parallel computers. One scheme for a parallel Genetic Algorithm is shown in Figure 7.1. The sorting is split between two processors. In this scheme, the host processor does the job of crossover, mutation, etc.

## 7.5 SIMULATED ANNEALING

Simulated annealing is another nature-inspired optimization technique that draws its inspiration from the metallurgical process of annealing, wherein a material is heated to a high temperature and then slowly cooled to achieve a desired crystalline structure with minimum energy. This physical process allows atoms to move freely at high temperatures and gradually settle into optimal positions as the temperature decreases, ultimately reaching a state of minimal internal energy and maximum structural stability. The algorithm was first introduced by Kirkpatrick et al. [2] who demonstrated

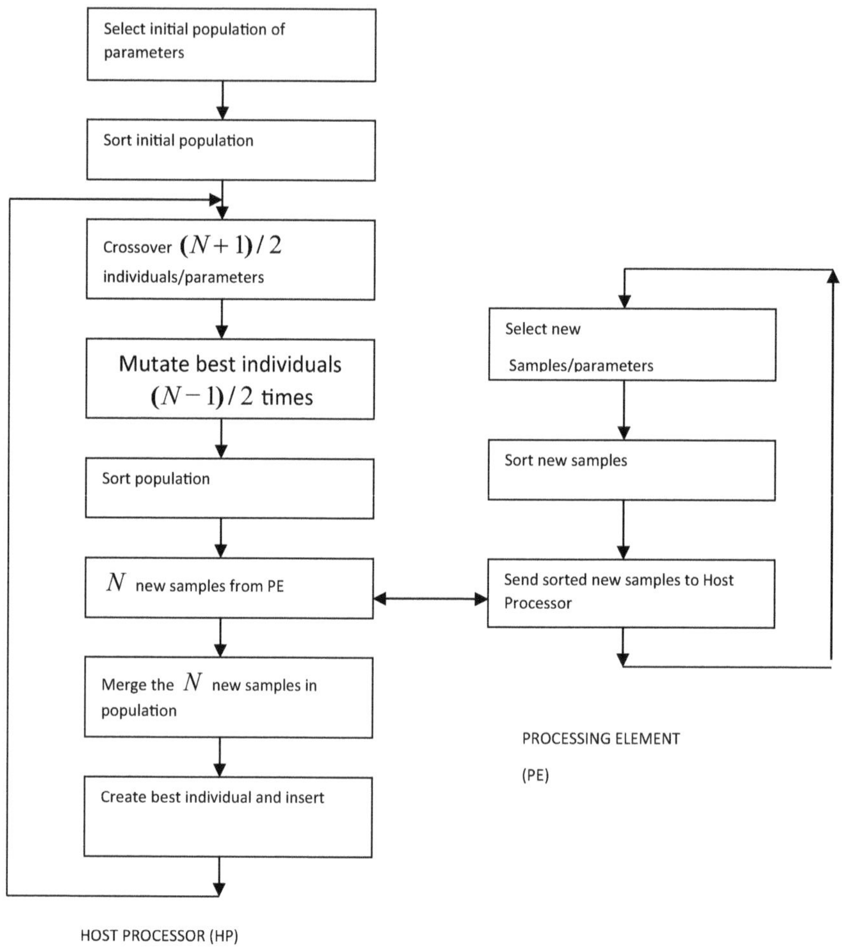


**FIGURE 7.1** Schematic of parallel genetic algorithm

its effectiveness in solving complex optimization problems that traditional methods found difficult to handle.

The simulated annealing algorithm mimics this thermal process by introducing a temperature parameter that controls the acceptance of solutions during the optimization process. At high temperatures, the algorithm accepts both better and worse solutions with relatively high probability, allowing extensive exploration of the solution space. As the temperature gradually decreases according to a predetermined cooling schedule, the algorithm becomes increasingly selective, accepting fewer inferior solutions and focusing on exploitation of promising regions. This mechanism enables the algorithm to escape local optima, which is particularly valuable in multimodal optimization landscapes commonly encountered in parameter estimation problems.

The core mechanism of simulated annealing involves generating a new solution by making a small perturbation to the current solution through a neighbourhood function. If the new solution is better than the current one, it is always accepted. If the new solution is worse, it may still be accepted with a probability that depends on both the magnitude of the deterioration and the current temperature. This acceptance probability is typically calculated using the Boltzmann probability distribution, expressed as $P = \exp(-\Delta E/T)$, where ΔE represents the change in the objective function value and T is the current temperature.

The algorithm begins with an initial solution and a high initial temperature. At each iteration, a neighbouring solution is generated through a predefined neighbourhood function, which could involve simple modifications like changing one or more parameters within specified bounds. The difference between the cost functions of the new and current solutions is calculated. If this difference indicates improvement, the new solution becomes the current solution. If the difference indicates deterioration, the acceptance probability is computed and compared with a randomly generated number between zero and one to determine whether to accept the inferior solution.

The temperature reduction follows a cooling schedule, which is crucial for the algorithm's performance. Common cooling schedules include exponential cooling, where the temperature is multiplied by a factor slightly less than one at each iteration, linear cooling, where a constant value is subtracted from the temperature, and logarithmic cooling, which provides slower temperature reduction. The choice of cooling schedule significantly affects the balance between exploration and exploitation and consequently the quality of the final solution. In parameter estimation applications, the cooling schedule must be carefully tuned to ensure adequate exploration of the parameter space while maintaining computational efficiency.

Simulated annealing possesses several attractive properties that make it suitable for complex optimization problems encountered in intelligent estimation. Unlike genetic algorithms, it operates on a single solution at a time, making it computationally efficient in terms of memory requirements. The algorithm can escape local optima due to its probabilistic acceptance of inferior solutions, particularly at higher temperatures. This characteristic makes it particularly effective for parameter estimation in dynamic systems where the objective function exhibits multiple local minima due to measurement noise and model uncertainties.

The algorithm's simplicity in implementation is another significant advantage, as it requires only a neighbourhood function, a cooling schedule, and an initial solution. Furthermore, simulated annealing provides theoretical guarantees of convergence to the global optimum under certain conditions, particularly when the cooling schedule is sufficiently slow. In practical implementations, faster cooling schedules are often employed to achieve reasonable computational times, though this may compromise the theoretical convergence guarantees.

Applications of simulated annealing in parameter estimation have been extensively documented in literature. The method has been successfully applied to estimate parameters in dynamic systems where conventional gradient-based methods fail due to the presence of multiple local minima. In aerospace applications, simulated annealing has proven effective for estimating aerodynamic parameters from flight test data, where the optimization landscape is often characterized by numerous local optima. The algorithm's ability to accept temporarily inferior solutions makes it particularly suitable for problems where the parameter space is highly discontinuous or where the objective function contains significant noise.

The performance of simulated annealing in parameter estimation problems is significantly influenced by the choice of initial temperature, cooling schedule, and neighbourhood function. The initial temperature should be set high enough to allow adequate exploration of the parameter space, while the cooling schedule should provide sufficient time for the algorithm to locate and converge to the global optimum. The neighbourhood function defines the local search mechanism and should be designed to generate feasible perturbations that maintain the physical constraints of the estimation problem.

## 7.6 ADAPTIVE RANDOM SEARCH AND GENETIC ALGORITHMS FOR OPTIMIZATION

The integration of adaptive random search techniques with genetic algorithms represents a significant advancement in evolutionary optimization methodologies. While genetic algorithms provide a robust framework for population-based search through their inherent mechanisms of selection, crossover, and mutation, the incorporation of adaptive random search elements enhances their exploration capabilities and convergence characteristics, particularly in complex parameter estimation problems.

Adaptive random search algorithms modify their search strategy based on the success or failure of previous iterations, thereby learning from the optimization history to improve future search directions [3]. These algorithms typically maintain a step size or search radius that is adjusted based on the recent performance of the optimization process. When consecutive iterations yield improvements in the objective function, the step size is increased to accelerate convergence. Conversely, when several consecutive iterations fail to produce improvements, the step size is reduced to focus the search in more promising regions.

The combination of adaptive random search with genetic algorithms can be implemented through several approaches. One effective strategy involves using adaptive random search to enhance the mutation operation in genetic algorithms. Instead of applying fixed mutation rates and magnitudes, the mutation parameters are adapted based on the population's diversity and the rate of improvement in fitness values. When the population diversity is high and improvements are being achieved, larger mutation steps can be employed to maintain exploration. As the population converges and improvements become marginal, smaller mutation steps help fine-tune the solutions.

Another approach involves incorporating adaptive random search as a local search mechanism within the genetic algorithm framework. After the standard genetic operations of selection, crossover, and mutation, selected individuals from the population undergo local improvement through adaptive random search. This hybrid approach combines the global search capabilities of genetic algorithms with the local search efficiency of adaptive random search, resulting in improved convergence characteristics and solution quality.

In parameter estimation applications, this hybrid methodology proves particularly beneficial when dealing with high-dimensional parameter spaces where genetic algorithms alone might require excessive computational time to achieve fine convergence. The adaptive random search component provides rapid local convergence once genetic algorithms have identified promising regions in the parameter space. This combination is especially effective for estimating parameters in complex dynamic systems where the objective function exhibits multiple local optima separated by regions of poor fitness.

The adaptive mechanism in random search can also be extended to control the population size in genetic algorithms dynamically. Based on the convergence rate and population diversity measures, the algorithm can increase the population size when diversity is low or decrease it when computational efficiency is prioritized. This dynamic population sizing ensures that computational resources are utilized optimally throughout the optimization process.

The performance of the combined approach depends on several factors including the adaptation strategy for the random search component, the balance between genetic operations and local search, and the criteria for triggering adaptive mechanisms. The adaptation can be based on various metrics such as the rate of improvement in the best fitness value, the diversity of the population measured through variance of fitness values, or the success rate of recent search operations.

For practical implementation in parameter estimation problems, the hybrid algorithm typically begins with standard genetic algorithm operations to establish a diverse population and identify promising regions in the parameter space. As the optimization progresses, adaptive random search mechanisms are progressively emphasized to achieve fine-tuned convergence. The termination criteria can combine traditional stopping conditions such as maximum number of generations with adaptive criteria based on the convergence rate and parameter uncertainty estimates.

The integration of adaptive random search with genetic algorithms has shown significant improvements in convergence speed and solution quality across various parameter estimation problems in engineering systems. This methodology represents a natural evolution of optimization techniques that leverages the strengths of both population-based and single-solution methods to achieve robust and efficient parameter estimation in complex dynamic systems.

## EXERCISES

1. Explain the analogy between the natural genetic system and genetic algorithms. Use the terms chromosome, gene, allele, locus, genotype, and phenotype to illustrate how biological evolution concepts map onto GA operations.
2. Describe the roles of selection, crossover, and mutation in a genetic algorithm. Why is mutation considered essential, even though it introduces randomness?
3. A genetic algorithm encodes integers in the range 0 to 63 as binary strings of length 6. a) Encode the number 45 as a binary string of length 6. b) If a crossover takes place between the binary strings of 45 and 23 at the 3rd bit, write down the new offspring strings.

4. Compare genetic algorithms (GAs) and simulated annealing (SA) in terms of:
    - The number of solutions handled at a time
    - Mechanisms used to escape local optima
    - Stopping criteria
5. Discuss how adaptive random search can enhance the performance of genetic algorithms. In what way does integrating local search (adaptive random search) with global search (GA) improve convergence and robustness?

## REFERENCES


1. Raol, J.R., Girija, G., and Singh, J., *Modelling and Parameter Estimation of Dynamic Systems. IEE/IET Control Series* Vol. 65, The Institution of Electrical Engineers (IEE), London, United Kingdom, Aug. 2004.
2. Kirkpatrick, S., Gelatt Jr., C.D., and Vecchi, M.P., "Optimization by Simulated Annealing", *Science*, Vol. 220, No. 4598, pp. 671–680, 1983.
3. Gheorghies, O., Luchian, H., and Gheorghies, A., "A Study of Adaptation and Random Search in Genetic Algorithms", *2006 IEEE International Conference on Evolutionary Computation*, Vancouver, BC, Canada, 2006.

# 8 Estimation with Genetic Algorithms (GAs)

## 8.1 INTRODUCTION

Parameter estimation using GA involves using the GA's evolutionary approach to find the OIFR (optimal in fact robust) values of parameters in a system's model, by iteratively refining a population of candidate solutions (chromosomes/samples) based on their fitness (how well they fit the data via a chosen cost function).

Parameter estimation is the process of finding the best values of the parameters of a mathematical model that best explain or predict a set of observed data [1]. These parameters could represent the coefficients of a regression model, the rates in a kinetic model, or any other unknown values that influence the model's behaviour.

Procedure of using GA for parameter estimation has the steps: i) **Encoding Parameters:** Each chromosome/sample (of parameter vector); CSPV represents a set of parameter values, and the GA uses genetic operators (crossover and mutation) to create new CSPV, exploring the parameter space; ii) **Fitness Function:** A fitness function (inverse of the cost function; this is often minimized, and hence the fitness function is maximized)) is used to evaluate how well each chromosome (CSPV) fits the observed data; iii) **Iteration and Refinement:** The GA iteratively generates new populations, with the fittest chromosomes having a higher chance of being selected for reproduction and contributing to the next generation; and iv) **Convergence:** The process continues until the GA converges to a set of parameter values that represent a good fit for the data, or until a predefined stopping criterion is met.

Advantages of Using GAs for Parameter Estimation are: a) **Handles Complex Models:** GAs can be particularly effective for parameter estimation in complex models where traditional optimization methods (like gradient-based methods) might struggle; b) **Global Optimization:** GAs are designed to find the global optimum (the best solution) rather than getting stuck in local optima; c) **Robustness:** GAs are relatively robust to noisy data and can handle non-smooth or discontinuous fitness functions/landscapes; and d) **Parallelization:** The GA's evolutionary nature lends itself well to parallelization, allowing for faster computation.

Some application examples are: i) **Building Thermal Network Model:** GAs can be used to estimate the parameters of a building thermal network model, which helps predict building cooling/heating loads; ii) **Cosmological Parameter Estimation:** GAs can be used to estimate parameters in cosmological models, like those related

DOI: 10.1201/9781003616580-8

to dark energy; iii) **Multiple Linear Regression:** GAs can be used to obtain maximum likelihood estimates of parameters in multiple linear regression models; and iv) **PID Control Parameter Tuning:** GAs can be used to tune the parameters of a Proportional-Integral-Derivative (PID) controller

## 8.2 PARAMETER ESTIMATION USING GENETIC ALGORITHM

Most of the parameter estimation methods are based on the minimization of the cost function of the estimation error, resulting in utilization of the gradient of the cost function (with respect to parameter vector). The application of the genetic algorithm to a parameter estimation problem does not need utilization of the gradient of the cost function.

Consider the problem of parameter estimation as follows:

$$z = H\beta + v; \quad \hat{z} = H\hat{\beta} \tag{8.1}$$

The cost function is formulated as

$$E = \frac{1}{2}\sum (z-\hat{z})^T (z-\hat{z}) = \frac{1}{2}\sum (z - H\hat{\beta})^T (z - H\hat{\beta}) \tag{8.2}$$

Now in the gradient-based method, the minimum is obtained by differentiation of the cost function with respect to parameter vector, $\partial E / \partial \beta$, and then the resulting equation is solved to get the expression for the parameter vector; instead we can use genetic algorithm.

**Example 8.1** Consider the third-order system described by:

$$\begin{bmatrix} \dot{x}_1 \\ \dot{x}_2 \\ \dot{x}_3 \end{bmatrix} = \begin{bmatrix} -2 & 0 & 1 \\ 1 & -2 & 0 \\ 1 & 1 & -1 \end{bmatrix} \begin{bmatrix} x_1 \\ x_2 \\ x_3 \end{bmatrix} + \begin{bmatrix} 1 \\ 0 \\ 1 \end{bmatrix} u \tag{8.3}$$

Here, $u$ is the doublet input to the system. The output is described by

$$z = \begin{bmatrix} 2 & 1 & -1 \end{bmatrix} \begin{bmatrix} x_1 \\ x_2 \\ x_3 \end{bmatrix} \tag{8.4}$$

Obtain the doublet response of the system and use u and z in the genetic algorithm to estimate all the 15 parameters.

**Solution 8.1**

The system is simulated with doublet input and has total simulation time of 20 seconds (sampling interval $\Delta t = 0.1$ sec; number of data points=200). Figure 8.1 (a)

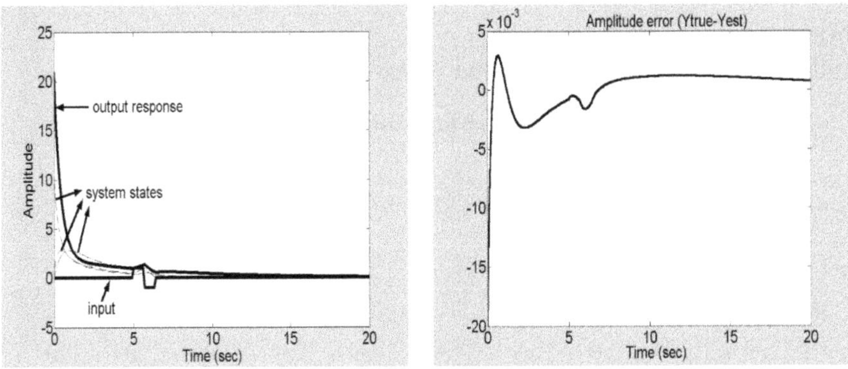


**FIGURE 8.1** (a) System response and doublet input (Example 8.1). (b) Output error w.r.t true data (SNR = ∞) (Example 8.1)

shows the doublet input $u$ and system response $z$. Figure 8.1 (b) shows the response error for the case with no noise. The estimation of parameters is accomplished by using file '**parestga.m**' placed in folder **'Ch8GAex1'**. The initial state of the system, $x(0) = [10\ 1\ 0.1]$; with: i) POPSIZE = 100 (sets of parameters/ population size); and ii) MAXITER = 100 (number of GA iterations).

The initial population of parameters and fitness values are given in Table 8.1 and the estimated parameters for various noise levels are given in Table 8.2.

**Example 8.2** Find the minimum of the function $f(b) = b^2 - 64b + 1025$ using the genetic algorithm, where $b$ varies from 0 to 63; see Figure 8.2.

**Solution 8.2** From Figure 8.2 we see that the minimum for f($b$) is at $b = 32$. Using the genetic algorithm the minimum was found at $b = 32$. Figure 8.3 shows the plot of '$b$' versus genetic algorithm iterations. The estimation of parameter '$b$' is accomplished by using file '**parestga.m**' placed in folder **'Ch8GAex2'**: i) POPSIZE = 10 (sets of parameter/ population size); and ii) MAXITER = 20 (number of iterations); we see that the convergence is reached in less than 10 iterations.

**Example 8.3** Find the global minimum of the function (see Figure 8.4) $f(b) = b^3 - 45b^2 + 600b + v$ using the genetic algorithm, where '$b$' varies from 1 to 25, and 'v' is the measurement noise.

**Solution** 8.3 The data simulation is carried out using function $f(b)$ with '$v$' as an additive white Gaussian noise. In the Figure 8.4, the global minimum for $f(b)$ is at $b = 1$. Using the genetic algorithm the global minimum was found to be at $b = 1.005$. Figure 8.5 shows the plot of '$b$' versus genetic algorithm iterations. The estimation of parameter '$b$' is accomplished by using file '**parestga.m**' placed in folder **'Ch8GAex3'**: i) POPSIZE = 100 (sets of parameter – population size); and ii) MAXITER = 250 (number of GA iterations). The estimates of parameter $b$ are presented in Table 8.3.

**TABLE 8.1**
**Initial population of parameters and fitness (Example 8.1)**

| | Initial 10 populations of parameters | | | | | | | | | |
|---|---|---|---|---|---|---|---|---|---|---|
| | 1 | 2 | 3 | 4 | 5 | 6 | 7 | 8 | 9 | 10 |
| $a_{11}$ | -1.1 | -2.19 | -2.97 | -1.99 | -2.01 | -1.60 | -2.73 | -2.55 | -2.17 | -2.18 |
| $a_{12}$ | 0 | 0 | 0 | 0 | 0 | 0 | 0 | 0 | 0 | 0 |
| $a_{13}$ | 1.11 | 1.42 | 0.95 | 0.93 | 1.32 | 0.95 | 1.39 | 1.26 | 1.37 | 0.77 |
| $a_{21}$ | 0.99 | 0.91 | 1.43 | 0.81 | 1.15 | 1.20 | 0.70 | 1.03 | 0.52 | 0.94 |
| $a_{22}$ | -1.22 | -1.21 | -2.07 | -2.62 | -1.36 | -1.76 | -2.40 | -1.72 | -1.46 | -1.13 |
| $a_{23}$ | 0 | 0 | 0 | 0 | 0 | 0 | 0 | 0 | 0 | 0 |
| $a_{31}$ | 0.96 | 0.85 | 1.35 | 1.18 | 0.84 | 1.46 | 0.78 | 0.88 | 1.49 | 0.71 |
| $a_{32}$ | 0.52 | 1.31 | 1.03 | 0.80 | 0.79 | 1.02 | 0.97 | 1.28 | 1.29 | 1.34 |
| $a_{33}$ | -0.68 | -1.49 | -1.30 | -0.96 | -1.16 | -0.62 | -1.44 | -0.82 | -1.06 | -0.87 |
| $b_{11}$ | 0.95 | 0.64 | 1.17 | 0.65 | 1.03 | 0.67 | 1.49 | 0.96 | 1.0 | 0.63 |
| $b_{21}$ | 0 | 0 | 0 | 0 | 0 | 0 | 0 | 0 | 0 | 0 |
| $b_{31}$ | 1.29 | 0.70 | 0.52 | 0.88 | 0.81 | 0.77 | 0.92 | 1.29 | 1.14 | 1.11 |
| $c_{11}$ | 2.84 | 2.21 | 2.36 | 2.72 | 2.68 | 1.51 | 2.03 | 1.12 | 1.64 | 2.26 |
| $c_{12}$ | 1.24 | 0.77 | 0.88 | 1.35 | 1.07 | 1.38 | 0.83 | 1.10 | 1.46 | 0.87 |
| $c_{13}$ | -1.32 | -1.30 | -0.67 | -0.91 | -1.13 | -0.76 | -1.07 | -1.45 | -0.77 | -0.93 |
| Fitness* | 0.0006 | -0.015 | -0.026 | 0.0082 | 0.0035 | 0.0007 | 0.024 | 0.0028 | 0.0018 | -0.008 |

* Fitness value $= [\frac{1}{2}\sum_{k=1}^{N}(z(k)-\hat{z}(k))^T \hat{R}^{-1}(z(k)-\hat{z}(k)) + \frac{N}{2}\ln(|\hat{R}|)]^{-1}$

$\hat{R} = \frac{1}{N}\sum_{k=1}^{N}(z(k)-\hat{z}(k))(z(k)-\hat{z}(k))^T$

**TABLE 8.2**
**Estimated parameters with GA (Example 8.1)**

| | True values | Estimated values | |
|---|---|---|---|
| Parameters | ---- | SNR = ∞ | SNR = 10 |
| $a_{11}$ | -2 | -2.0055 | -2.0401 |
| $a_{12}$ | 0 | 0.0000 | 0.0000 |
| $a_{13}$ | 1 | 1.0012 | 1.0208 |
| $a_{21}$ | 1 | 1.0033 | 1.0235 |
| $a_{22}$ | -2 | -2.0121 | -2.0459 |
| $a_{23}$ | 0 | 0.0000 | 0.0000 |
| $a_{31}$ | 1 | 1.0028 | 1.0185 |
| $a_{32}$ | 1 | 1.0027 | 1.0194 |
| $a_{33}$ | -1 | -1.0009 | -1.0215 |
| $b_{11}$ | 1 | 1.0011 | 1.0198 |
| $b_{21}$ | 0 | 0.0000 | 0.0000 |
| $b_{31}$ | 1 | 1.0078 | 1.0202 |
| $c_{11}$ | 2 | 2.0015 | 2.0505 |
| $c_{12}$ | 1 | 1.0043 | 1.0246 |
| $c_{13}$ | -1 | -0.9979 | -1.0145 |
| PEEN | ---- | 0.3730 | 2.1879 |

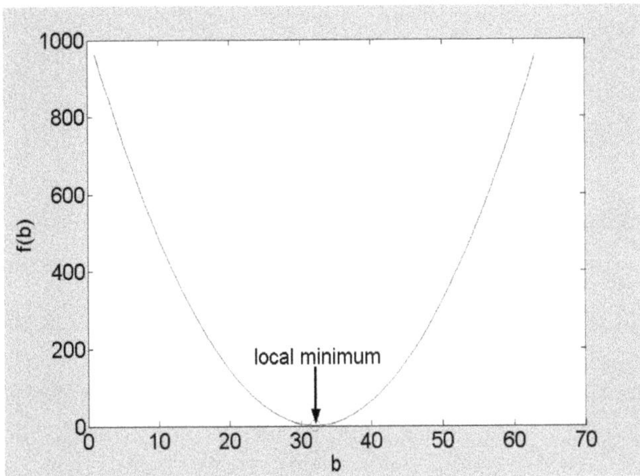


**FIGURE 8.2** Cost function f($b$) w.r.t parameter $b$ (Example 8.2)

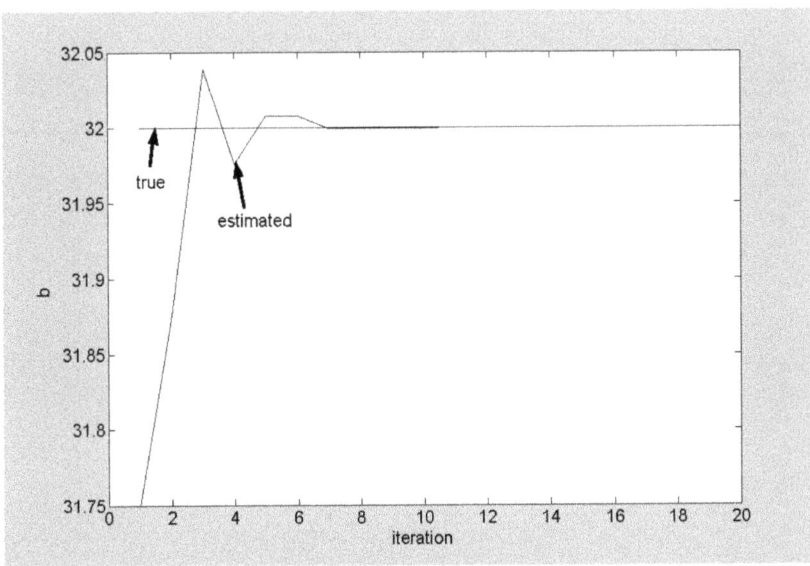


**FIGURE 8.3** Parameter $b$ vs. GA iterations (Example 8.2)

## 8.3 HIERARCHICAL GENETIC ALGORITHMS

A hierarchical genetic algorithm (HGA) is a refined genetic algorithm that uses a hierarchical chromosome-sample structure, allowing for flexible representation of both model structure (e.g., number and placement of knots) and model parameters (e.g., B-spline coefficients) in tasks like curve fitting and system identification [2]. HGAs are based on the principles of GAs, which are optimization techniques inspired

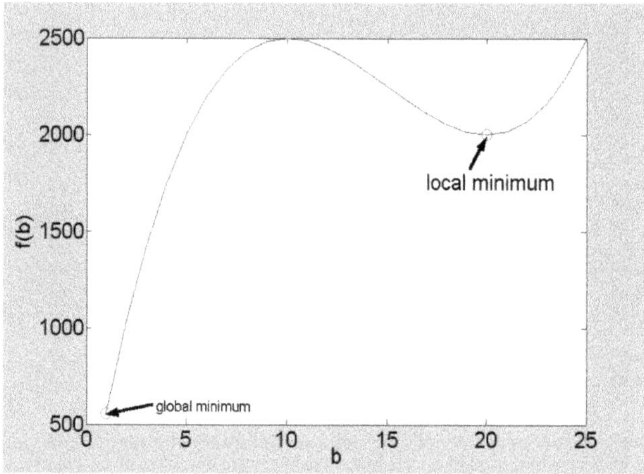


**FIGURE 8.4** Cost function f($b$) w.r.t parameter $b$ (Example 8.3)

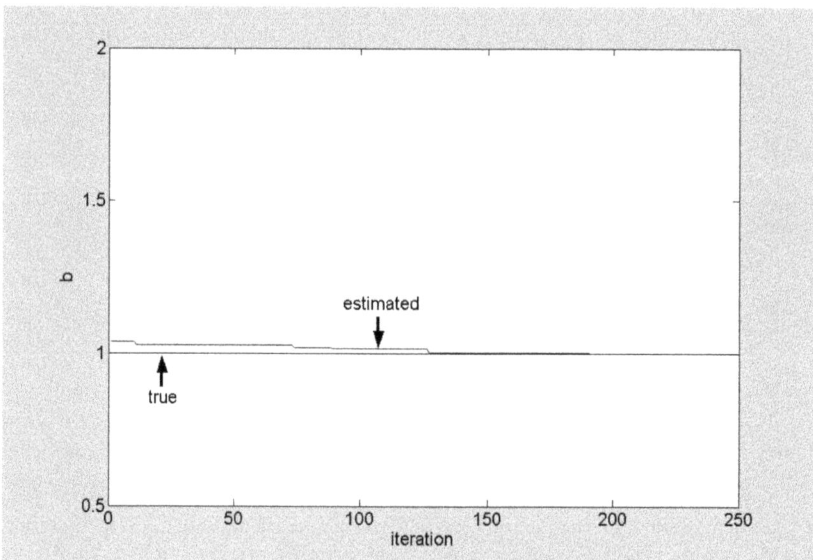


**FIGURE 8.5** Parameter $b$ versus GA iterations for SNR=100 (Example 8.3)

**TABLE 8.3**
**Estimation of 'b' with genetic algorithm**

| Parameters | True values | Estimated values | | |
|---|---|---|---|---|
| | | SNR = ∞ | SNR = 100 | SNR = 50 |
| $b$ | 1 | 1.00081 | 1.00015 | 1.0045 |

by biological evolution. GAs use concepts like selection, crossover, and mutation to evolve a population of potential solutions.

The key features are: i) **Hierarchical Chromosome:** Unlike traditional GAs with fixed chromosome structures, HGAs employ a hierarchical chromosome structure. This structure allows for representing complex problems with multiple levels of information; ii) **Control and Parametric Genes:** The hierarchical chromosome typically consists of two types of genes: control genes and parametric genes; iii) **Control Genes:** These genes, often encoded as binary digits, determine the structure of the model (e.g., the number and placement of knots in curve fitting); and iv) **Parametric Genes:** These genes, usually encoded as real numbers, represent the parameters of the model (e.g., B-spline coefficients). The activation of these genes depends on the values of the associated control genes.

The HGAs find applications in system identification and curve fitting: i) **System Identification:** HGAs can be used to identify the mathematical models of dynamic systems from measured data. This involves finding a function that best represents the relationship between the system's inputs and outputs; and ii) **Curve Fitting:** HGAs can be used to find the best-fitting curve or function for a set of data points. This can involve finding the optimal parameters of a model to minimize the difference between the model's predictions and the actual data.

Advantages of HGAs are: i) **Automatic Model Structure Determination:** HGAs can automatically determine the number and placement of knots (or other model structure parameters) during the optimization process; ii) **Automatic Parameter Optimization:** They can also automatically optimize the model parameters (e.g., B-spline coefficients); and iii) **Flexibility:** The hierarchical structure allows for representing complex relationships between variables.

An example is curve fitting with B-splines: Given a set of scattered data points, find the best-fitting B-spline curve.

HGA steps are: i) **Hierarchical Chromosome:** Represent both the number and placement of knots (control genes) and the B-spline coefficients (parametric genes) in a hierarchical structure; ii) **Fitness Function:** Define a fitness function that measures the goodness of fit of the B-spline curve to the data points; and iii) **Optimization:** Use the HGA to evolve the population of B-spline models, searching for the best combination of knots and coefficients that minimizes the fitness function

### 8.3.1 A Hierarchical Genetic Algorithm for System Identification and Curve Fitting

The hierarchical GA concept can be implemented through various encoding strategies depending on the application. While control genes and parametric genes provide one approach for specific applications like B-spline curve fitting, a more general framework uses flexible chromosome structures that can represent diverse functional forms and model architectures.

The hierarchical genetic algorithm approach for system identification represents an advancement over standard genetic algorithms by addressing both model structure selection and parameter optimization simultaneously. While conventional GAs optimize parameters within a fixed model structure, hierarchical GAs can search through different structural configurations while also optimizing the parameters for each configuration. This capability is particularly valuable in system identification problems where the appropriate functional form or model structure is not known in advance.

In traditional system identification, the analyst must first select a model structure based on prior knowledge, physical principles, or preliminary analysis, and then estimate the parameters of that chosen structure. If the initial structural choice is incorrect, even optimal parameter values will not yield a good model. Hierarchical GAs address this limitation by incorporating structural decisions into the evolutionary search process, allowing the algorithm to explore both structural alternatives and parametric variations across generations [3].

The concept of hierarchy in genetic algorithms refers to the organization of genetic information at multiple levels within the chromosome representation. At the highest level, certain genes control structural or configurational aspects of the solution, determining which components or features are active or present in the model. At lower levels, other genes encode the numerical parameters or coefficients associated with the active components. This multi-level organization mirrors the natural hierarchy found in biological systems, where regulatory genes control the expression of structural genes, which in turn determine organism characteristics.

The encoding of hierarchical information can be accomplished through various chromosome representations depending on the application domain. One approach uses a two-part chromosome structure where the first part contains discrete or binary genes indicating the presence or absence of model components, and the second part contains real-valued genes specifying the parameters of the active components. The length of the parametric portion may vary depending on how many components are activated by the structural genes, requiring specialized genetic operators that can handle variable-length chromosomes.

For system identification problems, consider the task of identifying a mathematical function from input-output data. The hierarchical chromosome must encode both which functional terms or basis functions should be included in the model and what coefficient values those terms should have. If the candidate model is expressed as a weighted sum of basis functions:

$$y = \sum_{j=1}^{M} \theta_j \varphi_j(x) \tag{8.5}$$

where $y$ is the output, $x$ represents the input variables, $M$ is the total number of candidate basis functions available, $\varphi_j(x)$ represents the j-th basis function which could be polynomials, trigonometric functions, radial basis functions, or other functional forms and $\theta_j$ is the coefficient for the j-th basis function. The hierarchical chromosome would contain structural information determining which subset of the M candidate basis functions should be included in the final model and parametric information specifying the coefficient values $\theta_j$ for the selected functions.

The fitness evaluation in hierarchical GAs must balance two competing objectives: accuracy of fit to the data and simplicity or parsimony of the model structure. A fitness function that considers only prediction error will tend to favour increasingly complex models that may overfit the training data. To prevent this, a complexity penalty is typically incorporated into the fitness calculation:

$$F = -\left[\sum_{i=1}^{N}\left(y_i - \hat{y}_i\right)^2 + \lambda C(S)\right] \tag{8.6}$$

where $F$ is the fitness value to be maximized, $N$ is the number of data points, $y_i$ is the actual output for the i-th data point, $\hat{y}_i$ is the predicted output from the model, $\lambda$ is a regularization parameter that controls the tradeoff between accuracy and complexity, and $C(S)$ is a complexity measure that depends on the model structure $S$. The complexity measure $C(S)$ could be the number of active basis functions, the total number of parameters, or other metrics quantifying model complexity. The negative sign converts the minimization problem into a maximization problem suitable for genetic algorithm fitness-based selection.

The selection mechanism in hierarchical GAs operates on complete chromosomes that encode both structural and parametric information. Chromosomes representing simpler models with good predictive accuracy will receive high fitness values due to the low error and low complexity penalty, while chromosomes representing complex models must demonstrate substantially better prediction accuracy to justify their complexity. This fitness-based selection pressure drives the population toward models that achieve an appropriate balance between fitting the data and maintaining structural simplicity.

Crossover operations in hierarchical GAs must handle the interaction between structural and parametric genes appropriately. When two parent chromosomes undergo crossover, the structural genes may be exchanged using standard crossover operators such as uniform crossover or single-point crossover. But, the parametric genes require careful treatment because the meaning of a particular parameter depends on which structural components are active. If crossover of structural genes results in offspring with different active components than either parent, the parametric genes must be adjusted accordingly. One approach performs crossover on structural genes first, then applies crossover to parametric genes only for those components that are active in both parents, while initializing parameters for newly activated components randomly or using inheritance rules.

Mutation in hierarchical GAs serves distinct purposes at different levels of the genetic hierarchy. Mutation of structural genes introduces or removes model components, enabling the algorithm to explore different structural configurations. The mutation rate for structural genes is typically set relatively low to prevent excessive disruption of well-performing structures. Mutation of parametric genes introduces variations in coefficient values, providing local search capability within a given model structure. The mutation rate for parametric genes can be set higher than for structural genes because parameter adjustments generally have less dramatic effects on overall fitness than structural changes.

The evolutionary dynamics of hierarchical GAs differ from standard GAs due to the interplay between structural and parametric evolution. In early generations, when the population contains diverse structural configurations with randomly initialized parameters, structural selection tends to dominate as markedly superior structures emerge. As evolution progresses and the population converges toward promising structural regions, parametric optimization becomes increasingly important as the algorithm fine-tunes coefficient values for the prevailing structures. This natural progression from structural exploration to parametric exploitation enables hierarchical GAs to efficiently search the combined space of structures and parameters.

Alternative hierarchical approaches implement the structural and parametric optimization as distinct evolutionary processes. In these two-level frameworks, an upper-level GA searches through possible model structures, while a lower-level GA optimizes the parameters for each structure proposed by the upper level. The upper-level GA maintains a population of structural configurations and uses genetic operators designed for discrete or combinatorial optimization. For each structural configuration in the upper-level population, the lower-level GA is invoked to find optimal parameter values, treating the structure as fixed during this parametric optimization. The fitness of each structural configuration in the upper-level population is determined by the best fitness achieved by the lower-level GA for that structure, possibly with a complexity penalty added.

This two-level approach provides a clear separation of concerns between structural search and parametric optimization, allowing each level to use genetic operators and search strategies appropriate for its task. The lower-level GA can use real-valued chromosomes and operators such as arithmetic crossover and Gaussian mutation that work well for continuous parameter optimization. The upper-level GA can use representations and operators suited to the specific structural search problem, whether that involves selecting features, choosing basis functions, or determining network architectures.

The computational cost of hierarchical GAs, particularly two-level implementations, can be substantial because each structural evaluation requires a complete parametric optimization. This cost can be managed through several strategies. Early termination of lower-level optimization for clearly inferior structures saves computation by not fully optimizing parameters for structures that will not survive selection. Parallel evaluation of different structures distributes the computational load across multiple processors when available. Incremental evaluation reuses parameter values from similar structures in previous generations as starting points for optimization, reducing the number of generations needed for parametric convergence.

Hierarchical genetic algorithms have been successfully applied to various system identification problems including nonlinear function approximation, time series prediction and dynamic system modelling. The ability to simultaneously optimize structure and parameters makes them particularly valuable when domain knowledge is insufficient to specify the model structure confidently. The increased complexity of implementation and higher computational cost compared to standard GAs means they are most appropriate for problems where structural uncertainty is a genuine concern and where the benefits of automatic structure selection justify the additional computational effort.

## 8.4 STATE ESTIMATION USING GAS

State estimation is a fundamental problem in control systems and signal processing that involves determining the internal states of a dynamic system based on noisy measurements and a mathematical model of the system. While traditional methods like the Kalman filter work well for linear systems, nonlinear systems present significant challenges that genetic algorithms can effectively address.

Consider a general discrete-time nonlinear system described by:

$$x(k+1) = F(x(k), u(k), \theta) \tag{8.7}$$

$$y(k) = G(x(k), u(k)) \tag{8.8}$$

where $x(k)$ is the state vector at time $k$, $u(k)$ is the input vector, $y(k)$ is the output (measurement) vector, $\theta$ is the parameter vector that characterizes the system, and $F$ and $G$ are nonlinear functions representing the system dynamics and measurement equation, respectively.

The objective of state estimation is to find the best estimate $\hat{x}(k)$ of the true state $x(k)$ given the available measurements $y(1)$, $y(2)$, … ., $y(k)$ and inputs $u(1)$, $u(2)$, … ., $u(k)$. This is particularly challenging when the system parameters $\theta$ are unknown or time-varying.

The genetic algorithm approach to state estimation differs fundamentally from conventional methods by treating the parameter estimation and state estimation as an integrated optimization problem. Instead of requiring gradient information or linearization assumptions, the GA evolves a population of parameter estimates to minimize the estimation error [4].

The GA-based state estimator operates on an estimated system model:

$$\hat{x}(k+1) = F(\hat{x}(k), u(k), \hat{\theta}(k)) \tag{8.9}$$

$$\hat{y}(k) = G(\hat{x}(k), u(k)) \tag{8.10}$$

where $\hat{\theta}(k)$ represents the parameter estimates at time $k$, $\hat{x}(k)$ is the estimated state, and $\hat{y}(k)$ is the predicted output.

The key insight is that each chromosome (or sample) in the GA population represents a complete set of parameter estimates $\hat{\theta}$. The population thus represents multiple candidate systems and the GA evolutionary process selects the system that best explains the observed data. This concept is illustrated in Figure 8.6, which shows the block diagram of a general GA-based state estimator. In this configuration, the estimated system block uses the current best parameter estimates $\hat{\theta}(k)$ from the GA

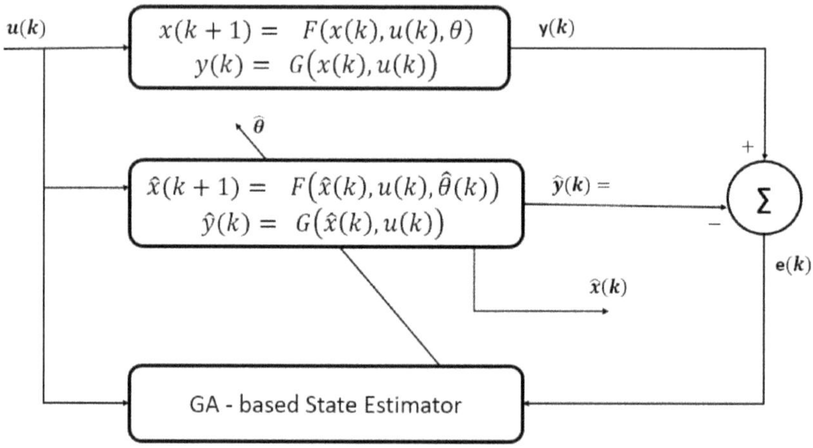


**FIGURE 8.6** Block diagram for a general GA-based state estimator

**Source: Adapted/modified from Gremling and Passino, 2000**

population, while the genetic adaptive state estimator continuously evolves these parameters based on the estimation error $e(k) = y(k) - \hat{y}(k)$.

The fitness function plays a crucial role in guiding the evolution toward accurate parameter estimates. A commonly used fitness function is based on the windowed prediction error:

$$J = a - \frac{1}{2} E^T E \tag{8.11}$$

where

$$E = \begin{bmatrix} e_{k-W} \\ e_{k-W+1} \\ \vdots \\ e_k \end{bmatrix} \tag{8.12}$$

with $e_j = y(j) - \hat{y}(j)$ for $j > 0$ and $e_j = 0$ for $j < 0$. The parameter $W$ defines the window size over which errors are accumulated and $a$ is chosen as the highest $\frac{1}{2} E^T E$ value in the current population to ensure all fitness values remain positive.

This fitness function formulation ensures that the GA maximizes fitness by minimizing the prediction error between the actual system output and the estimated system output. The windowed approach provides a balance between responsiveness to recent data and stability against noise.

The genetic adaptive state estimator operates through the following process. At each time step k, the current population of parameter estimates is used along with past and present values of input u(k) and output y(k) to generate estimated system states $\hat{x}(k)$. When actual measurement data y(k') is available for time k', it is used instead of the predicted output $\hat{y}(k')$ to improve estimation accuracy.

A window of predicted outputs $\hat{y}(k-W),\hat{y}(k-W+1),\ldots,\hat{y}(k)$ is generated for each candidate parameter set in the population. Each candidate is then assigned a fitness value according to eqns. (8.11) and (8.12). Using these fitness values, the GA selection process determines which candidates will serve as parents for generating the next population of parameter estimates.

The GA employs modified selection, crossover, and mutation operators optimized for real-time state estimation. The selection mechanism chooses the D best chromosomes from the population based on their fitness values rather than using probabilistic selection. This ensures that the most accurate parameter estimates have the highest probability of contributing to future generations.

For crossover operations, mating occurs at the parameter level rather than at the bit level. This approach recognizes that each parameter has physical meaning and should be treated as a coherent unit during genetic operations. The crossover process exchanges parameter values between parent chromosomes to create offspring that potentially combine favourable characteristics from both parents.

Mutation is applied with probability $p_m$ to introduce variations in parameter estimates. The mutation rate must be carefully balanced: higher rates provide faster exploration of the parameter space but may prevent convergence to optimal values, while lower rates may lead to premature convergence to suboptimal solutions.

For real-time implementation, computational complexity becomes a critical consideration. The genetic adaptive state estimator requires a specific number of operations per generation. When examining the W+1 most recent data points, each population member must undergo W+1 addition operations to calculate the error between actual and estimated system outputs, followed by W+1 multiplication operations to compute squared errors. The squared errors over the window are then summed, requiring another W addition operations.

The total computational requirements for fitness evaluation are [4]:

$$n_a = 2N(W+1) \tag{8.13}$$

$$n_m = 2N(W+2) \tag{8.14}$$

where N is the population size, W is the window size, $n_a$ represents the number of addition operations, and $n_m$ represents the number of multiplication operations required per generation.

Beyond fitness calculation, the genetic operators require additional computations. Parent selection involves choosing chromosomes from the best D members, crossover operations occur for each trait in the selected parents, and mutation affects individual genes with probability $p_m$. The number of random integer generations $r_i$ needed per population follows:

$$r_i = \left(\frac{n_t}{2} + p_m n_g + 1\right)(N - \rho D) \tag{8.15}$$

where $n_t$ is the number of traits per chromosome, $n_g$ is the number of genes per chromosome, $\rho$ is the portion of parents surviving to the next generation, and D is the number of parents selected for mating. These computational requirements scale linearly with population size and window size, making the method computationally intensive but parallelizable for real-time implementation on modern hardware.

The genetic adaptive state estimation approach offers several significant advantages over conventional methods. Unlike gradient-based techniques such as extended Kalman filters, the GA does not require linearization of nonlinear system dynamics or computation of Jacobian matrices. This elimination of linearization requirements makes the method particularly suitable for highly nonlinear systems where linear approximations may be inadequate or fail entirely [4].

The population-based search strategy provides inherent robustness against local minima, which represents a major advantage in complex optimization landscapes that commonly arise in nonlinear state estimation problems. Traditional gradient-based methods can become trapped in local optima, leading to poor parameter estimates and subsequently poor state estimates.

Furthermore, the GA approach can simultaneously estimate both system parameters and states without requiring precise a priori knowledge of parameter bounds or initial estimates. This adaptability makes it especially valuable for systems with time-varying parameters or uncertain model structures, where traditional methods may struggle or fail.

The method demonstrates particular effectiveness in applications involving highly nonlinear dynamics. For example, in jet engine compressor systems where surge and stall phenomena create complex nonlinear behaviour, the genetic adaptive state estimator can successfully track unmeasurable states related to critical operating conditions. In such applications, traditional extended Kalman filter approaches often fail due to the severe nonlinearities and the breakdown of linearization assumptions. Comparative studies have shown that while conventional extended Kalman filters become unstable and fail for such highly nonlinear systems, the genetic adaptive state estimator maintains stability and provides useful state information throughout the operating range. This robustness stems from the GA's ability to handle discontinuous fitness functions and avoid the mathematical difficulties associated with linearization of highly nonlinear dynamics.

The genetic adaptive state estimation framework represents a powerful alternative to conventional estimation techniques, particularly for nonlinear systems where traditional methods encounter fundamental limitations. While computational complexity is higher than linear methods, the robustness and global optimization capabilities make it an attractive choice for challenging state estimation problems in complex dynamic systems where conventional approaches prove inadequate.

## 8.5 GA FOR IDENTIFICATION OF TIME DELAY IN SYSTEMS FROM STEP RESPONSES

Time delay systems are ubiquitous in industrial processes and present significant challenges for system identification and control. The presence of dead time in a system affects both performance and stability, making accurate identification crucial for effective control design. Traditional identification methods often struggle with time delay systems due to the complexity introduced by the dead time component and potential model mismatch problems.

Industrial systems commonly exhibit characteristics such as long dead times, large time constants, multi-variable interactions, nonlinear behaviour, and non-minimum phase dynamics. Among these constraints, the identification of systems with long-range dead times requires particular attention due to their impact on control system performance [5]. The step response test emerges as the most practical approach for industrial applications because it is simple to implement, requires minimal equipment, and can be performed manually or through programmable logic controllers (PLCs) and distributed control systems (DCS).

The genetic algorithm approach offers significant advantages for time delay system identification compared to conventional methods. Unlike gradient-based techniques, GAs can handle the complex optimization landscapes that arise in time delay parameter estimation without requiring derivative information or risking convergence to local minima.

Consider the First-Order Plus Dead-Time (FOPDT) model, which provides an effective approximation for many industrial processes with long dead times:

$$G(s) = \frac{Ke^{-Ls}}{Ts+1} \tag{8.16}$$

where K represents the process gain, T is the time constant, and L denotes the dead time. The identification challenge involves determining these three parameters from step response data.

For more complex systems, the Second-Order Plus Dead-Time (SOPDT) model extends the representation capability:

$$G(s) = \frac{Ke^{-Ls}}{T_1 s^2 + T_2 s + 1} \tag{8.17}$$

where $T_1$ and $T_2$ are time constants that characterize the second-order dynamics, providing four parameters to be identified: $K, T_1, T_2$, and $L$.

Traditional identification methods face significant limitations when applied to time delay systems. The area-based method, while providing better robustness against measurement noise compared to graphical methods, relies on computing the average residence time from the area under the step response curve:

$$T_{ar} = \frac{A_0}{K} = \frac{\int_0^{\infty}\left[y(\infty) - y(t)\right]dt}{K} \tag{8.18}$$

The time constant and dead time are then obtained through additional area calculations:

$$\frac{A_1}{K} = \frac{\int_0^{T_{ar}} y(t)\,dt}{K} = T_{ar} - T \tag{8.19}$$

$$L = T_{ar} - T = \frac{A_0}{K} - \frac{A_1}{K} \tag{8.20}$$

The accuracy of this method depends heavily on the area $A_0$, requiring long testing periods for the process to reach steady state completely.

Direct identification methods attempt to derive system parameters through linear regression equations. For the FOPDT system with zero initial conditions and step input of amplitude $h$ at t = 0, the transient output for t ≥ L is:

$$y(t) = hK\left(1 - e^{-\frac{t-L}{T}}\right) + \varepsilon(t) \tag{8.21}$$

where $\varepsilon(t)$ represents measurement noise. This equation can be rearranged and integrated to form a system of linear equations suitable for least-squares parameter estimation. These methods, however, remain sensitive to noise and modelling assumptions.

The genetic algorithm approach addresses these limitations through its robust global optimization capabilities. The GA formulation treats the identification problem as an optimization task where each chromosome represents a complete parameter set. For FOPDT systems, each chromosome contains three genes corresponding to K, T, and L, while SOPDT systems require four genes for K, $T_1$, $T_2$, and L.

Real-coded genetic algorithms prove particularly effective for this application because they eliminate the encoding and decoding overhead associated with binary representation while providing higher precision and faster convergence. The real-coded approach represents each parameter directly as a real number within specified bounds, making the chromosome structure:

$$S_i = \{x_{i1}\ x_{i2}\ x_{i3} \ldots x_{in}\} \tag{8.22}$$

where $S_i$ represents the i-th chromosome in the population and $x_{ij}$ denotes the j-th gene (parameter) of the i-th chromosome.

The fitness evaluation utilizes the mean square error between the actual system step response and the estimated model response:

$$\varepsilon = \frac{1}{n}\sum_{k=m}^{m+n}\left[y(kT_s) - \hat{y}(kT_s)\right]^2 \tag{8.23}$$

where $y(kT_s)$ is the actual system output, $\hat{y}(kT_s)$ is the estimated system response, $T_s$ is the sampling interval, and n+1 is the number of data points used in the evaluation.

The fitness function transforms the minimization problem into a maximization problem suitable for GA operation:

$$f(S_i) = \frac{1}{J(S_i) + \alpha} \tag{8.24}$$

where $J(S_i)$ is the cost function value for chromosome $S_i$ and $\alpha$ is a small positive constant (typically 0.1) to prevent division by zero and ensure positive fitness values.

A key innovation in the GA approach is the modified crossover operator that adapts the search range based on parent fitness values. For two parent chromosomes $S_1$ and $S_2$ with fitness values$\mathrm{f}(S_1) \geq \mathrm{f}(S_2)$, the offspring are generated using:

$$\begin{aligned} \hat{x}_1 &= \lambda x_1 + (1-\lambda)x_2 \\ \hat{x}_2 &= \lambda x_2 + (1-\lambda)x_1 \end{aligned} \tag{8.25}$$

where the multiplier $\lambda$ range is dynamically adjusted according to:

$$-\frac{f(S_2) - f(S_1)}{\mathrm{f}(S_2) + f(S_1)} \leq \lambda \leq 1 + \frac{f(S_2) - f(S_1)}{\mathrm{f}(S_2) + f(S_1)} \tag{8.26}$$

This modification extends the search range for chromosomes with higher fitness, enhancing the algorithm's ability to explore promising regions of the parameter space more thoroughly.

The initialization strategy for time delay system identification incorporates practical bounds based on the step response characteristics. For FOPDT systems, reasonable parameter ranges are established as:

$$\begin{aligned} 0.8 \times Y_{final} \leq\ & K \leq 1.2 \times Y_{final} \\ 0 \leq\ & T \leq \frac{Rise\ Time}{2} \\ 0 \leq\ & L \leq 0.1 \times Y_{final} \end{aligned} \tag{8.27}$$

where $Y_{final}$ is the final steady-state value of the step response and the rise time is determined from the response characteristics.

The GA algorithm implementation employs tournament selection to choose parents based on fitness comparison, ensuring that better chromosomes have higher reproduction probability while maintaining population diversity. The mutation operator introduces controlled perturbations to prevent premature convergence and maintain exploration capability throughout the evolution process.

The theoretical advantages of the GA-based identification method include superior robustness against measurement noise, ability to handle both FOPDT and SOPDT systems effectively, and insensitivity to initialization compared to traditional methods. The global optimization capability ensures that the algorithm can escape local minima that trap conventional optimization approaches.

The GA method proves especially valuable when dealing with non-minimum phase systems and multi-lag systems where traditional methods often fail or provide poor accuracy. The population-based search provides inherent robustness against modelling uncertainties and measurement noise, making it suitable for industrial applications where data quality may be compromised.

Computational requirements for the GA-based identification remain reasonable for practical applications. The population size typically ranges from 40–100 chromosomes, with generation counts of 100–200 sufficient for convergence in most cases. The real-coded representation eliminates encoding/decoding overhead, making the method suitable for online identification applications where computational efficiency is important.

The genetic algorithm approach to time delay system identification represents a significant advancement over traditional methods, offering improved accuracy, robustness against noise and reliability in parameter estimation. The method's ability to handle both FOPDT and SOPDT systems with equal effectiveness, combined with its insensitivity to initialization and noise, makes it an attractive choice for industrial system identification applications where accurate time delay characterization is critical for control system design and performance [5].

## 8.6 SYSTEM IDENTIFICATION USING GENETIC ALGORITHM AND PARTICLE SWARM OPTIMIZATION

System identification often involves complex scenarios where the model structure itself is unknown, leading to structure identification problems where both the functional form and parameters must be determined simultaneously. Both genetic algorithms and particle swarm optimization provide evolutionary approaches to address these challenging identification tasks.

Consider a general system identification problem for discrete-time systems described by an ARMAX model:

$$\mathrm{A}\left(q^{-1}\right)y(t)=B\left(q^{-1}\right)u(t-d)+\mathrm{C}\left(q^{-1}\right)e(t) \tag{8.28}$$

where $\mathrm{A}\left(q^{-1}\right)$, $B\left(q^{-1}\right)$, and $\mathrm{C}\left(q^{-1}\right)$ are polynomials in the backward shift operator $q^{-1}$ ,$y(t)$, $u(t)$, and $e(t)$ represent output, input, and noise, respectively, and d is the time delay. The objective is to estimate the polynomial coefficients or, alternatively, the poles and zeros of the system directly from input-output data.

For continuous-time systems, the identification problem can be formulated in terms of poles and zeros:

$$\mathrm{G}(s)=\frac{\prod_{i=1}^{m}\left(s-z_i\right)}{\prod_{j=1}^{n}\left(s-p_j\right)} \tag{8.29}$$

where $z_i$ and $p_j$ represent the system zeros and poles, respectively.

The genetic algorithm approach to system identification treats the parameter estimation as an optimization problem where each chromosome in the population represents a complete set of system parameters. The fitness function is typically based on minimizing the prediction error between the actual system output and the model output [6].

One significant advantage of GAs in system identification lies in their ability to directly identify poles and zeros without requiring linearity in parameters. The system can be parameterized in pole-zero form as:

$$\mathrm{A}\left(q^{-1}\right)=\prod_{k=1}^{n_a/2}\left(1-\left(\alpha_k \pm \beta_k\right)q^{-1}\right) \tag{8.30}$$

$$\mathrm{B}\left(q^{-1}\right)=b_0\prod_{k=1}^{n_b/2}\left(1-\left(\gamma_k \pm \delta_k\right)q^{-1}\right) \tag{8.31}$$

where $\alpha_k$ and $\gamma_k$ represent the average values of pole and zero pairs, respectively, $\beta_k$ and $\delta_k$ represent deviations from the average, $b_0$ is the system gain, $n_a$ is the order of polynomial $\mathrm{A}\left(q^{-1}\right)$, $n_b$ is the order of polynomial $\mathrm{B}\left(q^{-1}\right)$. In this parameterization, when $\beta_k > 0$, the pair $\left(\alpha_k \pm \beta_k\right)$ represents two real poles; when $\beta_k < 0$, the notation represents a complex conjugate pole pair $\left(\alpha_k \pm j\beta_k\right)$ where $j=\sqrt{-1}$. This parameterization allows GAs to search for both real and complex poles using a unified representation.

Particle swarm optimization provides an alternative evolutionary approach for system identification. In PSO, each particle represents a candidate parameter set and the swarm evolves according to the fundamental equations:

$$v_{ij}(t+1)= w.v_{ij}(t)+c_1r_1\left(p_{ij}(t)-x_{ij}(t)\right)+c_2r_2\left(g_j(t)-x_{ij}(t)\right) \tag{8.32}$$

$$x_{ij}(t+1)= x_{ij}(t)+v_{ij}(t) \tag{8.33}$$

where $w$ is the inertia weight that controls the influence of previous velocity on current velocity, $v_{ij}(t)$ is the velocity of particle i in dimension j at time t, $x_{ij}(t)$ is the position of particle i in dimension j at time t, $p_{ij}(t)$ is the personal best position found by particle i in dimension j, $g_j(t)$ is the global best position found by the entire swarm in dimension j, $c_1$ and $c_2$ are acceleration coefficients that control the influence of personal and social learning respectively and $r_1$, $r_2$ are random numbers uniformly distributed between 0 and 1.

The PSO algorithm balances exploration and exploitation through the interaction between individual particle experience (personal best) and collective swarm knowledge (global best). The first term in eqn. (8.32) represents the particle's inertia, the second term represents cognitive learning from the particle's own experience, and the third term represents social learning from the swarm's collective experience [7].

For system structure identification problems where the model structure is unknown, both GA and PSO can be applied to combinatorial optimization problems. Consider a system model composed of mathematical sub-models:

$$y = p_0 + \sum_{k=1}^{N_s} K_k \cdot M_k\left(x, p_{k1}, p_{k2}, \dots, p_{km_k}\right) \tag{8.34}$$

where $N_s$ is the total number of available sub-models, $K_k$ is a binary variable indicating whether sub-model k is selected ($K_k = 1$) or not ($K_k = 0$), $M_k$ represents the k-th sub-model function, $p_0$ is a constant term, $p_{kj}$ are the parameters of the k-th sub-model, and $m_k$ is the number of parameters in sub-model k. For PSO application to structure identification problems, the binary selection variables $K_k$ require special handling since PSO is inherently designed for continuous optimization. The binary variables can be managed through discrete mutation operations that reverse their values (0↔1) during the particle update process, while the continuous parameters $p_{kj}$ are updated using the standard PSO velocity and position equations (8.32) and (8.33)

The fitness function for both GA and PSO in structure and parameter identification is typically based on the mean squared error:

$$f = \sum_{i=1}^{n} \left(y_i - \hat{y}_i\right)^2 \tag{8.35}$$

where n is the number of data samples, $y_i$ is the actual system output for sample I, and $\hat{y}_i$ is the model output for sample i.

Both genetic algorithms and particle swarm optimization offer significant advantages over traditional gradient-based approaches by avoiding local minima and handling non-differentiable objective functions effectively. For system identification with purely continuous parameters such as pole-zero estimation or coefficient optimization, PSO often exhibits faster convergence due to its velocity-based search mechanism and simpler algorithmic structure with fewer parameters requiring tuning. However, for problems combining structure selection (discrete binary variables) with parameter optimization (continuous variables), GAs provide more natural handling of mixed variable types through their genetic operators such as crossover and mutation. The choice between GA and PSO depends on problem characteristics: GA's population-based genetic operators excel in exploring discrete combinatorial spaces and maintaining diversity in multimodal landscapes, while PSO's continuous particle movement with velocity updates is particularly effective for smooth, unimodal parameter spaces where exploitation of gradient-like information accelerates convergence [6,7].

## EXERCISES

1. Why are genetic algorithms particularly effective for parameter estimation in complex, nonlinear, or noisy systems compared to traditional gradient-based methods?

2. Describe the main steps of parameter estimation using a genetic algorithm. How do encoding, fitness evaluation, and genetic operators contribute to finding optimal 3.
3. What distinguishes a hierarchical genetic algorithm (HGA) from a standard genetic algorithm and what advantages do the hierarchical structure provide for system identification?
4. Explain how genetic algorithms can be used for state estimation. What are the key benefits of this approach over conventional methods such as the Kalman filter?
5. Compare genetic algorithms and particle swarm optimization for system identification. What are the relative strengths of each method and what factors should guide the choice between them?

## REFERENCES

1. Raol, J.R., Girija, G., and Singh, J., "Modelling and Parameter Estimation of Dynamic Systems", *IET/IEE Control Series* Vol. 65, The Institution of Electrical Engineers (IEE), London, UK, Aug. 2004.
2. Garcia, C.H., Cuevas, F.J., Trejo-Caballero, G., and Rostro-Gonzalez, H., "A Hierarchical GA Approach for Curve Fitting with B-Splines", *Genetic Programming and Evolvable Machines*, Vol. 16, No. 2, pp. 151–166, 2015.
3. Gulsen, M., and Smith, A.E., "A Hierarchical Genetic Algorithm for System Identification and Curve Fitting with a Supercomputer Implementation", In L. Davis, K. DeJong, M. Vose, and D. Whitley (Eds.), *IMA Volumes in Mathematics and its Applications – Evolutionary Algorithms*, Springer-Verlag, New York, 1998.
4. Gremling, J.R., and Passino, K.M., "Genetic Adaptive State Estimation", *Engineering Applications of Artificial Intelligence*, vol. 13, no. 6, pp. 611–623, 2000.
5. Shin, G.-W., Song, Y.-J., Lee, T.-B., and Choi, H.-K., "Genetic Algorithm for Identification of Time Delay Systems from Step Responses", *International Journal of Control, Automation, and Systems*, vol. 5, no. 1, pp. 79–85, Feb. 2007.
6. Kristinsson, K., and Dumont, G.A., "System Identification and Control Using Genetic Algorithms", *IEEE Transactions on Systems, Man, and Cybernetics*, vol. 22, no. 5, pp. 1033–1046, Sep/Oct. 1992.
7. Deng, X., "System Identification Based on Particle Swarm Optimization Algorithm", In: *Proceedings of the 2009 International Conference on Computational Intelligence and Security*, pp. 259–263, 2009.

# 9 Hybrid Methods-NN-FL-GA triad Paradigm

In the previous chapters we have discussed several estimation techniques that use either ANNs, or FL, or GAs. Mainly we have considered the problem of System Identification, and Parameter Estimation, and briefly the state estimation. In this chapter we discuss hybrid methods, where the combination of two subcomponents of SC is used for estimation with a view to deriving some benefits from two such subcomponents taken at a time.

## 9.1 NN-FL, FL-GA, AND GA-NN DUOS FEASIBILITIES FOR SIPEST

We briefly summarize some important features of each of these subcomponents of SC. ANNS have learning abilities and need to be trained using some observed data of the system under estimation exercise, and a training algorithm. FL allows to incorporate the domain experts' knowledge in the form of If,...,Then,... rules; and as such FL does not have learning ability except via using ANNs as is done in ANFIS. GAs can be used for optimization of the solution by several iterations (of generations) of an evolutionary process, and mainly would result in a robust solution rather than (only) the optimal one. It would be prudent here to revisit Figure 1.6 (Chapter 1) to get an overall picture of what these subcomponents can contribute to the estimation process when you use any two subcomponents of SC at a time [1]-[4].

### 9.1.1 NN-FL Method

Combining ANNs and FL, called neuro-fuzzy systems (NFS) allows for robust and adaptable estimation by deriving the benefits from both [1]. ANNs learn complex relationships from the presented data, while fuzzy logic deals with uncertainty and imprecision, i.e., vagueness. This hybrid approach is useful when we are dealing with complex and uncertain modelling problems. The benefits of the combination are: i) improved accuracy of estimated parameters, ii) robustness to uncertainty, and iii) interpretability: The FL rules provide insights into the system's reasoning, making it more understandable. The applications of this NN-FL duo are: i) reconfiguration control strategies, ii) decision-making, iii) pattern recognition, and iv) prediction and forecasting.


DOI: 10.1201/9781003616580-9

### 9.1.2 FL-GA Method

Combining FL and GAs offers a powerful approach for estimation by taking the advantages of both techniques [2]. FL handles uncertainty and imprecision, and GA optimizes parameters to achieve the best possible results. The hybrid approach can be applied in various fields: control systems, data analysis, and optimization problems. Fuzzy logic is used to create a model of the system: the model can include fuzzy rules, membership functions, and other fuzzy logic components that define the relationships between input and output variables. GAs are then used to optimize the parameters of the FL model: by adjusting the form of membership functions, the fuzzy rules, or other parameters to achieve the best possible performance.

The applications are: i) control systems – GAs can be used to tune the parameters of fuzzy logic controllers for optimal performance, e.g., in robotics; ii) data analysis – combining FL and GAs can be used to analyze complex datasets and identify patterns or relationships that might not be apparent otherwise, useful for situations where data are noisy or/and incomplete; and iii) optimization problems – GAs can be used to optimize the parameters of a FL model to find the best possible solution to a given optimization problem.

### 9.1.3 GA-NN Method

Combining GAs and ANNs for estimation involves using GAs to optimize the parameters of an ANN (may be in learning algorithm) by using the GA's ability to search for optimal solutions and the ANN's capability to learn complex/nonlinear relationships: can lead to more accurate and robust estimation models [3]. Specific aspects are: i) parameter optimization – NNs require proper tuning of parameters like learning rate, momentum, and the number of hidden layers. GAs would efficiently search for the optimal combination of these parameters, improving the NN's performance; ii) GAs and NNs excel in solving complex, non-linear problems; and iii) GAs can help find more robust solutions that are less sensitive to variations in the input data, making the resulting ANN model more reliable. How the combination would work: i) The parameters of the ANN (e.g., weights, biases, network architecture) are encoded as genes in a GA chromosome; and ii) A fitness function evaluates the performance of the ANN based on its estimation accuracy.

### 9.1.4 NN-FL-GA Approach

Combining ANN, FL, and GA in estimation has some benefits: ANN excels at learning complex relationships from data, FL provides a way to represent and reason in presence of vagueness, and GA can fine-tune the parameters of the hybrid system [4]. Hybrid ANN-FL-GA models have been used to estimate ore grade, which is crucial for mining project feasibility. Hybrid models can be more robust to noisy data and uncertainties. Fuzzy logic can provide insights into the reasoning process of the hybrid system, which can be valuable for understanding the model's predictions. In CANFIS (Combining ANFIS with GA) GA is used to optimize the parameters of an ANFIS, leading to a more accurate and robust model.

## 9.2 NN-FL: NEURAL IDENTIFICATION OF FUZZY ANOMALIES IN PRESSURIZED WATER SYSTEM

Water supply systems represent critical infrastructure that conveys treated water to consumers through pressurized networks of pipes. The operational management of these systems relies on continuous or periodic measurements taken by meters and gauges, which are transmitted via telemetry systems to control and operation centres for network monitoring. However, the complexity of these systems, combined with the limited number of available measurements, presents significant challenges for accurate state assessment and anomaly detection. Traditional approaches that depend on human operators to interpret measurement data and diagnose system anomalies often prove inadequate, particularly when dealing with subtle or gradually developing problems such as progressive leaks or valve malfunctions.

The fundamental challenge in water supply system management lies in performing accurate state estimation with incomplete and uncertain information. While telemetry systems provide real-time data from various points in the network, these measurements represent only a small fraction of the total system state. The sparse nature of measurements, coupled with inherent uncertainties in both sensor readings and demand predictions, complicates the task of assessing network health and detecting anomalies. Furthermore, as water becomes increasingly recognized as a scarce and valuable resource, particularly for human consumption where significant quality-related investments have already been made, the need for more efficient management and suitable water audits becomes paramount for sustainable development and integrated water management.

Neural networks offer a promising approach to address these challenges by encapsulating the necessary knowledge to assess network state and identify anomalies. The integration of fuzzy logic with neural network architectures provides a mechanism to handle the inherent uncertainties in measurement data while performing pattern recognition and classification tasks. This combination enables the development of intelligent systems that can diagnose network anomalies based on patterns of fuzzy state estimates derived from uncertain measurements.

The mathematical foundation for water supply system analysis rests on solving a large system of nonlinear equations that describe the hydraulic behaviour of the network. For a pressurized water system with N demand junctions and L connecting lines, the governing equations consist of continuity equations at network junctions and energy equations along the pipelines.

The continuity equations express flow conservation at each junction i:

$$\sum_{j \in A_i} q_{ij} = Q_i, \; i = 1,2,\ldots,N-1 \tag{9.1}$$

where $A_i$ represents the set of lines directly connected to node i, $q_{ij}$ denotes the flowrate in line j, and $Q_i$ is the demand at node i.

The energy equations for each line k relate the head loss to the flowrate:

$$H_{k_1} - H_{k_2} = R_k q_k \left| q_k \right|, \; k = 1,\ldots,L \tag{9.2}$$

where $H_{k_1}$ and $H_{k_2}$ are the piezometric heads at the two end nodes, $q_k$ is the flowrate through line k, and $R_k$ represents the line resistance, which depends on $q_k$ through the Reynolds number. The complete system of equations can be assembled in block matrix form and the solution yields the state vector $x = (q, H)^t$ representing the complete hydraulic state of the system comprising flowrates through all lines and piezometric heads at all demand nodes.

State estimation in water supply systems integrates this mathematical model with specific measurements obtained through telemetry systems. The measurements provide additional constraints beyond those imposed by the conservation and energy balance equations, resulting in an expanded overdetermined system. Since measurements are contaminated by noise and errors from telemetry devices and uncertainty in demand predictions, this system must be solved using optimization techniques.

For large water systems involving thousands of equations and unknowns, the solution typically employs an over-relaxation iterative process applied to a linearized version of the expanded system. At each iteration k, the linearized system can be written compactly as:

$$A_k \Delta x_k = b_k \tag{9.3}$$

where $A_k$ is the coefficient matrix of size $(L + N_p + N_t) \times (L + N_p)$ with $N_p$ being the number of demand nodes and $N_t$ being the number of telemetry measurements, $\Delta x_k$ is the correction to the state estimate, and $b_k$ is the right-hand side vector containing the residuals between measured and predicted values. The iterative solution procedure uses the normal equations approach with weighted least squares:

$$A_k^t W A_k \Delta x_k = A_k^t W b_k \tag{9.4}$$

where W is a diagonal weighting matrix that accounts for different levels of confidence in various measurements, allowing more reliable measurements to have greater influence on the estimated state. The state estimate is updated according to $x_{k+1} = x_k + \alpha \Delta x_k$, where α is an over-relaxation factor and the process continues until convergence is achieved.

The reliability of estimated states depends fundamentally on the precision of input data, including both telemetry measurements and demand predictions. Error limit analysis provides a mechanism to determine uncertainty bounds for state estimates that originate from the lack of precision in measurements. This representation of uncertainty through bounded errors produces more meaningful results than probabilistic noise characterization because most telemetry systems do not employ suitable probabilistic models for measurement noise.

The central question in error limit analysis concerns the reliability of the estimated state x∗ when measurement vectors y are not crisp values but may vary within bounded regions $[y - \delta y, y + \delta y]$. The relationship between measurement uncertainties and state estimate uncertainties can be derived through sensitivity matrix analysis applied to the linearized state estimation equations. From the linearization process, it can be shown that the variation in the estimated state, denoted $\delta x^*$, is related to variations in measurements δy through:

$$A_k^{*t} W A_k^* \delta x^* = A_k^{*t} W \delta y \tag{9.5}$$

where the subscript asterisk indicates evaluation at the converged state estimate x∗. This relationship allows computation of component-by-component bounds for the state estimate uncertainty. A bound vector e∗ for δx∗ can be obtained by:

$$e^* = \left| \left( A_k^{*t} W A_k^* \right)^{-1} A_k^{*t} W \right| \left| \delta y \right| \tag{9.6}$$

where the vertical bars indicate absolute values of all matrix and vector entries. Since this relationship is linear, the computed bounds are symmetrical and the uncertainty in the state estimate can be expressed as a multidimensional interval [x∗] in the state space. The maximum and minimum values for the state variables are given by:

$$x_{inf}^* = x^* - e^*, \qquad x_{sup}^* = x^* + e^* \tag{9.7}$$

and the complete fuzzy state estimate is represented as:

$$\left[ x^* \right] = \left[ x_{inf}^*, x_{sup}^* \right] \tag{9.8}$$

The result of this error limit analysis is a fuzzy state estimate where each component of the state vector is represented by an interval rather than a single value. This fuzzy representation captures the uncertainty inherent in the estimation process and provides a more complete picture of system state than a single point estimate.

Neural networks provide an effective mechanism for learning to recognize patterns in fuzzy state estimates and associating these patterns with specific network anomalies. The neural network architecture for this classification task employs a three-layer structure that grows and adapts during training to accommodate the characteristics of the problem. The network performs clustering and classification based on multidimensional cells, which are regions of the pattern space defined by intervals in each dimension.

A cell C in an n-dimensional pattern space is defined as the intersection of n pairs of half-spaces. Each dimension i has associated minimum and maximum values $m_i$ and $M_i$, such that the cell comprises all points x satisfying $m_i \le x_i \le M_i$ for all dimensions i=1,2,…,n. The vectors $m = \left( m_1, m_2, \ldots, m_n \right)$ and $M = \left( M_1, M_2, \ldots, M_n \right)$ are called the min and max points of cell C and completely determine the cell.

For fuzzy patterns represented by intervals $P = \left[ P_{inf}, P_{sup} \right]$, such as the fuzzy state estimates obtained from error limit analysis, membership of a pattern to a cell is defined through a membership function. This function evaluates how well the fuzzy pattern fits within the cell boundaries. The membership function is given by:

$$c(P) = 1 - \max_{i=1,2,\ldots,n} \left\{ \max \left\{ \phi_i \left( m_i - P_{inf_i} \right), \phi_i \left( P_{sup_i} - M_i \right) \right\} \right\} \tag{9.9}$$

where $\phi_i$ is a non-decreasing function for each coordinate i that maps real numbers to the interval zero to one. This function takes the value zero for negative arguments, monotonically increases for small positive arguments and approaches one for large positive arguments. The shape and sensitivity of $\phi_i$ are controlled by sensitivity parameters and a common choice is a ramp function $\phi_i = \mu_i x$ for $x \epsilon \left[0, \frac{1}{\mu_i}\right]$ where $\mu_i$ is a positive parameter. The membership function evaluates the maximum extent to which the fuzzy pattern extends beyond the cell boundaries in any dimension, with values near one indicating good containment within the cell and values near zero indicating poor fit.

The neural network implementing the classification process consists of three layers: an input layer, a hidden layer and an output layer. The input layer has 2n neurons, two for each dimension of the fuzzy pattern $P = \left[P_{inf}, P_{sup}\right]$. When a fuzzy pattern is presented to the network, the components of vectors $P_{inf}$ and $P_{sup}$ are compared with the minimum and maximum points of all existing cells in the hidden layer. For each cell j with min point m and max point M, the differences $m_i - P_{inf_i}$ and $P_{sup_i} - M_i$ are calculated for all dimensions i and then mapped through the functions $\phi_i$.

The hidden layer contains J neurons, one for each cell that has been created during training. Each hidden neuron computes the membership value of the current pattern to its associated cell using the membership function. The membership functions serve as the activation functions for the hidden neurons and the values of the minimum and maximum points of the cells, which are adjusted during training, constitute the synaptic weights between the input and hidden layers.

The output layer has p neurons, one for each possible class or diagnosis that the system can exhibit, such as leaks at specific locations, incorrect valve positions, or normal operation. The connection between hidden and output layers is governed by binary weights $n_{jk}$ where $n_{jk} = 1$ if cell $C_j$ belongs to class $e_k$ and $n_{jk} = 0$ otherwise. The output of neuron k representing class $e_k$ is computed as:

$$g_k = \max_{j=1,2,\ldots,J} n_{jk} c_j \tag{9.10}$$

where $c_j$ is the membership value of the current pattern to cell j. This output provides either a fuzzy membership assignment where $g_k$ indicates the grade of membership to each class, or a crisp classification by selecting the class with the highest membership value.

During training, patterns represented as ordered pairs (P,e) are presented sequentially, where P is a fuzzy state estimate and e is a label describing the anomaly type. For each pattern, the network either creates a new cell or expands an existing cell. A new cell is created when all existing cells either are too large or belong to different classes. The criterion for a cell being too large is:

$$\max_{i=1,2,\ldots,n} \left\{ \max \left\{ \left(m_i - P_{inf_i}\right), \left(P_{sup_i} - M_i\right) \right\} \right\} > \sigma \tag{9.11}$$

where $\sigma$ is a threshold parameter limiting maximum cell size. New cells are initialized with $m = P_{inf}$ and $M = P_{sup}$ and assigned to the appropriate class. Existing cells belonging to the same class as the current pattern are expanded by updating their boundaries:

$$m_i \to \min\left(m_i, P_{inf_i}\right),\ M_i \to \min\left(M_i, P_{sup_i}\right) \tag{9.12}$$

for all dimensions. After creation or expansion, the network checks for overlapping between cells of different classes and applies contraction procedures to eliminate ambiguity while maintaining pattern coverage.

The effectiveness of classification depends on appropriate selection of sensitivity parameters $\mu_i$ and the threshold $\sigma$. Since different state variables have different importance for anomaly detection, the error limit analysis applied to nominal system operation provides guidance for parameter selection. Variables with larger uncertainty bounds under normal conditions are assigned lower sensitivity, while variables with tight error bounds that change significantly during anomalies receive higher sensitivity [5].

The neural identification approach offers several practical advantages for water supply system management. The network learns from simulated or historical data, capturing complex relationships between measurement patterns and anomalies that would be difficult to encode explicitly. The fuzzy pattern representation naturally accommodates measurement uncertainties, providing robust classification with imprecise data. The graduated output provides both qualitative anomaly identification and quantitative confidence assessment. The modular architecture allows specialized sub-networks for different operating regimes such as peak, valley, or seasonal demand patterns, with appropriate sub-networks selected based on easily identifiable characteristics like total system inflow. As new conditions are encountered, additional sub-networks can be incorporated without retraining the entire system. This capability enables early identification of progressive anomalies such as gradually worsening leaks or partially stuck valves, allowing preventive maintenance before failures occur, thereby reducing water losses, minimizing service disruptions and enabling more efficient resource allocation for improved integrated water management.

## 9.3 NN-FL: A NEURO-FUZZY APPROACH FOR PARAMETERS ESTIMATION IN OIL WELLS

The production of hydrocarbons from oil wells involves complex interactions between reservoir properties, well completion configurations, and surface facilities. A critical challenge in oil well operations is the estimation of bottom-hole parameters, particularly the bottom-hole flowing pressure, which is essential for monitoring well performance, optimizing production rates, and making informed operational decisions. Traditional approaches require expensive instrumentation at the well bottom, with devices that have short useful life and high maintenance costs. This motivates the development of estimation techniques that can infer bottom-hole parameters from readily available surface measurements using intelligent computational methods [6].

The neuro-fuzzy approach for bottom parameter estimation combines neural networks and fuzzy logic through a specialized architecture called the neo-fuzzy neuron model. This hybrid method takes advantage of neural networks' ability to learn from data while incorporating fuzzy logic's capability to handle vague and imprecise information. The neo-fuzzy neuron represents a simpler alternative to conventional multilayer neural networks, consisting of a single neuron structure where synaptic weights are not constant values but nonlinear functions of inputs represented through fuzzy logic models. This architecture uses collections of IF-THEN rules with approximate reasoning in the inference process, making it particularly suitable for modelling complex systems with limited structural complexity.

The neo-fuzzy neuron differs fundamentally from traditional artificial neural networks in its design philosophy. While conventional neural networks require careful selection of the number of layers, neurons per layer and activation functions to achieve good performance, the neo-fuzzy neuron simplifies this process by requiring only the adjustment of fuzzy partitions in the input variables. This makes finding a suitable structure much easier and more straightforward. The neuron does not use a typical activation function but instead employs a summing point that aggregates outputs from fuzzy logic models associated with each input variable. The input variable spaces are divided into segments that constitute fuzzy subgroups, with each segment characterized by triangular complementary membership functions that provide smooth transitions between fuzzy regions.

The learning process in neo-fuzzy neurons involves modifying synaptic weights to minimize errors between desired and actual outputs. Since synapses are represented by fuzzy logic models with IF-THEN rules having constant weights as consequents and typically two rules are activated for any single input signal, the learning process focuses on adjusting the two weights of each synapse corresponding to activated rules. This is accomplished through a gradient descent algorithm that iteratively updates weights according to:

$$w_{ik}(T+1) = w_{ik}(T) + \Delta w_{ik}(T) \tag{9.13}$$

where the weight adjustment is computed based on the error between desired and actual outputs. The output y of the neo-fuzzy neuron is given by:

$$y = f_1(x_1) + f_2(x_2) + \ldots + f_m(x_m) = \sum_{i=1}^{m} f_i(x_i) \tag{9.14}$$

where f(x) represents the defuzzification function for each input. During training, patterns are presented sequentially, each activating only two rules per input and the corresponding weights are modified until convergence to acceptable error levels is achieved.

For oil well applications, the approach focuses on wells using the artificial gas lift method, which is a common technology for producing oil from wells with low reservoir pressure. In this method, gas is injected into the production tubing to reduce the hydrostatic pressure of the fluid column, thereby decreasing bottom-hole pressure and increasing reservoir inflow. The injected gas mixes with reservoir fluids and reduces the density of the mixture in the tubing, making it easier for the well to flow. But, gas lift wells can exhibit unstable behaviour with oscillatory flow patterns that reduce production efficiency. The relationship between gas injection rate and oil production follows a characteristic curve where production initially increases with gas injection but eventually reaches a maximum, after which additional gas injection causes production to decline.

The estimation system uses a configuration with two neo-fuzzy neurons working in sequence. The first neuron estimates the tubing head pressure using three surface measurements as inputs: gas lift pressure, differential pressure of injected gas and casing head pressure. These variables are readily available through standard instrumentation at the wellhead and can be measured continuously without requiring bottom-hole equipment. The output from this first neuron provides an estimated tubing head pressure. The second neo-fuzzy neuron then uses this estimated value along with the measured tubing head pressure to estimate the bottom-hole flowing pressure. This cascaded structure allows the system to build up from surface measurements to bottom-hole conditions through a logical progression of estimations.

The validation mechanism for the estimated bottom pressure relies on comparing the measured tubing head pressure with the estimated tubing head pressure from the first neuron. If the absolute difference between these values is less than a specified threshold $\beta$, it indicates that the bottom-hole pressure estimation is reliable. This validation approach provides a practical way to assess estimation quality without requiring direct bottom-hole measurements during normal operation. When the difference exceeds the threshold, it signals that operating conditions may have changed significantly from those represented in the training data, suggesting the need for model retraining or further investigation.

The training process begins with sampling input and output variables simultaneously to create training patterns. Variables are normalized to facilitate fuzzy partitioning, typically using triangular membership functions. For the first neo-fuzzy neuron with three inputs and one output, each input variable is partitioned into fuzzy sets, with membership functions defining the degree to which input values belong to each fuzzy region. Consider an input value that falls between two fuzzy regions: the membership degrees are calculated based on the triangular function geometry, providing values that sum to unity for adjacent regions and zero for non-adjacent regions. These membership degrees determine which rules are activated and influence the output calculation.

The synapse output for each input is computed by multiplying the corresponding weights by membership degrees and summing the results. For example, if an input activates two fuzzy regions with membership degrees $\mu_1$ and $\mu_2$ and the associated weights are $w_1$ and $w_2$, the synapse contribution is $f(x) = w_1\mu_1 + w_2\mu_2$. The total neuron output is the sum of all synapse contributions from all inputs. During training, after each pattern presentation, weights associated with activated fuzzy sets are updated using:

$$w_{ik}(T+1) = w_{ik}(T) + \alpha\mu_i(y - y_d)x_{ik} \tag{9.15}$$

where $\alpha$ is the learning rate, $\mu_i$ is the membership degree, y is the actual output, $y_d$ is the desired output, and $x_{ik}$ is the input contribution. This update rule adjusts weights in proportion to both the output error and the degree of membership activation, providing targeted learning that focuses on the most relevant fuzzy rules for each pattern.

The second neo-fuzzy neuron follows a similar training procedure but uses the estimated tubing head pressure from the first neuron as one of its inputs, along with the measured tubing head pressure. The target output for training is the actual bottom-hole pressure measured using temporary fibre optic systems. These optical fibre-based measurement systems can record bottom-hole pressure and temperature but are expensive to operate continuously, which is why the estimation approach is valuable for routine monitoring. Once trained, the neo-fuzzy system can provide continuous bottom-hole pressure estimates using only surface measurements, eliminating the need for permanent bottom-hole instrumentation.

The practical benefits of this approach include low implementation cost compared to bottom-hole instrumentation, and the ability to use existing data acquisition systems and historical databases for training and continuous monitoring capability without well intervention. The estimated bottom-hole pressure allows operators to assess well productivity, determine operational state, and detect conditions such as water or sediment accumulation at the well bottom. Results from field applications show that the neo-fuzzy system can estimate both surface and bottom variables with quadratic errors below one percent, demonstrating effective tracking of the dynamic behaviour of actual measured pressures. The validation threshold $\beta$ typically remains below five psi throughout operation, confirming the reliability of surface pressure estimation and consequently the bottom-hole pressure inference.

The neo-fuzzy approach is flexible and can be adapted to different wells and production methods by retraining the model with data representative of specific operational conditions. Each well has unique characteristics including reservoir properties, completion configuration and production method, so the fuzzy partitions and trained weights must reflect these specific conditions to obtain reliable estimates. The simplicity of the neo-fuzzy neuron structure, requiring only adjustment of fuzzy partitions rather than complex network architectures, makes this adaptation process more manageable than with conventional neural networks. This combination of learning capability from neural networks and linguistic rule representation from fuzzy logic provides an interpretable yet powerful tool for parameter estimation in oil well operations.

## 9.4 FL-GA

Fuzzy logic controllers are rule-based systems that are particularly useful for controlling complex processes that can be managed by skilled human operators without requiring detailed mathematical models of the underlying dynamics. The performance of a fuzzy logic controller depends critically on two main components: the control rules expressed as IF-THEN statements and the membership functions that define the meaning of linguistic terms used in these rules. While domain experts can provide initial control rules based on their experience, adjusting these parameters to achieve optimal controller performance remains a challenging task. This is where the combination of fuzzy logic and genetic algorithms becomes valuable for parameter estimation and system tuning [7].

The challenge in tuning fuzzy controllers arises from the interference between different tuneable parameters. A skilled operator's control behaviour can be described through training data consisting of input-output pairs, where the inputs represent system states and the outputs represent desired control actions. In a fuzzy controller, each control action results from the combined effect of multiple rules, with each rule contributing based on how well its antecedent conditions match the current input. Therefore, adjusting any single membership function typically affects multiple rules and each rule may influence multiple control actions across the operating range. The key problem is determining how to adjust these interconnected parameters so that the controller's behaviour matches the desired behaviour as closely as possible across all operating conditions.

The FL-GA approach addresses this tuning problem by using genetic algorithms to optimize the membership functions of fuzzy control rules. The objective is to minimize the difference between the controller's output and the desired output specified by training data, taking into account the inference system and defuzzification strategy being used. This optimization process operates offline, tuning the controller before deployment by fitting it to a set of reference data that represents expert or desired behaviour. The method treats the complete set of fuzzy control rules as a system to be optimized, where the membership function parameters are the decision variables that the genetic algorithm adjusts.

A fuzzy logic controller operates through several stages. First, linguistic control rules relate input state variables to output control variables through IF-THEN statements. For example, a rule might state 'IF temperature is high AND pressure is low THEN valve opening is medium'. Second, the linguistic labels such as high, low, and medium are defined by membership functions, typically having trapezoidal or triangular shapes. These shapes are characterized by parameters that define the function's support and core regions. Third, when actual input values are provided, the fuzzy inference process evaluates how well these inputs match each rule's antecedent conditions using fuzzy conjunction operations. Fourth, all activated rules are combined using a disjunctive operation to produce an overall fuzzy output. Finally, a defuzzification method converts this fuzzy output into a crisp control action that can be applied to the process.

The genetic algorithm formulation for this tuning problem requires careful consideration of several components. The representation scheme must encode the membership function parameters in a way that allows the genetic algorithm to search effectively. Each candidate solution represents a complete set of fuzzy control rules and within each rule, both the antecedent and consequent membership functions are defined by their parameters. For trapezoidal membership functions, four parameters specify each function: the left and right endpoints of the support and the left and right endpoints of the core where membership equals one. A complete rule base with multiple rules and multiple variables per rule involves many parameters that must be optimized simultaneously.

Traditional binary encoding of parameters can become problematic for this application because of the large number of parameters and the precision required. For instance, with multiple input variables, several rules, and requirements for high numerical precision, binary encoding would create extremely long chromosomes and vast search spaces that genetic algorithms would explore inefficiently. Instead, real-valued encoding provides a more natural representation where each gene directly represents a parameter value as a floating-point number. This approach maintains better precision, creates more compact chromosomes and allows the search space to remain manageable while still capturing the necessary detail in the membership function shapes.

The initial population for the genetic algorithm is constructed from the expert-provided fuzzy rules. These initial rules represent the starting point and the genetic algorithm explores variations around this baseline to find improved configurations. For each parameter in the initial membership functions, an interval of permissible variation is defined based on the structure of the membership function. For example, if a trapezoidal function is defined by four consecutive points, the left support point can vary within limits that maintain proper ordering with adjacent points, ensuring that the resulting functions maintain valid membership functions. This constrained initialization ensures that the genetic algorithm searches within a meaningful region of the parameter space while maintaining the basic structure intended by the expert.

The fitness function evaluates how well a particular set of membership functions performs by comparing the controller's output to desired outputs from training data. Given a set of input-output training pairs where inputs represent system states and outputs represent desired control actions, the controller is evaluated on each training instance. The squared error between the controller's output and the desired output is computed for each instance and these errors are summed to produce the overall fitness measure. The genetic algorithm seeks to minimize this total squared error, meaning that better solutions have lower fitness values. This fitness function directly reflects the tuning objective of making the controller behave as closely as possible to the training data.

The genetic operators used must be appropriate for real-valued representations. One effective mutation operator is nonuniform mutation, which modifies selected parameter values by adding or subtracting amounts that depend on the current generation number. Early in the search, this operator makes relatively large changes, allowing broad exploration of the parameter space. As generations progress, the magnitude of changes decreases, causing the search to become more localized and fine-tuned. This adaptation helps the genetic algorithm balance exploration and exploitation phases naturally over the course of the run. For a parameter selected for mutation, the operator adds or subtracts a value drawn from a range that shrinks as the generation count increases, providing global search initially and local refinement later.

The crossover operators can take several forms with real-valued encoding. Simple crossover operates similarly to traditional crossover by selecting a crossover point and exchanging parameter segments between two parent solutions to create offspring. However, arithmetic crossover operators that are specific to real-valued representations can be more effective. Max-min-arithmetical crossover generates multiple offspring from two parents by combining their parameter values in different ways: taking weighted averages with different weights, taking the minimum value at each position and taking the maximum value at each position. The best offspring from these alternatives are retained, allowing the operator to explore different combinations of parental characteristics. This approach can be particularly effective when the optimal parameter values lie between those found in different good solutions.

The training data used for tuning represent the desired control behaviour and are critical to the success of the approach. These data typically come from observations of skilled operators controlling the process, from simulations of ideal control behaviour, or from previously successful controller configurations. Each training instance consists of values for all input state variables and the corresponding desired output control variable. The size and coverage of the training set affect how well the tuned controller generalizes to situations not explicitly represented in the training data. Adequate coverage of the operating range with representative examples helps ensure that the optimized membership functions produce good performance across all conditions the controller will encounter.

The iterative optimization process proceeds over many generations, with the genetic algorithm maintaining a population of candidate solutions. In each generation, solutions are evaluated according to the fitness function, better solutions are selected preferentially for reproduction and genetic operators create new candidate solutions that form the next generation's population. Over successive generations, the population evolves toward regions of the parameter space that produce lower errors on the training data. The process continues until convergence criteria are met, such as reaching a maximum number of generations, achieving a target fitness level, or observing that improvement has stagnated over recent generations.

The effectiveness of this FL-GA tuning approach has been demonstrated in control applications such as the inverted pendulum problem, which is a standard benchmark in control engineering. An inverted pendulum involves balancing a pole in an upright position by applying forces to its base, requiring rapid and precise control actions. The controller must consider both the angle of the pole from vertical and its angular velocity to determine appropriate corrective forces. Initial fuzzy rules provided by experts define the basic control strategy, but the membership functions require tuning to achieve optimal performance. After applying genetic algorithm optimization to adjust the membership functions, the resulting controller shows substantially reduced error compared to the initial expert-provided rules. The optimized controller stabilizes the pendulum more quickly and maintains better control across different initial conditions and disturbances.

Different defuzzification methods convert the fuzzy output of the rule base into crisp control actions and the optimal membership functions may vary depending on which method is used. Common defuzzification strategies include centre of area, which computes the centroid of the fuzzy output set; mean of maximum, which averages the points where membership is highest; and various maximum-based methods. The genetic algorithm can be run separately for each defuzzification method, finding the membership functions that work best with that particular strategy. Results show that some defuzzification methods are inherently more effective than others for particular control problems and that genetic algorithm tuning improves performance regardless of which method is used, though the degree of improvement varies.

The FL-GA combination provides several advantages for controller tuning. The genetic algorithm's population-based search explores multiple regions of the parameter space simultaneously, reducing the risk of becoming trapped in poor local optima. The approach does not require gradient information, making it applicable even when the relationship between parameters and performance is complex or discontinuous. The method is flexible and can accommodate different types of membership functions, inference mechanisms, and defuzzification strategies without fundamental changes to the optimization procedure. Once tuned, the resulting controller retains the interpretability of fuzzy rule-based systems while achieving performance that closely matches the training data. This combination of optimization power and interpretability makes the FL-GA approach valuable for developing high-performance fuzzy controllers that can be understood and trusted by human operators and engineers.

## 9.5 GA-NN: GENETICALLY EVOLVED DYNAMICALLY FUNCTIONAL-LINK NEURAL NETWORKS

System identification involves developing mathematical models that describe the behaviour of dynamic processes based on observed input-output data. For complex industrial processes where underlying physical principles are difficult to model or where processes exhibit significant nonlinear behaviour, data-based modelling approaches using artificial neural networks have become valuable tools. However, the effectiveness of neural network models depends critically on selecting an appropriate network architecture, including the number of layers, neurons, types of connections and dynamic elements. Manual selection of these architectural parameters through trial and error is difficult and time-consuming, particularly for models that must capture dynamic behaviour over time. The combination of genetic algorithms with neural networks addresses this challenge by automating the architecture selection process through evolutionary optimization [8].

Neural networks for system identification require some form of memory to capture the dynamic relationships between inputs and outputs over time. Two main approaches exist for providing this memory capability. The first approach uses static neural networks equipped with external dynamic elements such as tapped delay lines, which feed past values of inputs and outputs as additional inputs to the network. This increases the dimensionality of the input space and requires the network to receive multiple time-delayed versions of each process variable. The second approach provides the network with internal dynamics through feedback connections and filters within the network structure itself. This internal dynamics approach is generally more efficient because it processes current input values directly without requiring explicit presentation of past measurements, with the network's internal state maintaining necessary historical information.

The functional-link neural network represents an alternative architecture to the traditional multilayer perceptron. Instead of using multiple hidden layers with many neurons, the functional-link approach expands the input space through nonlinear transformations of the original inputs before feeding them to a simpler network structure. Common functional expansions use trigonometric functions such as sine and cosine applied to the inputs with different frequencies. This approach reduces computational cost during training while maintaining good approximation performance, making it attractive for practical applications. The functional-link network essentially moves complexity from the network's hidden layer structure to the input transformation stage, trading architectural complexity for enhanced input representation.

The generalized dynamic functional-link neural network extends the basic functional-link concept by incorporating internal dynamic elements in a flexible manner. The network architecture includes several components that can be selectively activated or deactivated. First, each network input can be functionally expanded using trigonometric functions, with the order of expansion determining how many frequency components are included. Different inputs may use different expansion orders, allowing the model to capture varying degrees of nonlinearity in different variables. Second, the network can include autoregressive moving average filters at different locations within the structure. A filter can be placed before the nonlinear activation unit, creating local activation feedback, or on the feedback path from the output back to the input, creating local output feedback, or both filters can be used simultaneously. These filter structures are characterized by numerator and denominator polynomial orders that determine their dynamic properties.

The design problem for such a network involves selecting which functional expansions to use for each input, determining the expansion orders, deciding which dynamic filter structures to activate and choosing the filter orders. These architectural decisions create a large discrete search space where exhaustive evaluation would be computationally prohibitive. Additionally, multiple objectives must be balanced: the model should fit the training data accurately, generalize well to new data not used in training and maintain reasonable complexity to avoid overfitting and ensure practical implementation. This multi-objective optimization problem with discrete and continuous parameters makes genetic algorithms a natural choice for finding good solutions.

The genetic algorithm formulation uses a hierarchical chromosome structure to represent candidate network architectures. The highest level contains binary control genes that activate or deactivate functional expansions for each input and the dynamic filter structures. When a functional expansion or filter is inactive, it simply does not contribute to the network's operation. The lower level contains parametric genes that specify the orders of functional expansions and filters when these features are active. During the genetic search, inactive genes remain in the chromosome and can become active in future generations through genetic operations, maintaining the potential for these features to be explored. This hierarchical encoding efficiently represents the variable-structure nature of the network architecture.

The genetic operators work with this encoding scheme to explore the architecture space. Standard crossover and mutation operations are applied independently to each level of the chromosome. Binary control genes undergo typical bit-flip mutation and crossover, while integer parametric genes use operations appropriate for integer values. The search space is bounded by specifying maximum values for expansion orders and filter orders, preventing the genetic algorithm from considering unreasonably complex structures. The practical maximum values depend on the application and available computational resources, but are typically chosen to allow sufficient flexibility without excessive complexity.

Multiple objectives guide the evolutionary search toward good architectures. The primary objectives relate to model accuracy, measured by the sum of squared errors between network outputs and target values on both training and validation data sets.

Using separate training and validation sets helps identify models that generalize well rather than simply memorizing training data. Secondary objectives relate to model complexity, counting the number of functional expansion terms and filter coefficients in the active architecture. The objectives are organized by priority, with accuracy objectives having higher priority than complexity objectives. This means the genetic algorithm first seeks architectures with good accuracy and among comparably accurate models, it prefers simpler ones.

The multi-objective optimization produces not a single solution but a set of solutions representing different trade-offs between objectives. This solution set is called the Pareto-optimal front, where each solution is optimal in the sense that improving one objective would require degrading another. For the network design problem, the Pareto front might contain simple models with modest accuracy, complex models with excellent accuracy and intermediate models balancing these extremes. The designer can examine this range of solutions and select one that best matches the application requirements, whether those emphasize accuracy, simplicity, or some balance.

During each generation of the genetic algorithm, candidate architectures must be evaluated for fitness. This evaluation process involves more than just examining the chromosome structure. For each candidate architecture, the network parameters including connection weights, filter coefficients, and bias terms, must be determined through training. This training uses a dynamic backpropagation algorithm that extends standard backpropagation to handle the internal dynamic elements. The algorithm minimizes the sum of squared errors between network outputs and training targets over multiple training epochs. The trained network is then evaluated on both training and validation data to compute the accuracy objectives, while the complexity objectives are computed directly from the architecture specification. This nested optimization, where parameter training occurs within the architecture search, makes each fitness evaluation computationally expensive, but is necessary to properly assess each architecture's potential.

The complete design process uses three sets of data with different purposes. The training set is used to adjust network parameters during the dynamic backpropagation learning process. The validation set guides the genetic algorithm's architecture search by providing an independent measure of generalization performance. The test set, which is not used at all during the design process, provides a final assessment of the selected model's performance on completely new data. This three-way split helps ensure that the resulting model captures genuine process behaviour rather than peculiarities of any particular data set. The correlation coefficient between network outputs and target values provides a convenient measure of fit quality, with values near one indicating excellent agreement.

The practical implementation of this approach requires careful consideration of data preparation and model validation. Process measurements often contain noise from sensors and disturbances, requiring preprocessing through techniques like spectral analysis and appropriate filtering to extract meaningful patterns. The correlation coefficient between network outputs and target values serves as a useful measure of model quality, with values approaching one indicating strong agreement between predicted and actual behaviour. The three-way data split into training, validation, and test sets ensures that selected models genuinely capture process dynamics rather than overfitting to particular datasets.

The advantages of using genetic algorithms to evolve neural network architectures include automation of the design process, exploration of a wide range of architectural possibilities and principled handling of multiple competing objectives. Rather than requiring the designer to manually try different architectures and compare results, the genetic algorithm systematically searches the architecture space guided by the defined objectives. The population-based search explores multiple regions simultaneously, reducing the risk of settling on mediocre solutions. The multi-objective formulation makes trade-offs explicit and provides decision-makers with a range of solutions to choose from based on application-specific priorities. The user must still specify parameters like maximum expansion orders and genetic algorithm settings, but these high-level specifications are generally easier to determine than detailed architecture choices.

The combination of genetic algorithms with functional-link neural networks that have internal dynamics provides a powerful tool for nonlinear system identification. The genetic algorithm handles the discrete architectural decisions and structure optimization, while the neural network training algorithm handles the continuous parameter optimization. Together, these methods produce models that balance accuracy and complexity, generalize well to new data and capture both nonlinear relationships and dynamic behaviour. This approach demonstrates how genetic algorithms and neural networks complement each other effectively in estimation problems, with each technique contributing its strengths to solve aspects of the overall problem that would be difficult to address with either method alone.

## 9.6 NN-FL-GA

While the previous sections have explored how pairs of soft computing technologies can be combined for estimation problems, the integration of all three components – neural networks, fuzzy logic, and genetic algorithms – offers even greater potential for addressing complex system identification and parameter estimation challenges. Each technology brings distinct strengths to the estimation process, but also has inherent limitations. Neural networks excel at learning from data and approximating complex functions, but suffer from slow convergence and lack interpretability. Fuzzy logic provides a framework for reasoning with uncertain and imprecise information, but struggles with parameter tuning and lacks effective learning mechanisms. Genetic algorithms perform robust global optimization without requiring gradient information, but are difficult to tune and have no clear convergence criteria. By merging all three technologies, we can create hybrid systems that capitalize on complementary strengths while compensating for individual weaknesses [9].

The fundamental motivation for combining these three technologies lies in recognizing that estimation problems often require capabilities that no single approach can provide adequately. Consider the challenge of building a model for a complex nonlinear system where available data contains uncertainty, the relationships between variables are not fully understood and the model structure itself must be determined. Neural networks can learn the nonlinear relationships from data, but determining the appropriate network architecture and parameters is difficult. Fuzzy logic can represent the uncertain relationships in interpretable rule form, but constructing the membership functions and rules manually is challenging. Genetic algorithms can search for optimal structures and parameters, but need a proper representation scheme and fitness evaluation method. When these technologies work together, neural networks can provide the learning capability, fuzzy logic can handle uncertainty and provide interpretability, and genetic algorithms can optimize the overall system configuration.

The integration of neural networks and fuzzy logic creates neuro-fuzzy systems where fuzzy logic components control or modify neural network behaviour. One approach involves making the inputs or weights of neural networks fuzzy rather than crisp. In a fuzzy neural network with fuzzy weights, each connection is represented not by a single weight value but by a fuzzy membership function that describes the weight's uncertainty. For any given input, multiple adjacent membership functions are activated and the effective weight becomes a combination of the activated values. This representation allows the network to naturally handle uncertain or imprecise training data. Similarly, fuzzy inputs can be processed through the network using interval arithmetic or fuzzy extension principles, producing fuzzy outputs that reflect the uncertainty in the input data and model parameters.

Another way fuzzy logic enhances neural networks is by adaptively controlling learning parameters during training. The learning rate and momentum coefficient in backpropagation determine how quickly the network adjusts its weights, but fixed values for these parameters throughout training can be inefficient. A fuzzy system can monitor training progress through variables like total error and change in error between iterations, then apply fuzzy rules to adjust learning parameters dynamically. For example, if the error is large and decreasing, the learning rate might be increased to accelerate progress. If the error is small but oscillating, the momentum might be adjusted to stabilize convergence. This adaptive control typically reduces training time substantially compared to fixed parameter values while improving the quality of the final solution.

Genetic algorithms enhance neural networks by automating architecture selection and parameter optimization. Rather than manually determining the number of hidden layers, neurons per layer, connectivity patterns, learning rates, and other architectural decisions through trial and error, these choices can be encoded in a genetic algorithm chromosome. A straightforward encoding uses binary strings where different bit patterns represent different architectural options. For instance, several bits might specify the number of hidden neurons, other bits might encode the learning rate from a set of discrete values, and additional bits might specify the momentum coefficient. Each chromosome represents a complete network configuration that can be constructed, trained on data, and evaluated for performance.

The fitness function for evaluating network configurations typically combines training accuracy with generalization performance and sometimes includes complexity penalties. A network is constructed according to the chromosome specification, trained using backpropagation or another learning algorithm for a fixed number of iterations, then tested on validation data. The fitness might be computed as the inverse of the mean squared error on the validation set, possibly adjusted for network complexity to favour simpler models among equally accurate alternatives. The genetic algorithm population evolves over generations, with better-performing configurations having higher probability of reproduction and passing their characteristics to offspring through crossover and mutation operations.

An important consideration when using genetic algorithms with neural networks is the computational cost of fitness evaluation. Since each candidate network must be trained and evaluated and a genetic algorithm population might contain dozens or hundreds of individuals evolving over many generations, the total computational effort can be substantial. However, this cost is often justified by the quality of the resulting network and the difficulty of finding good architectures manually. The genetic algorithm's population-based search explores multiple regions of the architecture space simultaneously, reducing the risk of settling on poor local optima that might trap manual or gradient-based design approaches.

Genetic algorithms also optimize fuzzy inference systems by adapting membership functions and learning rule bases. The parameters defining membership functions, such as the positions and shapes of triangular or trapezoidal functions, can be encoded in chromosomes. For a fuzzy system with multiple inputs and outputs, each having several membership functions, the chromosome becomes a concatenation of all these parameters. The genetic algorithm searches for membership function configurations that minimize the error between the fuzzy system's outputs and desired targets on training data. Beyond just tuning membership functions with fixed rules, genetic algorithms can also determine the rules themselves by encoding which input membership functions appear in each rule's antecedent and which output membership function appears in the consequent.

A particularly effective approach represents fuzzy rules using a hierarchical chromosome structure. At the highest level, binary control genes activate or deactivate functional expansions for inputs and select which dynamic elements to include in the system. At lower levels, parametric genes specify the orders and parameters of activated components. During genetic search, inactive genes remain in the chromosome but do not affect the current system behaviour, though they might become active in future generations through genetic operations. This representation allows the genetic algorithm to explore not just parameter values but also structural variations in the fuzzy system, determining which inputs are relevant and how they should be combined.

When all three technologies are merged together, the design becomes more complex but also more powerful. A typical architecture might use genetic algorithms to determine both the structure of a neuro-fuzzy system and its parameters. The chromosome would encode the fuzzy membership functions for network inputs and possibly weights, the network architecture including number and types of layers and any dynamic elements like feedback connections. The genetic algorithm would evolve a population of these complete system specifications, with fitness evaluation requiring construction of each specified system, training it using an appropriate learning algorithm and testing its performance on validation data.

The learning algorithm for the neuro-fuzzy system might itself incorporate adaptive features controlled by fuzzy rules. As the system trains on data to determine connection weights and other adjustable parameters, fuzzy logic could monitor progress and adjust learning rates, momentum coefficients, or other training parameters to improve efficiency.

This creates a multi-level optimization where genetic algorithms handle high-level structural decisions and learning algorithm selection, neural network training handles continuous parameter optimization for a given structure, and fuzzy logic provides adaptive control to improve training efficiency and handle uncertain data.

Several practical challenges arise when implementing these merged systems. The increased number of parameters and design choices requires careful consideration of computational resources and time constraints. Training a population of complex neuro-fuzzy systems through multiple genetic algorithm generations demands significant computation, though this can be mitigated through parallel processing where multiple individuals are evaluated simultaneously. The complexity also makes it harder to understand which design choices are most important, though this can be addressed through careful experimental design and sensitivity analysis. There is also the risk that the expanded design space might lead to overfitting if not properly constrained through validation procedures and complexity penalties in the fitness function.

Despite these challenges, the merged NN-FL-GA approach offers compelling advantages for difficult estimation problems. The combination provides simultaneous optimization of structure and parameters, handles uncertain and imprecise data naturally, produces interpretable models through fuzzy rule representation, and achieves robust solutions through the genetic algorithm's global search capability. The neural network component ensures the system can learn complex relationships from data, the fuzzy logic component provides transparency and approximate reasoning capabilities, and the genetic algorithm component automates the difficult design decisions that would otherwise require extensive trial and error.

The practical implementation typically proceeds in stages. First, the problem is formulated by identifying input and output variables, collecting training and validation data, and defining appropriate performance measures. Second, the representation scheme is designed by deciding how to encode network architectures, fuzzy membership functions, rules, and other components in genetic algorithm chromosomes. Third, the fitness evaluation procedure is established, including how candidate systems are constructed from chromosomes, trained on data, and assessed for performance. Fourth, genetic algorithm parameters like population size, crossover, and mutation probabilities and selection mechanisms are chosen. Fifth, the evolutionary process is executed over multiple generations until satisfactory solutions emerge. Finally, the best solutions are analyzed, validated on test data, and selected for deployment based on the specific requirements of the application.

The success of the merged approach depends on several factors including the quality and quantity of available training data, the appropriateness of the chosen representations and operators, and the computational resources available for the evolutionary search. When these factors are favourable, the NN-FL-GA combination can produce estimation systems that significantly outperform those designed using any single technology. The resulting systems combine accuracy in capturing complex relationships with interpretability through fuzzy rules and robustness through genetic optimization. This makes the merged approach particularly valuable for applications where estimation accuracy is critical, model transparency is important for user acceptance and the problem structure is too complex for manual design to be practical.

## EXERCISES

1. Explain how the three soft computing technologies (ANNs, FL, and GAs) complement each other in estimation problems. What are the key limitations of each individual technology and how does combining them address these limitations?
2. In the water supply system anomaly detection example, explain how fuzzy state estimates are generated from uncertain measurements and how the neural network uses these fuzzy patterns for classification. What is the advantage of using fuzzy intervals rather than crisp values in this application?
3. Compare and contrast the neo-fuzzy neuron architecture used in oil well parameter estimation with traditional multilayer neural networks. Why is the neo-fuzzy approach considered simpler to design and what role do the fuzzy membership functions play in the learning process?
4. In the FL-GA approach for controller tuning, explain why adjusting membership functions is challenging when done manually. How does the genetic algorithm address the problem of parameter interference and what is the purpose of using real-valued encoding instead of binary encoding?
5. In the GA-NN approach for evolving dynamic functional-link neural networks, explain the concept of multi-objective optimization and the Pareto-optimal front. Why is it valuable to obtain multiple solutions representing different trade-offs between accuracy and complexity rather than a single 'best' solution?

## REFERENCES

1. Jang, J.S.R., "ANFIS: Adaptive-Network-Based Fuzzy Inference System", *IEEE Transactions on Systems, Man, and Cybernetics*, Vol. 23, No. 3, pp. 665–685, 1993.
2. Cordon, O., Herrera, F., Hoffmann, F., and Magdalena, L., Genetic Fuzzy Systems: Evolutionary Tuning and Learning of Fuzzy Knowledge Bases, World Scientific Publishing, Singapore, 2001.
3. Montana, D.J., and Davis, L., "Training Feedforward Neural Networks Using Genetic Algorithms", Proceedings of the 11th International Joint Conference on Artificial Intelligence (IJCAI-89), Vol. 1, pp. 762–767, Morgan Kaufmann Publishers, San Mateo, California, USA, 1989.
4. Tahmasebi, P., and Hezarkhani, A., "A Hybrid Neural Networks–Fuzzy Logic–Genetic Algorithm for Grade Estimation", *Computers & Geosciences*, Vol. 42, pp. 18–27, 2012.
5. Izquierdo, J., Pérez, R., López, P.A., and Iglesias, P.L., "Neural Identification of Fuzzy Anomalies in Pressurized Water Systems", *International Congress on Environmental Modelling and Software, Burlington,* Vermont, USA, July 2006.
6. Camargo, E., Aguilar, J., Rios, A., and Rivas, F., "A Neo-Fuzzy Approach for Bottom Parameters Estimation in Oil Wells", *WSEAS Transactions on Systems and Control,* Vol. 4, pp. 445-454, 2009.
7. Herrera, F., Lozano, M., and Verdegay, J.L., "Tuning Fuzzy Logic Controllers by Genetic Algorithms", *International Journal of Approximate Reasoning*, Vol. 12, pp. 299–315, 1995.

8. Marcu, T., and Köppen-Seliger, B., *"Dynamic Functional-Link Neural Networks Genetically Evolved Applied to System Identification",* In Proceedings of European Symposium on Artificial Neural Networks (ESANN), Bruges, Belgium, pp. 115–120, 2004.
9. Shapiro, A.F., "The Merging of Neural Networks, Fuzzy Logic, and Genetic Algorithms", *Insurance: Mathematics and Economics*, Vol. 31, pp. 115–131, 2002.

# Appendix A Key Theorems in System Identification, Filtering, and State & Parameter Estimation

In this Appendix, we briefly present certain properties of signals, matrices, estimators, and estimates. Most of these are also valid for the theory of Intelligent Estimation that is based on Soft Computing.

A good estimator should possess certain properties in terms of errors in parameter estimation and/or errors in the predicted measurements or responses of the mathematical model thus determined. Since the measured data used in the estimation process are noisy, the parameter estimates can be considered to have some random nature. In fact, the estimates that we would have are the mean of the probability distribution, and hence the estimation error would have some associated covariance matrices (dispersion uncertainty). Thus, due to the stochastic nature of the errors, one would desire that the probability of the estimate being equal to the true value be 1. We expect an estimator to be unbiased, efficient, and consistent – not all of which might be achievable. Here, we collect several properties of signals, matrices, estimators, and estimates that would be useful in judging the properties and 'goodness of fit' of the parameter/state estimates and interpreting the results [A.1]. Many of these definitions, properties and other useful aspects are used or indicated in the various chapters of the book and are compiled in this appendix.

## A1 AUTOCORRELATION

For a random signal $x(t)$, it is defined as

$$R_{xx}(\tau) = E\{x(t)x(t+\tau)\}; \ \tau \text{ is the 'time-lag'}.$$

Here, $E$ stands for a mathematical expectation operator.

For stationary process, $R_{xx}$ is dependent on $\tau$ and $x$ only and not on '$t$'. Its value is maximum when $\tau = 0$, then it is the variance of the signal $x$ (assuming the mean of the signal is removed). As the time tends to be large, if the $R_{xx}$ shrinks then it means that the nearby values of the process $x$ are not correlated and hence not dependent on each other. Autocorrelation of the white noise/process is an impulse function. Autocorrelation of discrete-time residuals is given as:

$$R_{rr}(\tau) = \frac{1}{N-\tau}\sum_{k=1}^{N-\tau} r(k)\, r(k+\tau);\ \tau = 0,\ldots,\tau_{max} \text{ are the discrete time lag).}$$

In order that residuals are white, the normalized values $R_{rr}$ should lie within $(\pm 1.97/\sqrt{N})$ band; and only 5% of $R_{rr}$ values are allowed out of the band. This property is used for checking the performance of state/parameter estimation algorithms. In practice, about 30 to 50 autocorrelation values are obtained and checked if at least 95% of these values fall within the band. Then it is assumed that practically, these autocorrelation values are zero and hence, the residuals are white, thereby signifying that they are not 'autocorrelated'. This means that complete information has been extracted out of the measurement data for state/parameter estimation.

## A2 ALIASING OR FREQUENCY FOLDING

According to Shannon's sampling theorem, if continuous time signal is sampled at least at 2 times the Nyquist frequency, the information content in the signal is preserved and original continuous-time signal can be recovered from the sampled signal by reverse process. Now usually, the measured signal contains noise, which is believed to be of high frequency. For a white noise, the frequency spectrum is flat of constant (power) magnitude. For a band-limited noise, it extends up to a certain frequency. If such a continuous-time measurement is sampled, then aliasing or frequency folding (Figure A.1) is likely to occur.

Let $\omega_N$ be the Nyquist or cut off frequency, $\omega_s$ the sampling frequency, and $\Delta t$ the sampling interval. For any frequency in the range $0 \le f \le f_N$, the higher frequencies that are aliased with $f$ are: $(2f_N \pm f)$, $(4f_N \pm f)$, …, $(2nf_N \pm f)$;

$$\text{Let } t = \frac{1}{2f_N} = \frac{1}{f_s}$$

$$\text{Then } \mathbf{cos}(2\pi f t) \cong \mathbf{cos}\left\{2\pi(2nf_N \pm f)\,\frac{1}{2f_N}\right\}$$

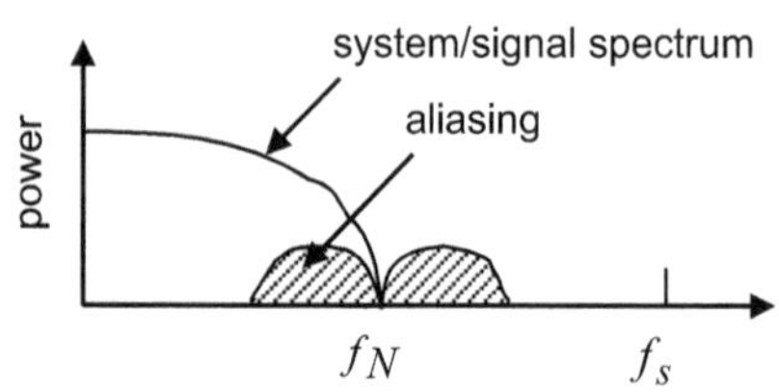


**FIGURE A.1** Effect of aliasing

$$\cos\left(2\pi f\left(\frac{1}{2f_N}\right)\right) \cong \cos\left\{2\pi(n) \pm \frac{\pi f}{f_N}\right\}$$

$$\cos\left(\pi\frac{f}{f_N}\right) = \cos\left\{\pi\frac{f}{f_N}\right\}$$

This shows that the noise spectra would alias with signal spectra under certain conditions. This means that all data at frequencies $(2nf_N \pm f)$ will have the same cosine function as the data at the frequency $f$ when sampled at points $1/f_N$ apart. If $f_N = 100Hz$, then data at $f = 30Hz$ would be aliased with data at frequencies 170, 230, etc. Similarly, power will also be aliased.

There are two approaches to overcome the problem of aliasing:

i) Sample the original signal at 4 to 6 times the Nyquist frequency. Then apparently, the (new) Nyquist frequency will be $f_N' = \frac{1}{2}f_s$ where the $f_s = 6f_N$, and hence we get

$$f_N' = \frac{1}{2}f_s = \frac{1}{2}(6f_N) = 3f_N$$

Now, the frequency folding will occur around $f_N' = 3f_N$ and not around $f_N$. This pushes the folding further away from the actual $f_N$, and hence, essentially minimizing the aliasing of the power spectrum below $f_N$ (thereby not affecting the frequency range of interest).

ii) Filter the continuous-time signal to reduce substantially the effect of noise. However, this will introduce time lag in the signal because of the low pass filter (lag).

Often the signals are collected at 200 samples/second and then digitally filtered down to 50 samples/second.

## A3 BIAS AND PROPERTY OF UNBIASED ESTIMATES

It is the difference between the true value of the parameter $\beta$ and expectation value of its estimate: Bias $(\beta) = \beta - E(\hat{\beta})$. Bias, in general, cannot be determined since it depends on the true value of the parameter that is in practice unknown! Often the estimates would be biased, if the noise were not zero mean. We use a large amount of data to estimate a parameter, then we expect an estimate to centre closely on the true value. The estimate is called unbiased if $E\{\hat{\beta} - \beta\} = 0$. This property means that on the average the expected value of the estimate is the same as the true parameter. One would expect the bias to be small. Unbiased estimates are always sought and preferable. Unbiased estimates may not exist for certain problems. If an estimate is unbiased as the number of data points tends to infinity, then it is called an asymptotically unbiased estimate.

## A4 BAYES' THEOREM

If A and B are two events, then the formula for the Bayes theorem is:

P(A/B)={P(B/A)P(A)}/P(B); where P(A|B) is the probability of condition when event A is occurring while event B has already occurred. One of the many applications of Bayes' theorem is Bayesian inference, an approach to statistical inference. Bayesian inference has found application in various activities, including medicine, science, philosophy, engineering, sports, law, etc. For example, we can use Bayes' theorem to define the accuracy of medical test results by considering how likely any given person is to have a disease and the test's overall accuracy. Bayes' theorem relies on consolidating prior probability distributions to generate posterior probabilities. In Bayesian statistical inference, prior probability is the probability of an event before new data is collected.

### A4.1 BLUE Estimator

It stands for Best Linear Unbiased Estimator, which refers to an estimator in statistics that is considered the best among all linear and unbiased estimators, meaning it has the smallest variance while still providing an unbiased estimate of the population parameter; in linear regression, the Ordinary Least Squares (OLS) estimator is typically considered BLUE under certain conditions, as stated by the Gauss-Markov theorem.

Key points about BLUE estimators:

- Best: This means it has the smallest variance among all linear and unbiased estimators, making it the most efficient choice.
- Linear: The estimator is a linear combination of the observed data points.
- Unbiased: On average, the estimator will provide an accurate estimate of the true population parameter.

## A5 CENTRAL LIMIT PROPERTY/THEOREM

Assume a collection of random variables that are distributed individually according to some different distributions. Let $y = x_1 + x_2 + \ldots + x_n$; then the central limit theorem states that the random variable $y$ is approximately Gaussian (normally) distributed, if $n \to \infty$ and $x$ should have finite expectations and variance. Often $n$ is even 6 or 10 and the distribution of $y$ would be almost similar to the theoretical normal distribution. This property helps in making a general assumption that noise processes are Gaussian, since one can say that they arise due to the sum of various individual noise processes of different types. The central limit theorem says that the sampling distribution of the mean will always be normally distributed, as long as the sample size is large enough. Regardless of whether the population has a normal, Poisson, binomial, or any other distribution, the sampling distribution of the mean will be normal.

## A6 CHI-SQUARE TEST

Let $x_i$ be normally distributed and mutually uncorrelated variables around mean $m_i$ and with variance $\sigma_i$. Form the normalized sum of squares:

$$s = \sum_{i=1}^{n} \frac{(x_i - m_i)^2}{\sigma_i^2}$$

Then $s$ follows the $\chi^2$ distribution with $n$ DOF. In estimation practice often, the $\chi^2$ test is used for hypothesis testing.

## A7 CONFIDENCE LEVEL

In parameter/state estimation, requirement of high confidence in the estimated parameters/states is imperative; without which the results cannot be trusted. Often this information is available from the estimation results. A statistical approach and judgment are used to define the confidence interval within which the true parameters/states are assumed to lie with 95% of confidence, signifying high probability with which truth lies within the upper and lower intervals. This signifies that the estimation error, e.g. $\tilde{\beta}_{LS}$, should be within a certain interval band. In that case, one can define:

$$P\{\, l < \beta < u \,\} = \alpha$$

It means that $\alpha$ is the probability that $\beta$ is constrained in the interval $(l,u)$. In other words, the probability that the true value, $\beta$, is between $l$ (the lower bound) and $u$ (the upper bound) is '$\alpha$'. As the interval becomes smaller, the estimated value $\hat{\beta}$ can be taken, more confidently, as the value of the true parameter.

## A8 CONSISTENCY OF ESTIMATES

One can study the behaviour of an estimator with increased amount of data. An estimator is called asymptotically unbiased, if the bias approaches zero as the number of data tends to infinity. An asymptotically efficient estimator is obtained if the equality in CRI (Chapter 3) is approached as the number of data tends to infinity (see definition of an efficient estimator). It is very reasonable to postulate that as the number of data used increases, the estimate tends to the true value. This property is called 'consistency'. This is a stronger property than asymptotic unbiasedness, since it has to be satisfied for a single realization of estimates and not 'on the average' behaviour. It means that the strong consistency is defined in terms of the convergence of the individual realizations of the estimates and not in terms of the average properties of the estimates. Hence, all the consistent estimates are unbiased asymptotically.

The convergence is required to be with probability 1 (one) and is expressed as

$$\lim_{N\to\infty} P\left\{\left|\hat{\beta}(z_1, z_2, \ldots, z_n) - \beta\right| < \delta\right\} = 1 \qquad \forall \delta > 0$$

This means that the probability that the error in estimates (w.r.t the true values) is less than a certain small positive value is one, as the number of data used in the estimation process tends to infinity.

## A9 CORRELATION COEFFICIENT

$$\rho_{ij} = \frac{\mathbf{cov}(x_i, x_j)}{\sigma_{x_i}\ \sigma_{x_j}} - 1 \le \rho_{ij} \le 1$$

Here, $\rho_{ij} = 0$ for independent variables $x_i$ and $x_j$.

For certainly correlated process, $\rho = 1$. Thus $\rho$ defines the degree of correlation between two random variables. This test is used in the model error method for parameter estimation. For example, in KF theory, often an assumption is made that the state error and measurement error or residuals are uncorrelated. If a variable $d$ is dependent on several $x_i$, then the correlation coefficient for each of $x_i$ can be utilized to determine the degree (extent) of this correlation with $d$ as:

$$\rho(d, x_i) = \frac{\sum_{k=1}^{N} (d(k) - \underline{d})\ (x_i(k) - \underline{x}_i)}{\sqrt{\sum_{k=1}^{N} (d(k) - \underline{d})^2}\ \sqrt{\sum_{k=1}^{N} (x_i(k) - \underline{x}_i)^2}}$$

Here, 'under bar' represents mean of the variable.

If $|\rho(d, x_i)|$ is nearly equal to 1, then d can be considered to be linearly related to a particular $x_i$. In that case, the $x_i$ terms with higher correlation coefficient can be included in the model.

## A10 COVARIANCE

It is defined as

$$\text{Cov}(x_i, x_j) = E\left\{ \left[x_i - E(x_i)\right] \left[x_j - E(x_j)\right] \right\}$$

For the independent variables $x_i$ and $x_j$, the covariance matrix is null. But if the matrix is zero, it does not mean that $x_i$ and $x_j$ are independent. The covariance matrix is supposed to be symmetric and positive semi-definite by definition. However, in practice, when the estimation (iteration) proceeds the matrix may not retain these properties. The covariance matrix plays a very important role in Kalman filter time-propagation and measurement data update equations. It provides theoretical prediction of the state-error variance and covariance-matching concept can be used for judging the performance/consistency of the filter (tuning) (Chapter 4). A similar concept is also used in the method of model error for tuning the deterministic state estimator. The square roots of the diagonal elements of this matrix give standard deviations of the errors in estimation. It must be also emphasized here that the inverse of the covariance matrix gives the indication of the information content in the signals about the parameters. Thus, a large covariance matrix signifies higher uncertainty and low information and low confidence in the state/parameter estimation results.

## A11 CONVERGENCE TEST OF ESTIMATION

A mathematical method used to determine whether a sequence of estimates generated by an estimation algorithm will eventually approach a true value as the number of iterations increases, essentially checking if the estimations are getting closer and closer to the correct answer over time; this is often done by analyzing the behaviour of the error between the estimated value and the true value as the number of iterations grows.

Limit test: This basic test involves checking if the limit of the difference between successive estimations approaches zero as the number of iterations increases.

Ratio test: Used when the estimation process involves a ratio between values, this test checks if the ratio of consecutive errors is less than 1, indicating convergence.

## A12 CR LOWER BOUND (CRLB)

It gives a lower estimate for the variance of an unbiased estimator. Estimators that are close to the CLRB are more unbiased (i.e. more preferable to use) than estimators further away. It is theoretical; Sometimes a perfectly unbiased estimator (i.e. one that meets the CRLB) doesn't exist. Additionally, the CRLB is difficult to calculate unless you have a very simple scenario.

Let $X_1, X_2, \ldots X_n$ be a random sample with probability density function (pdf) $f(x,\Theta)$. If $\hat{\Theta}$ is an unbiased estimator for $\Theta$, then:

$$\operatorname{var}(\hat{\Theta}) \geq \frac{1}{nI(\theta)}$$

$$I(\theta) = E\left[\frac{\partial}{\partial\theta}\ln f(X;\theta)\right]^2 = -E\left[\frac{\partial^2}{\partial\theta^2}\ln f(X;\theta)\right]$$

Here, I(,) is the Fisher Information.

## A13 ERGODICITY

Assume a number of realizations of a random process are present. For an ergodic process, any statistic computed by averaging over all the members of this ensemble (realizations) at a fixed time point can also be calculated (and will be identical) by averaging over all times on a single representative member of the ensemble. Ergodicity implies stationarity, but stationary processes need not be ergodic. Often the assumption of ergodicity is implicit in the parameter estimation process. This assumption allows one to handle only one realization of the process, e.g., data collected from only one experiment. However, from a point of view from consistency of results, it will be desirable to have at least three repeat experiments at the same operating condition. Then these data sets can be used for system identification and parameter estimation purposes, either by averaging the data or by using two sets of data for estimation and the third for model validation purposes.

## A14 EFFICIENCY OF AN ESTIMATOR

We have seen in Chapter 2 that we can obtain covariance of the estimation error. This covariance, which is theoretical in nature, can be used as measure of the quality of an estimator. Assume that $\hat{\beta}_1$ and $\hat{\beta}_2$ are the unbiased estimates of the parameter vector $\beta$. We compare these estimates in terms of error covariance matrices. We form the inequality:

$$E\left\{(\beta-\hat{\beta}_1)(\beta-\hat{\beta}_1)^T\right\} \leq E\left\{\ (\beta-\hat{\beta}_2)(\beta-\hat{\beta}_2)^T\right\}$$

From this, we notice that the estimator $\hat{\beta}_1$ is said to be superior to $\hat{\beta}_2$ if the inequality is satisfied. If it is satisfied for any other unbiased estimator, then it is called an efficient estimator. Another useful measure is the mean square error. Since, the mean square error and the variance are identical for unbiased estimators, such optimal estimators are also called minimum variance unbiased estimators. The efficiency of an estimator can be defined in terms of the so-called Cramer-Rao inequality. It obtains a theoretical limit to the achievable accuracy, irrespective of the estimator used:

$$E\left\{[\hat{\beta}(z)-\beta]\ [\hat{\beta}(z)-\beta]^T\right\} \geq \mathrm{M}^{-1}(\beta)$$

The matrix $M$ is the Fisher information matrix (see Eqn. (3.7) of Chapter 3). The inverse of $M$ is a theoretical covariance limit. It is assumed that the estimator is unbiased. Such an estimator with equality valid is called an efficient estimator. Thus, the Cramer-Rao inequality means that for an unbiased estimator, the variance of parameter estimates cannot be lower than its theoretical bound $\mathrm{M}^{-1}(\beta)$. However, one can get an estimator with lower variance, but it would be the biased estimate. Therefore, a compromise has to be struck between acceptable bias and variance. The $\mathrm{M}^{-1}(\beta)$ gives Cramer-Rao lower bounds for the estimates and is very useful in judging the quality of the estimates. Mostly these Cramer-Rao bounds are used in defining uncertainty levels around the estimates obtained by using maximum likelihood/output error method (see Chapter 3).

## A15 ENTROPY

It is a measure of some disorder in the system. Here, the system could be a plant or some industrial process. Always in a system, there could be some disorder and if the disorder is reduced, some regularization will set in the system. Let $p$ be the probability of the state of a system, then

$$E_s = k\,\mathbf{log}(p) + k_0$$

Let each state of the system be characterized by probability $p_i$, then

$$S = -\sum_{i=1}^{n} p_i \,\mathbf{log}\, p_i$$

In the information theory concept, if new measurements are obtained, then there is a gain in information about the system's state and the entropy is reduced. The concept of entropy is used in model order/structure (Chapter 6) determination criteria. The idea here is that first a low order model is fitted to the data. The entropy is evaluated. Then a higher order model is fitted in succession and a reduction in the entropy is sought. Physical interpretation is that when a better model is fitted to the data the model is the refined one and the fit error is substantially reduced. The disorder is reduced, and hence the entropy.

## A16 EXPECTATION VALUE

Let $x_i$ be the random variable, then the mathematical expectation $E$ is given as

$$E(x) = \sum_{i=1}^{n} x_i P(x = x_i); E(x) = \int_{-\infty}^{\infty} x\, p(x) dx$$

Here, $P$ is the probability distribution of variables $x$ and $p$ the pdf of variable $x$. The usual definition of mean of a variable does not take into account the probability of the (favourable) occurrence of the variables and just gives a conventional average value of the variables. The expectation concept plays important role in many parameter estimation methods. It can be considered as a weighted mean, where the weights are individual probabilities. In general, it can also be used to get average properties of squared quantities or two variables like $x_i, y_i$.

## A17 FIT ERROR CRITERIA

a) **Fit error criterion (%)**: Also called prediction fit error (PPE), it can be used to judge the suitability of the model fit:

$$FEC = \frac{1/N\sum_{k=1}^{N}[z_k - z_k(\hat{\beta})]^2}{1/N\sum_{k=1}^{N} z_k^2} \times 100(\rightarrow \text{Percentage FEC})$$

Replacing N with N-n can correct the criterion for the degrees of freedom in the numerator. Essentially, this criterion compares models based on the reduction of residuals to the signal power ratio of successive models.

b) **Residual sum of squares (RSS)**: Often, sum of residuals is used for judging the model adequacy:

$$RSS = \sum_{k=1}^{N}[z_k - \hat{z}_k(\hat{\beta})]^2$$

If any new parameter enters the model, then there should be significant reduction in RSS, otherwise it is not included in the model.

## A18 GAUSSIAN PROBABILITY DENSITY FUNCTION

The Gaussian pdf is given as:

$$p(x) = \frac{1}{\sqrt{2\pi}\,\sigma}\,\mathbf{exp}\left(-\frac{(x-m)^2}{2\sigma^2}\right)$$

Here, $m$ is the mean and $\sigma^2$ is the variance of the distribution. For the measurements, given the state $x$ (or parameters), the pdf is given by:

$$p(z \mid x) = \frac{1}{(2\pi)^{n/2}\left|R\right|^{1/2}}\,\mathbf{exp}\left(-\frac{1}{2}(z-Hx)^T R^{-1}(z-Hx)\right)$$

In the above, $R$ is the covariance matrix of measurement noise. The variable $x$ can be replaced by $\beta$, the parameter vector. The maximization of $p(z \mid x)$ is equivalent to minimization of the term in the parenthesis.

## A19 GAUSS-MARKOV PROCESS

Assume a lumped parameter linear system of the first order driven by a white Gaussian noise. Then the output will be a Gauss-Markov process of the first order. This assumption is used in KF theory. A continuous process $x(t)$ is first order Markov if, for every $k$ and $t_1 < t_2 < \ldots < t_k$,

$$P\{x(t_k) \mid x(t_{k-1}),\ldots,x(t_1)\} = P\{x(t_k) \mid x(t_{k-1})\}$$

This means that the probability distribution of $x(t_k)$ is dependent on the value at point $k-1$ only.

## A20 HESSIAN

The symmetric matrix of dimension $n$ x $n$ of second partial derivatives of a cost function $f$ is termed the Hessian of the cost function. Let the cost function be dependent on the components of $\beta$, then

$$H_f = \begin{bmatrix} \dfrac{\partial f}{\partial \beta_1 \beta_1} & \dfrac{\partial f}{\partial \beta_1 \beta_2} & \cdots \\ \vdots & & \\ \dfrac{\partial f}{\partial \beta_n \beta_1} & \cdots & \dfrac{\partial f}{\partial \beta_n \beta_n} \end{bmatrix}$$

The positive Hessian indicates the minimum of the function $f$ and the negative Hessian indicates the maximum of the cost function $f$. This property is useful in optimization/estimation problems. For the LS method, $H_f = H^T H$ (see Chapter 2), and it indicates the minimum of the cost function.

## A21 IDENTIFIABILITY

Given the input/output data of a system and chosen form of the model (which when operated upon by the input, produces the output), one must be able to identify the coefficients/parameters of the model, with some statistical assumptions on the noise processes (acting on measurements). The identification methods (e.g., least squares) then yield the numerical values of these coefficients. The term 'system identification' is used in the context of identification of transfer function and time-series models. One important assumption is that the input should be persistently exciting, in order to be able to capture the modes of the system from its output. This roughly means that the spectrum of the input signal should be broader than the bandwidth of the system (that generates a time-series).

## A22 MEASUREMENT NOISE COVARIANCE MATRIX

This matrix for discrete-time noise, given as *R(k),* is called the noise covariance matrix. For continuous-time measurement noise, the covariance matrix *R(t)* is called the spectral density matrix.

In the limit $\Delta t \to 0$, $R(k) = \frac{R(t)}{\Delta t}$, such that the discrete noise sequence tends to the infinite valued pulses of zero duration. This ensures that the area under the 'impulse' autocorrelation function $R_k \Delta t$ = the area $R$ under the continuous white noise impulse autocorrelation function.

## A23 MODE

In parameter estimation, we use data affected by random noise, etc. Hence, the estimate of the parameter vector is some measure or quantity related to the probability distribution. It could be mode, median, or mean of the distribution. The mode of the distribution defines the value of $x$ (here $x$ could be parameter vector) for which the probability of observing the random variable is a maximum. Thus mode signifies the argument (i.e., $x$ or parameter vector) that gives the maximum of the probability distribution. The distribution could be unimodal or multi-modal. In practical situations multi-modal distribution could occur.

## A24 MONTE-CARLO METHOD

For a dynamic system, assume that the simulated data are used for parameter estimation. Therefore, for one set of data, we get one set of estimated parameters. Next, we change the seed number for random number generator, add these data to measurements as noise, and again estimate the parameters with the new data set. In the new data set, the original signal remains the same. Thus, we can formulate a number of such data sets with different seed numbers and obtain parameters to see the variability of the estimates across different realizations of the data, mimicking the practical real life situation. Then we can obtain the mean value and the variance of the parameter estimates using all the individual estimates from different realizations. This will help in judging the performance of the estimation method. The mean of the parameters should converge to the true values. If we take two estimation procedures/methods, then the one that gives estimates (mean value) closer to the true value and less variance will be the better choice. This approach can be used for linear or nonlinear systems. A similar procedure can be used for state estimation methods also. This procedure is numerical and could become computationally intensive. Depending upon the problem and its complexity often 400 or 500 simulation runs are required. However, as few as 20 runs are also often used to generate average results.

## A25 NORM OF A VECTOR

We need to have a measure of a vector or matrix (of a signal) in order to have knowledge of their magnitudes and strengths. This will also help in judging the magnitude of state error or measurement error or residuals. Let $x$ be a vector. Then the distance measure or norm is defined as:

$$L_p = \|x\|_p = \left( \sum_{i=1}^{n} |x_i|^p \right)^{1/p} ; \quad p \geq 1$$

We have three possibilities:

i) if $p = 1$ then the length of vector $x$ is $\|x\|_1 = |x_1| + |x_2| + \ldots + |x_n|$. Then the centre of a probability distribution estimated using $L_1$ norm is the median of the distribution.
ii) if $p = 2$, then it is called the Euclidean norm and it gives a length of the vector. We see that it is the square root of the inner product of the vector $x$ with itself. In addition, it is equal to the square root of sum of the squares of the components of $x$. This leads to the Schwarz inequality:

$$|x^T y| \le \|x\| \cdot \|y\|$$

Here $y$ is another vector. Also for $p = 2$, the centre of a distribution estimated using $L_2$ norm is the mean of the distribution and is the chi-square estimator. This norm is used in many state/parameter estimation problems to define the cost functions in terms of state error or measurement error. The minimization problems with this norm are mathematically highly tractable. This leads to the least squares or maximum likelihood estimator, as the case may be.
iii) If $p = \infty$, then it gives the Chebyshev norm. It signifies the maximum of the absolute v alue of $x_i$

$$\|x\|_{p=\infty} = \max |x_i|$$

It looks as if this norm is related to H-Infinity norm.

## A26 NORM OF MATRIX

The measure of strength of a matrix can be determined in terms of its determinant or eigenvalues (e.g., the largest or the smallest eigenvalue). One measure is given as:

$$\|A\| = \underset{\|x\| = 1}{\text{Sup}} \{\|Ax\|\}$$

Often a singular value is used as a norm of a matrix.

## A27 OBSERVABILITY

This generally applies to state observability. It means that if the system (its representation) is (controllable and) observable, then given the input/output responses of the system, one must be able to determine/observe the states of the system (also given the model information, essentially its structure). Often certain assumptions on statistics of the noise processes are made.

## A28 PARAMETER ESTIMATION ERROR NORM (PEEN)

$$\text{PEEN} = \frac{\text{norm}(\beta - \hat{\beta})}{\text{norm}(\beta)} \times 100$$

### A28.1 Projection Theorem

It refers to a fundamental concept in statistics and linear algebra, stating that within a linear model, the best linear unbiased estimator (BLUE) of a parameter is the orthogonal projection of the observed data onto the subspace spanned by the design matrix, meaning the error vector is orthogonal to the space of possible linear combinations of the explanatory variables; essentially, the estimator minimizes the squared error by being as close as possible to the data while remaining within the model space, with the projection aspect signifying that the error vector is perpendicular to this space. Let y be the vector of observed data, X be the design matrix, β be the vector of parameters to be estimated, and ε be the error vector; then the projection theorem states that the best linear unbiased estimator of β, denoted as β̂, satisfies the condition: (y – Xβ̂) is orthogonal to the columns of X.

## A29 ROOT SUM SQUARE ERROR (RSSE)

Let $x_t, y_t, z_t$ be the true trajectories and $\hat{x}, \hat{y}, \hat{z}$ be the estimated/predicted trajectories. Then

$$RSSE(t) = \sqrt{(x_t(t) - \hat{x}(t))^2 + (y_t(t) - \hat{y}(t))^2 + (z_t(t) - \hat{z}(t))^2}$$

This is valid also for the discrete time signals.

$$\text{Percentage RSSE} = \frac{RSSE(t)}{\sqrt{x_t^2(t) + y_t^2(t) + z_t^2(t)}} \times 100$$

## A30 ROOT MEAN SQUARE ERROR (RMSE)

$$RMSE = \frac{1}{N}\sqrt{\frac{(x_t(t) - \hat{x}(t))^2 + (y_t(t) - \hat{y}(t))^2 + (z_t(t) - \hat{z}(t))^2}{3}}$$

Percentage RMSE can also be defined.

## A31 RAO-BLACKWELL THEOREM

It provides a process by which a possible improvement in efficiency of an estimator can be obtained by taking its conditional expectation with respect to a sufficient statistic. The **Rao-Blackwell theorem** (called the Rao-Blackwell-Kolmogorov theorem or Rao-Blackwellization) is a way to improve the efficiency of initial estimators. Estimators are observable random variables used to estimate quantities. For example, the (observable) sample mean is an estimator for the (unknown) population mean. More formally, the theorem is defined as follows:

Suppose g(X) is an unbiased estimator of a scalar parametric function h(θ) and $T_i(X)$, i = 1,2,…,p, are jointly sufficient for (θ); then there exists an estimator u(T), depending on the data only through the sufficient statistics, such that E[u(T)] = h(θ) and Var[u(T)] ≤ Var[g(X)].

## A32 SINGULAR VALUE DECOMPOSITION (SVD)

A matrix $A(m \text{ x } n)$ can be factored into:

$$A = USV^T$$

Here, $U$ and $V$ are orthogonal matrices with dim. $(m,m)$ and $(n,n)$ respectively. $S$ is an $(m,n)$ diagonal matrix. Its elements are real and non-negative and are called singular values, $\rho_i$, of the matrix $A$. The concept of singular values is used in control system analysis and design as well as in the determination of model order of the system when significant SVs are retained to reduce the complexity of the identified model. Also, SVD is used in parameter/state estimation problems to obtain numerically stable algorithms.

## A33 SINGULAR VALUES (SV)

Singular values $\sigma$ are defined for a matrix $A$ as:

$$\sigma_i(A) = \sqrt{\lambda_i\{A^T A\}} = \sqrt{\lambda_i\{AA^T\}}$$

Here $\lambda_i$ are the eigenvalues of the matrix $A^T A$.

The maximum SV of a matrix A is called the spectral norm of A:

$$\sigma_{\max}(A) = \mathbf{max}_{x \neq 0} \frac{\|Ax\|_2}{\|x\|_2} = \|A\|_2$$

For a singular matrix $A$, one can use $\sigma_{\min}(A) = 0$.

Thus, for a vector, the Eucledian norm is

$$l_2 = \left( \sum_i |x_i|^2 \right)^{1/2}$$

For a matrix $A$, $\sigma_{\max}(A)$ can be used.

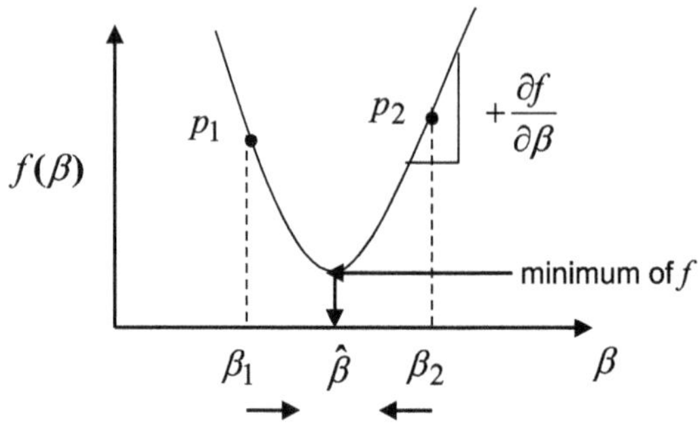


**FIGURE A.2** Cost function

## A34 STEEPEST DESCENT METHOD

The simplest form is explained below.

Let $f$ be the function of a variable, say, parameter $\beta$, i.e., $f(\beta)$. We consider that $f(\beta)$ is a cost function with at least one minimum as shown in Figure A.2.

Then we use the parameter estimation rule as:

$$\frac{d\beta}{dt} = -\frac{\partial f(\beta)}{\partial \beta}$$

What this means is that the rate of change in the parameter (with respect to time) is in the negative direction of the gradient of the cost function with respect to the parameter.

We can discretize the above formulation as: $\beta(i+1) = \beta(i) - \mu \frac{\partial f}{\partial \beta}$

In the above expression, $\Delta t$ is absorbed in the factor $\mu$. We see from Figure A.2 that at point $p_2$, the slope of $f$ is positive and hence we get new value of $\beta$ (assuming $\mu = 1$) as: $\beta = \beta_2 -$ (positive value of the slope). Hence, $\beta < \beta_2$ and $\beta$ is approaching $\hat{\beta}$. Similarly, when the slope is negative, $\beta$ will approach $\hat{\beta}$, and so on. The method will have problems if there are multiple minima and there is high noise in the measurement data. Small values of $\mu$ will make the algorithm slow and large values might cause it to oscillate. Proper choice of $\mu$ should be arrived at by trials using the real data for the estimation purpose. The $\mu$ is called obviously the "step size" or "tuning parameter". The method is suitable also for a function of more than one variable. It is also known as the steepest ascent or "hill climbing" method.

## A35 SYSTEM IDENTIFICATION THEOREM

Persistence of Excitation: This states that for a system to be properly identified, the input signal must be rich enough, meaning it excites all the system dynamics by having sufficient frequency content in the excitation signal, to accurately estimate the system parameters. If this happens then there are some effects of the excitation/input signal in the output responses of the system under test experiment, and we would be able to correlate the output signals to the input signal and identify the system's parameters.

## A36 VARIANCE OF RESIDUALS

$\sigma_r^2 = \frac{1}{N-1}\sum_{k=1}^{N}(r(k)-\underline{r})^2$. Here, $\underline{r}$ is the mean of the residuals.

## REFERENCE

A.1 Raol, J.R., Girija, G., and Singh, J., *Modelling and Parameter Estimation for Dynamic Systems, IEE/IET Control Series Books*, Vol. 65, The Institution of Electrical Engineers (IEE), London, UK, 2004.

# Appendix B
# Theorem in Soft Computing

We present here some important theorems in the area of soft computing, including machine learning.

## THEOREMS IN ARTIFICIAL NEURAL NETWORKS

### B1 Universal Approximation Theorem

The Universal approximation theorem states that a neural network with at least one hidden layer and a non-linear activation function can approximate any continuous function to an arbitrary level of accuracy, essentially meaning a neural network can learn to represent any complex function with enough neurons and proper training; this is why neural networks are often called universal approximators [B.1].

### B2 No Free Lunch Theorem

This theorem states that no single machine learning algorithm can outperform all other algorithms on average across all possible problems, highlighting the importance of choosing the right model for the specific task.

### B3 Convergence Theorem

A model always converges to a stable state when operating in a serial mode and to a cycle of length at most 2 when operating in a fully parallel mode. This property is the basis for the potential applications of the model, such as associative memory devices and combinatorial optimization.

### B4 Generalization Property

It refers to the ability of a trained ANN to accurately predict outcomes on new, unseen data that was not part of the training set; essentially meaning, it can apply learned patterns to situations beyond the specific examples it was trained on – a key aspect of a well-performing model in real-world applications. Good generalization is important as it indicates that the ANN has learned underlying patterns and relationships in the data, not just memorized the training examples, allowing it to make accurate predictions on new data.

When an ANN overfits the training data, it performs very well on the training set but poorly on new data, indicating poor generalization.

**Factors affecting generalization:** i) Training data quality: a diverse and representative training set is vital for good generalization; ii) Network complexity: a network that is too complex relative to the amount of training data can lead to overfitting and poor generalization; and iii) Regularization techniques: techniques like dropout, weight decay, and early stopping can help prevent overfitting and improve generalization.

To assess how well the trained ANN-model generalizes, it is tested on a separate validation set (of data) which contains unseen data that was not used for the training,

## B5 Stability-plasticity Dilemma

This is a challenge in learning systems that involves balancing the need for stability and plasticity.

- **Stability**: The ability to remember past knowledge;
- **Plasticity**: The ability to learn new knowledge.
- In biological neural networks (BNNs), the brain needs to balance plasticity and stability to integrate new knowledge (plasticity) while avoiding forgetting old knowledge (stability).
- In ANNs, a model's ability to learn new knowledge comes at the expense of its ability to remember old knowledge. Some continual learning algorithms use a hyper-parameter to control the trade-off between stability and plasticity; some argue that the trade-off should be dynamically tuned instead of being kept constant.

## B6 Kolmogorov-Arnold Representation Theorem (Superposition Theorem)

It is a foundational theorem and demonstrates that any continuous function can be represented as a composition of simpler functions, which underlies the concept of neural networks with multiple layers.

## B7 Cover's Theorem

It is a statement in computational learning theory and is one of the primary theoretical motivations for the use of non-linear kernel methods in machine learning applications. It is so termed after the information theorist Thomas M. Cover who stated it in 1965, referring to it as counting function theorem.

## B8 Perceptron Algorithm Theorem

Perceptron convergence: For any finite set of linearly separable labelled examples, the Perceptron learning algorithm will halt after a finite number of iterations; the algorithm yields a vector w that classifies perfectly all the examples.

### B9 Vapnik-chervonenkis (VC) Dimension Theorem

This theorem defines the capacity of a classification model and its ability to generalize to unseen data. It provides a measure of complexity for hypothesis classes in statistical learning theory, where a higher VC dimension indicates a more complex model that can fit a wider range of patterns but may also lead to overfitting. The VC dimension helps determine the minimum number of training samples required for a model to generalize well.

## THEOREMS IN FUZZY LOGIC

### B10 Decomposition Theorem

It states that any fuzzy set can be represented as a union of 'alpha-cuts' (crisp sets based on a membership threshold).

### B11 Gödel's Fuzzy Completeness Theorem

Here, the provability degree of a formula in a fuzzy theory is equivalent to its truth degree, allowing for a direct relationship between proof and truth values within fuzzy logic. This generalization extends the classical completeness theorem to fuzzy logic, ensuring that a formula is provable in a fuzzy system if and only if it is true to a certain degree, providing a strong connection between the logical system and the fuzzy truth values.

### B12 Extension Principle

It allows for the calculation of fuzzy set operations on non-fuzzy data by utilizing the concept of membership functions; essentially, it defines how to fuzzify any function by assigning membership degrees to the output based on the membership degrees of the input variables.

### B13 De Morgan's Laws

These laws relate the complement of a fuzzy union and intersection, similar to classical set theory, but with the membership function operations of 'max' and 'min' applied instead of 'or' and 'and'.

### B14 Baire Category Theorem (Fuzzy Version)

It deals with the concept of fuzzy dense sets and has applications in fuzzy topology and decision-making under uncertainty.

### B15 Fuzzy Inference System Theorems

These relate to the process of deriving fuzzy outputs based on fuzzy rules, including the Mamdani Inference and Sugeno Inference methods.

### B16 Central Limit Theorem for Fuzzy Random Variables

It is shown that a sum or mean of independent fuzzy random variables converges in the limit to a fuzzy Gaussian random variable, thus providing a fuzzy analogue of the central limit theorem of classical probability theory.

### B17 Zadeh's Extension Principle

Zadeh's extension principle is a fundamental theorem in fuzzy logic that allows classical mathematical functions to be extended to fuzzy domains. It provides a method for defining functions on fuzzy sets by mapping fuzzy inputs to fuzzy outputs through membership functions. This principle is widely used in fuzzy control systems, decision-making, and approximate reasoning, enabling operations on imprecise and uncertain data.

## THEOREMS OF GENETIC ALGORITHMS

### B18 Holland's Schema Theorem

Also called the fundamental theorem of genetic algorithms, it is an inequality that results from coarse-graining an equation for evolutionary dynamics. The schema theorem says that short, low-order schemata with above-average fitness increase exponentially in frequency in successive generations.

### B19 Convergence Theorem

In the context of genetic algorithms, the convergence theorem primarily refers to Holland's schema theorem, which mathematically explains how a genetic algorithm tends to converge towards solutions that represent beneficial schemata (building blocks of good solutions) by selectively propagating them through the population over generations, effectively guiding the search towards the optimal solution.

## REFERENCE

**B.1** Jang, J-S.R., Sun, C-T., and Mizutani, E., *Neuro-Fuzzy and Soft Computing-a Computational Approach to Learning and Machine Intelligence*, Upper Saddle River, NJ, Prentice Hall, 1997.

# Appendix C Procedure of Building an AI System

In this appendix we describe a procedure of building an AI system, and some examples where one or more aspects of AI are utilized.

## C.1 INTRODUCTION

To build an AI system, one has to follow these steps: i) define the problem and goals, ii) gather and prepare data, iii) select an appropriate model, iv) train the model, v) evaluate the model's performance, and vi) deploy the AI system into a production environment [C.1].

This involves identifying the specific problem one wants the AI to solve, collecting relevant data, cleaning and structuring it, choosing the right machine learning (ML) algorithm, training the model on the collected data, testing the model's accuracy, and integrating it into the desired application. Interestingly this sounds very similar to the procedure of mathematical model building of/for any dynamic system.

## C.2 KEY STEPS IN BUILDING AN AI SYSTEM

- **Define the problem and goals:** Clearly articulate the problem one wants the AI to address and set specific, measurable goals for its performance.
  - **Data collection and preparation:** i) Gather data: identify and acquire relevant data sources to train the AI model; ii) Data cleaning: remove inconsistencies, missing values, and irrelevant data from your dataset; and iii) Data pre-processing: transform data into a suitable format for the chosen model, including normalization or feature engineering.
  - **Model selection/choose an algorithm:** i) Based on the problem type and data characteristics, select a suitable ML algorithm like decision trees, ANNs, or support vector machines (SVM); and ii) Consider frameworks: utilize ML libraries like Python, TensorFlow, PyTorch, or Scikit-learn to build and train your model.
  - **Model training:** i) Train the model: feed the prepared data into the chosen algorithm to allow it to learn patterns and relationships; and ii) Parameter tuning: adjust model hyper-parameters to optimize its performance.
  - **Evaluation and testing:** i) Validate performance: test the model on a separate hold-out dataset to assess its accuracy, precision, recall, and other

relevant metrics; and ii) Iterate and improve: if necessary, revisit data preparation, model selection, or training parameters to enhance performance.

- **Deployment:** i) Integrate with applications: integrate the trained model into the desired software or system, allowing it to make predictions or decisions in real time; and ii) Monitor and maintain: continuously monitor the model's performance in production and update it as needed to adapt to changing data or requirements.

Some important considerations while building an AI system are: i) **Ethical implications:** be aware of potential biases in your data and model, and take steps to mitigate them; ii) **Data quality:** high-quality data is crucial for training an accurate AI model; iii) **Scalability:** design the system to handle increasing data volumes and computational demands; and iv) **Explainability:** Consider the need to understand how your AI model makes decisions, especially in critical applications.

## C.3 EXAMPLES AND APPLICATIONS OF AI

Applications of AI systems spans a wide range of technical and applied domains. Despite the prevalent use of AI-powered services and devices across industries and everyday life, many users remain unaware of the extent to which AI drives the tools and platforms they interact with daily.

In healthcare, machine learning algorithms trained on large patient datasets have demonstrated the ability to detect diseases from medical imaging with accuracy comparable to and in some cases exceeding, that of trained clinicians. Specific implementations include algorithms that detect ocular diseases using optical coherence tomography scans and deep learning models that identify signs of diabetic retinopathy from retinal images [C.2]. Algorithms are also used in cancer detection, where advanced machine learning and deep learning models analyze large volumes of medical imaging data to detect abnormalities at early stages. This helps to improve diagnostic accuracy. Apart from diagnostics, these algorithms are applied to drug discovery, where models identify molecular compounds and flag potentially harmful drug interactions and to predictive modelling for disease surveillance and outbreak tracking.

In transportation, autonomous vehicle platforms have demonstrated the feasibility of using computer-controlled navigation in real-world conditions. Route optimization systems have achieved meaningful reductions in delivery mileage and operational costs [C.3]. Traffic management systems employ real-time data fusion from road sensors, GPS feeds and weather inputs to dynamically predict and reduce congestion. Predictive maintenance systems help to reduce vehicle downtime by identifying component failures before they occur [C.4].

In finance, fraud detection systems monitor transaction streams in real time, using anomaly detection algorithms to flag deviations from established customer behaviour patterns. Algorithmic trading platforms employ machine learning models to analyze market data and execute trades at timescales which are usually beyond the speed of human decision-making. Natural language processing models are increasingly used to parse regulatory documents and support compliance reporting.

In consumer applications, intelligent virtual assistants such as Apple Siri, Amazon Alexa, and Google Assistant demonstrate the extensive use of artificial intelligence in everyday life. These systems use speech recognition, natural language processing and machine learning techniques to interpret user commands, retrieve information, control smart devices and also provide personalized assistance.

These examples illustrate only a small sample of the wide range of domains in which AI systems are being successfully applied. Additional applications spanning agriculture, education, environmental monitoring, smart manufacturing, robotics and assistive technologies. These considerations, which cut across all application domains, reinforce the importance of following a disciplined and systematic approach when building and deploying such systems.

## REFERENCES

C.1 Alzubaidi, L., Zhang, J., Humaidi, A.J. et al., "Review of Deep Learning: Concepts, CNN Architectures, Challenges, Applications, Future Directions", *Journal of Big Data*, Vol. **8**, p. 53, 2021.

C.2 Mudgal, S.K., Agarwal, R., Chaturvedi, J., Gaur, R., Ranjan, N. Real-world application, challenges and implication of artificial intelligence in healthcare: an essay. *Pan African Medical Journal*, 43, p. 3, 2022.

C.3 Thoughtful AI. *Applications of Artificial Intelligence (AI) Across 7 Key Industries*. www.thoughtful.ai/blog/examples-of-artificial-intelligence-ai-in-7-industries. Accessed March 2026.

C.4 www.tableau.com/data-insights/ai/examples, accessed March 2026.

# Index

## J



## K



## L



## M



## N

## S



## T



## U



## V



## W

For Product Safety Concerns and Information please contact our EU
representative GPSR@taylorandfrancis.com
Taylor & Francis Verlag GmbH, Kaufingerstraße 24, 80331 München, Germany

www.ingramcontent.com/pod-product-compliance
Ingram Content Group UK Ltd.
Pitfield, Milton Keynes, MK11 3LW, UK
UKHW022217190926
475702UK00003B/396

* 9 7 8 1 0 4 1 0 1 8 2 3 0 *